ECO-INDUSTRIAL STRATEGIES

UNLEASHING SYNERGY BETWEEN ECONOMIC
DEVELOPMENT AND THE ENVIRONMENT

Edited by Edward Cohen-Rosenthal
with Judy Musnikow

DEDICATION

Edward Cohen-Rosenthal
1952–2002

I count myself among the lucky individuals who had the opportunity to work with, learn from, laugh with and be inspired by Edward Cohen-Rosenthal. I know I speak for many when I express my appreciation to Ed for offering tangibility to the numerous crazy thoughts and dreams I've contemplated about how the world should be and what a few really committed people can actually achieve.

A pillar in the foundation of progressive foresight, Ed was fundamental in advancing various socially and environmentally conscious ideals, theories and endeavours. In particular, he was instrumental in enriching and fostering eco-industrial development at every level, from its mere notions to its concrete manifestations. This book reveals only a glimpse of Ed's impact on the field. In many ways, his greater influence on eco-industrial development will become evident in how this book is used: specifically, how it triggers ideas and projects and how it drives seemingly marginal concepts to the forefront of thought and practice. Truly the best way to realise Ed's influence, however, may be simply to ask anyone who knew him to share an Ed story. For I am sure they have one.

Judy Musnikow
Assistant Editor

Eco-industrial Strategies

UNLEASHING SYNERGY BETWEEN ECONOMIC DEVELOPMENT AND THE ENVIRONMENT

EDITED BY EDWARD COHEN-ROSENTHAL
WITH JUDY MUSNIKOW

Greenleaf
PUBLISHING
2003

© 2003 Greenleaf Publishing Limited

Published by Greenleaf Publishing Limited
Aizlewood's Mill
Nursery Street
Sheffield S3 8GG
UK

Printed and bound, using acid-free paper from managed forests,
by The Bath Press, UK.
Cover by LaliAbril.com/Utter.

All rights reserved. No part of this publication may be reproduced, stored in a retrieval system, or transmitted, in any form or by any means, electronic, mechanical, photocopying, recording or otherwise, without the prior permission in writing of the publishers.

British Library Cataloguing in Publication Data:
   Eco-industrial strategies : synergy between economic
     development and the environment
      1. Industrial districts   2. Economic development -
     Environmental aspects   3. Industries - Environmental aspects
     4. Business networks
     I. Cohen-Rosenthal, Edward   II. Musnikow, Judy
     338.6'042

   ISBN 1874719624

CONTENTS

Foreword 9
    Michael Krause

Introduction 11
    Judy Musnikow, Simmons College Graduate School of
    Library and Information Science, USA

PART 1: *Introduction* 13

  1. What is eco-industrial development? 14
     Edward Cohen-Rosenthal, Work and Environment Initiative,
     Cornell University, USA

  2. Making sense of industrial ecology:
    a framework for analysis and action 30
     Edward Cohen-Rosenthal, Work and Environment Initiative,
     Cornell University, USA

  3. A walk on the human side of industrial ecology 51
     Edward Cohen-Rosenthal, Work and Environment Initiative,
     Cornell University, USA

PART 2: *Eco-industrial development for whom? The role of stakeholders* ... 67

4. **The role of government in eco-industrial development** 68
   Bracken Hendricks, *Institute for America's Future, USA*
   Suzanne Giannini-Spohn, *US Environmental Protection Agency*

5. **The role of local government in eco-industrial park development** 89
   Maile Takahashi, *Harvard University, USA*

6. **Community engagement in eco-industrial development** 100
   Mary Schlarb and Judy Musnikow, *Cornell University, USA*

7. **The developer's role in eco-industrial development** 112
   Mark Smith, *Pario Research, USA*

8. **The energy of eco-industrial development** 128
   Daniel K. Slone, *McGuireWoods LLP, USA*

PART 3: *Key issues in eco-industrial development* 137

9. **Legal aspects of eco-industrial development** 138
   Daniel K. Slone, *McGuireWoods LLP, USA*

10. **Environmental and resource issues: an overview** 148
    Raymond P. Côté, *School for Resource and Environmental Studies, Canada*

11. **Management of eco-industrial parks, networks and companies** 163
    Edward Cohen-Rosenthal, *Work and Environment Initiative,
    Cornell University, USA*

12. **Links to the ISO 14000 series at the park and company level** 186
    Astrid Petersen, *Cornell University, USA*

13. **Making it happen: financing eco-industrial development** 200
    Dennis Alvord, *US Department of Commerce*

14. **Real estate and eco-industrial development:
    the creation of value** 243
    Edward Cohen-Rosenthal, *Work and Environment Initiative,
    Cornell University, USA*
    Mark Smith, *Pario Research, USA*

15. Evaluating the success of eco-industrial development 258
Marian R. Chertow, Yale University, USA

PART 4: *Eco-industrial compendium of cases* 269

16. The industrial symbiosis in Kalundborg, Denmark:
an approach to cleaner industrial production 270
Noel Brings Jacobsen, The Symbiosis Institute, Denmark

17. The Green Institute Phillips Eco-enterprise Center in
Minneapolis, Minnesota 276
Michael Krause, The Green Institute, USA
Corey Brinkema, Trillium Planning & Development, USA

18. Cape Charles Sustainable Technology Park: the eco-industrial
development strategy of Northampton County, Virginia 288
Timothy Hayes, County of Northampton, Virginia, USA

19. Sustainable Londonderry 300
Peter C. Lowitt, AICP, Devens Enterprise Commission, USA

20. The Red Hills Industrial EcoPlex: a case study 307
Ron Forsythe, Pickering Inc., USA

21. Region-wide eco-industrial networking in North Carolina 317
Judy Kincaid, Triangle J Council of Governments, USA

22. A case study in eco-industrial development:
the transformation of Burnside Industrial Park 322
Raymond P. Côté, Dalhousie University, Canada
Peggy Crawford, Eco-efficiency Center, Burnside Industrial Park, Canada

23. Eco-industrial development as a defence conversion strategy:
a case study of the Louisiana Army Ammunition Plant Re-use 330
Joshua L. Tosteson, Columbia University, USA
Victor A. Guadagno, Northwest Louisiana Commerce Center, USA

24. Eco-industrial development in Asian developing countries 341
Ernest Lowe, RPP International, USA

Bibliography 353

List of abbreviations 364

Author biographies 369

Index 376

FOREWORD

It was a moment of insight and creation that prompted Ed Cohen-Rosenthal to suggest that the world of work and labour should be brought together with the study of the natural environment. He was saying what others were beginning to recognise as well, that the old paradigm of jobs versus the environment was rapidly giving way to a new order that saw the possibility of reducing pollution and waste while simultaneously creating jobs and improving chances for business success. In fact, it is not simply possible but essential that we change our linear thinking about industrial production if we are to set human civilisation on a sustainable development path.

With the creation of the Work and Environment Initiative at Cornell University in 1994, and the founding of the National Center for Eco-Industrial Development in 2000, Ed Cohen-Rosenthal established himself as a leader in the new field of industrial ecology, and he continued to shape it in his own unique ways until his death in January 2002.

Ed Cohen Rosenthal's reputation was not just national but international. He was a US delegate to the United Nations General Assembly meeting on the Earth Summit +5. The last time I saw him, in the summer of 2001, we had just spent two days on a tour of the Kalundborg eco-industrial park in Denmark. That evening, we had dinner back in Copenhagen, and Ed kept up his rapid-fire thinking about the many fronts on which to continue promoting eco-industrial development. Ed was also a member of the Eco-Industrial Park Task Force of the President's Council on Sustainable Development and was a member of the faculty at Cornell, in the School of Industrial and Labour Relations.

One of the primary strategies for building an understanding of this new approach to industrial development was Ed's creation and leadership of the Eco-Industrial Development Roundtable in 1995. The group met three times a year, twice in communities where eco-industrial activities were under way, and once in Washington, DC, where we interacted with federal policy-makers and supporters of industrial ecology.

Ed did not seem like an academic himself and, in fact, was always more focused on how to build a better life for average people, having grown up in a working-class neighbourhood of Baltimore and having spent time as a staff member for the Bricklayers Union. And no doubt his concern for the health of the planet was influenced by his own long battle with the cancer that eventually took his life.

However, this book is not a tribute to Ed, and he would not want it to be so. It is, rather, a collection of concepts, tools and case studies about eco-industrial development in the USA. Ed would have been pleased at its publication but he would have been more pleased with it as the work of many people at many levels and across several disciplines.

Industrial ecology, eco-industrial development, or industrial symbiosis had very little conscious application in the industrial nations of the world in 1994. The multidisciplinary approach to industrial development has since taken hold in Asia and across Europe, where it is being applied in ways that may create significant competitive advantages for businesses in these nations.

The importance of academic research in the field of industrial ecology cannot be underestimated, and it is reflected in these pages. Research will continue to develop measures of performance and refine analytical tools for use by practitioners. However, there is a recognition among the practitioners who were brought together by the programme at Cornell that the greatest contribution of eco-industrial development would be to make a real difference in communities that are struggling to remain competitive in a global marketplace. So, you will find an equal emphasis here on the application of academic thinking in real settings as an answer to actual problems.

It is also important to note that industrial ecology is much more than a network of waste and energy exchanges among businesses. The voluntary networks that distinguish eco-industrial development patterns from more traditional models of industrial activity will be most successful when they include ongoing dialogues and are conceived of in the context of their larger communities. And the real benefit of these networks may come through joint efforts addressing issues such as human resources and training, marketing or transportation logistics as well as materials and energy.

The fundamental principle of our approach is to think about industrial and related activities—not as separate and disconnected islands unto themselves but as an overall system much like an ecosystem that gives this field its name and defining mental model. The goals are for the industrial base to become stronger while it also becomes a better neighbour within the mixed land-use plan of a community.

The study and the practice of industrial ecology or eco-industrial development is still at an early stage, but it is advancing quickly in different ways in every corner of the planet. The contributors to this volume have much to offer in experience and ideas to those who are working to find those simultaneous benefits to businesses and communities. In memory of Ed Cohen-Rosenthal who was a pioneer and a teacher to many of us, we wish you well in your efforts.

Michael Krause
June 2002

INTRODUCTION

Judy Musnikow
Simmons College Graduate School of
Library and Information Science, USA

Eco-industrial development, born from the realisation that the places where we work waste too much and unnecessarily pollute the land, air and water, simply stated, demands a better way of working. It challenges the outcast view of industrial development and strives to transform these facilities into safe workplaces for community and employee health. This progressive perspective on economic development is taking shape in communities across the United States and around the world. *Eco-industrial Strategies: Unleashing Synergy between Economic Development and the Environment* explores the key issues involved in eco-industrial development and identifies the stakeholders and their roles in such projects. In addition, it offers a compendium of eco-industrial development case studies, each written by key players in the case's development process.

While an increasing number of handbooks and manuals focus on eco-industrial development, no other formal book containing process analysis, a breakdown of stakeholder responsibility and case-study assessment presently exists. What sets this work apart is its ability to pool together resources and knowledge from a wide array of sources within the eco-industrial field and therefore frame the concept from multiple angles. Eco-industrial development is based on the notion of interrelated and interworking systems. In the same vein, this work maintains an interdisciplinary nature and benefits from the synergies that occur among and between various topics.

Eco-industrial Strategies: Synergy between Economic Development and the Environment aims to accomplish a two-step 'inform and empower' process. First, it familiarises readers with eco-industrial development, its innovative proclivity and applicability to diverse circumstances. Second, it provides the fundamental tools and motivational creativity to implement independent eco-industrial projects. This method of educating and enabling has been the impetus for several eco-industrial endeavours throughout the years. Broadening accessibility, this book allows eco-industrial development to expand its impact and reach previously untapped audiences.

Briefly stated, the introductory chapters of this book, each written by Edward Cohen-Rosenthal, herald eco-industrial development onto the main stage, presenting several overarching concepts and perspectives of the field and paying particular attention to the technological, economic and social elements. In addition, Ed addresses the general imperative for industry to engage in environmental problem-solving, including how eco-industrial development can improve industry's environmental and health safety track records.

The second part of this book focuses on the role of the various stakeholders involved in eco-industrial development. Each chapter sets out to answer various questions relating to the stakeholders' place in the system. Specifically, they ask: what are the stakeholders' particular interests, in what ways can they participate in the process and how do they relate to other actors and stakeholders? These chapters also respond to questions regarding the relationship between stakeholders and eco-industrial development. Chiefly, they trace the flow of benefits, and various other impacts and repercussions, among and between the stakeholders and the development project. Since eco-industrial development involves countless groups and individuals, this book focuses on five distinct and comprehensive categories: namely, federal government, local government, the surrounding community, the development community and energy resources.

The third part of this edited collection outlines several matters related to conceptualisation, design, operation and assessment of eco-industrial projects. Its main objective is to help readers consider eco-industrial development from the specific perspective that each chapter highlights. Concentrating on the core legal, environmental, management, financial, real estate and evaluative aspects—incidentally, significant in both traditional and eco-industrial development projects alike—this part of the book presents the critical components of each issue and also provides an understanding of the unique attributes that eco-industrial development brings to the equation.

The fourth, case study, part of this book provides vignettes of actual work in progress. Each chapter details the key characteristics of the effort and the process undertaken in developing the eco-industrial project. The studies focus primarily on issues considered in the preceding parts, such as project funding, stakeholder engagement and environmental stewardship. In addition, they recount achievements, threats to success, the ways in which obstacles were overcome and details on the project's future. This book showcases development projects from around the world, including Asia, Canada, Denmark and the United States, situated in a variety of settings, such as army bases, industrial parks and virtual networks. This eclectic mix of development structures and contexts is indicative of the diversity apparent in eco-industrial projects overall and allows readers to glean functional and constructive lessons adaptable to their particular circumstances. Accordingly, this part of the book stands as a testament to the widespread applicability of eco-industrial development and provides inspiration for practitioners in both traditional and unconventional settings.

An idea and practice still in its infancy, eco-industrial development will undergo much evolution beyond what this collaborative work is able to capture. As a document of the concept's earliest theorists, *Eco-industrial Strategies: Unleashing Synergy between Economic Development and the Environment* provides current and future readership with an understanding of eco-industrial development's foundations, its beginnings and its aspirations. Most excitingly, policy-makers, industry professionals, community developers, grass-roots activists and all other readers yearning for a better way to work and live will experience a glimpse of the thoughts, concerns, ambitions, technological insight, communities and economies that embody eco-industrial development now and forever.

Part 1
INTRODUCTION

1
WHAT IS ECO-INDUSTRIAL DEVELOPMENT?

Edward Cohen-Rosenthal
Work and Environment Initiative, Cornell University, USA

Eco-industrial development presents an archway to a better future. For business, eco-industrial development offers new avenues for profitable companies. For communities, eco-industrial options lead to more rooted businesses, good jobs and a cleaner environment. For local and global ecosystems, eco-industrialism promises a lighter load on the environment. In some ways, eco-industrialism is the sunny side of the street from doom-and-gloom environmental scenarios. It seeks to uplift, not to commiserate; to connect rather than dismantle. Eco-industrialism doesn't solve all environmental or business challenges but instead deploys a systemic scan at multiple levels to find and re-find best possible solutions.

My goal in this chapter is to sketch out the broad themes of eco-industrial development—to broaden perspective on theory and practice. I'd like to describe this broader framework first and then relate it to what have become standard definitions of eco-parks and networks. I'll leave it to others in subsequent chapters to more fully articulate the role of stakeholders and actors, describe key parts of the implementation process and provide vignettes of actual work in progress.

A new way of thinking about economic development is taking shape in communities around the world. It comes out of a common-sense observation that most places where we work waste too much, dump too much waste in the land, air and water and wantonly pass along disposal of products to the next users. The costs of these are borne in higher prices, large sums for clean-up and in pollution of the commons. We know that there has to be a better way of working.

For those who have come to be intrigued by these issues as a powerful economic development approach, finding a way out of inflated controversies between the economy and the environment serves as a magnet to this concept. It promises better, faster, more responsible development; synergy, not enmity.

Those who see the issue through the lens of industrial ecology dream that, as in natural systems, waste equals food and that linking one company's 'throw-aways' to another's needs will provide better environmental and business outcomes. Finding and acting on those possibilities is a role for eco-industrial practice.

Those who seek a new age of environmental excellence recognise that incorporating environmental excellence at the system and process level leads to better and less expensive results. Problems with limited solutions at the department or firm level have new options presented when the field of play is a cluster of companies.

For business people bold enough to understand that environmental responsibility is a bottom-line concern for the future—one that insurance companies, banks and customers will increasingly demand—this is the path to take. For entrepreneurs who understand that business opportunities abound not solely in control technologies but even more so in the integration of environmental benefits into the full range of products, here is a new playing field.

For those who see cookie-cutter development in strip malls or call centres that look and act all the same, eco-industrial development provides an alternative that celebrates the possibility of place. It looks at the particular geography, business climate, human potential and other factors that make places special. It avoids a race to the bottom in terms of tax giveaways and instead emphasises what locally can lead to extraordinary success.

For communities, eco-industrial development rather than promising utopia challenges those already doing business in the community and those recruited to move in to ask 'Why not the best?' Significant and continuous improvement in business and environmental performance is the goal. Especially in areas reeling from environmental injustice, it turns the tables on the approach to development. Rather than focus the most energy on keeping out bad investment, it actively seeks the best possible businesses that fit into the local business, social and natural ecology. Instead of begging for a break, it proudly offers opportunity.

1.1 Sustainable development: the contributions of industry

Eco-industrial development is a subset of sustainable development but walks in largely uncharted territory. As we will demonstrate, it also reflects the three Es of the sustainability stool: economy, environment and equity. Sustainable development tends to focus on broad models of biodiversity, global warming, forest cover and oceans. Solutions look at overall fiscal policy, tax laws, tradable permits and so on. These can be valuable perspectives and scales to consider when combined with actionable strategies that make a real difference in a relatively short time.

When we take an ecosystemic approach to analysis and solutions, we begin quickly to understand that any large-scale environmental problem such as desertification, soil and water pollution, species preservation, air quality and population are manifestations of billions and billions and billions of point-sources. Together they exhibit themselves at a larger scale as critical environmental problems.

The role of industry in anthropogenic alteration of the environment writ large is given short shrift.[1] Even community-based sustainable development activities often ignore industry as part of the solution. Community-based indicator projects run out of steam when it comes to the real engagement of industry (Hart 1999).

Yet industry in each transaction—every purchase and sale—influences the larger commons. Each decision on what materials and energy to use and how to use them is a pixel in the picture of industry's contribution to environmental problems or to their resolution. Aggregating micro-level efforts are too often overlooked in addressing sustainability's conundrum. The Hoover Dam or Yellow River approach represents large-scale public works as the prime solution: allegedly big solutions to big problems. And these dam projects come packed with lots of unwanted side-effects for many species, including humans. I prefer, instead, to learn whether water conservation, crop selection and distribution systems could be more fully explored, often at lower cost and greater impact. Many little choices add up to major impact while adjusting more dynamically to local constraints. The same is true in industry, where broad and systematic approaches are more likely to have a larger and sustained aggregate impact than a single or limited set of flashy projects.

No one solution will resolve the blights caused by industry. Within these larger expressions of environmental damage are many particular subsystems, each with its own unique characteristics, often requiring special solutions. Carrying eggs requires an idea of a basket—not just running back and forth with a few eggs in hand. A framework or scale shift is the conceptual basket for carrying ideas to where they can be unpacked and deployed. Eco-industrialism is just such a basket. Particular responses reflect the system's core values, practices or patterns. In the case of eco-industrialism, its applications evolve from applying several basic principles:

- Always ask how to achieve business and environmental excellence in the same breath.
- Always look for mutually beneficial connections with and among businesses, materials, energy, natural systems, markets and the local community.
- Think systemically; experiment locally.

Eco-industrial answers don't arrive wrapped up neatly in a kit with instructions on how to glue it together. No one approach, one machine, one chemical process or one law is up to the task of systemic changes in time or for all time. This is true for people too when we look for who is responsible for creating and correcting environmental problems. No one manager, one worker, one inspector, one activist or one guru can or ought to bear that burden. Taken together, each person, all six billion of us, represents parts of a human ecosystem by which we are all affected and for which all bear responsibility. This accountability also must find expression in the many millions of businesses around the globe who are enjoined to exploit the opportunities that they have to influence our future for the better—not diminish our common legacy.

---

1 The International Human Dimensions Programme of the UN based in Germany attempts to deal with these issues around climate change, but its budget is limited and only one of its five major programmes is on industrial transformation.

Eco-industrial development results in uniqueness. None of the projects described in this book is exactly the same as another. Those who seek benchmarks to copy will do better to visit today's efforts for inspiration—not the Holy Grail. The reason is simple. All businesses (one can make a broader assertion about all things) represent niche players. Each business or set of businesses is nested within an environment that includes its market, suppliers, geography, community, materials availability, energy sources, weather, population, transportation and so on. By necessity, its particular identity is a singular mould. We can sort them into categories. However, category is categorically an abstraction. To succeed in the real world, the particular is pre-eminent. Companies that ignore this truth find that their dissonance with the world around them leads to fundamental conflicts. Eco-industrialism asks each person, company, community or country to find its niche, reinvent its niche if necessary and celebrate the power of its specialness. Some have tried to create the franchisable eco-park and have run into this obstinate truth.

1.2 Why focus on industry?

For many of my colleagues who look at ecological systems and natural resource management a focus on industry feels strange to them. For some, it is consorting with the devil, preferring petitions and regulations to harness the beast. For others in the policy arena, industry is just one of the many stakeholders at the table. Sustainable development theorists frequently ignore the role of industry.

On the contrary, the role of industry is central in how we address the environmental crisis. Industry is a much larger part of the problem than is normally recognised and can take much more a part of the solutions than it has been allotted. If what we are trying to do is to manage the impact humankind has on the environment, then it is in industry that we find fault and hope. The room for improvement is incredible: 90% of materials and energy that go into the making of a product are never incorporated in the final product. These products are discarded very briefly after being purchased and, in a single leap, wind up in the rubbish heap (Hawken *et al.* 1999). A modest improvement in these figures could go a long way. Industry is an instrument of human intention and, as such, can refocus itself in ways that profitably meet human and environmental needs.

When I use *industry* as a term it is not related just to private-sector manufacturing. The state-owned aluminium plants in the former Soviet Union were much more of a problem than similar plants privately owned in the United States. Transportation is an industry with enormous environmental impact. Further, the service sector constitutes an industry, as does the government in that they require particular patterns of resource and energy use. They too must change. When we ask the following questions, it provides an important focus:

- Who makes the products that make up the building and construction sector? Industry.

- Who makes the retail units that lead to household contribution to pollution? Industry.

- Who makes the pesticides and fertilisers that provide the basis for non-point-source pollution and for industrial agriculture? *Industry.*

I make these statements not because there is malice or premeditation among the overwhelming percentage of businesses. I also recognise that 'natural causes' contribute to environmental loading, as does the stress on overburdened natural cleansers of the environment. But, to really deal with household, construction, industrial or municipal waste, etc., go upstream and you find industries where product design, production processes and embedded energy and resource use has to be addressed. If you want to create solutions, these industries can keep materials and energy in play delaying depletion while seeking sustainable substitutes and, we hope, restoring new life to natural systems.

1.3 Eco-industrial parks and networks

Defining eco-industrial development (EID) is not an easy task. It has been used to describe a wide variety of applications. My perspective is that it can cover industrial parks and estates, specific EID networks and relations in industrial districts as well as conscious partnering at the enterprise level. In some ways, it is easier to describe what it is not. Ernie Lowe has observed (2001):

> Some developers and communities have used the term EIP in a relatively loose fashion. To be a real eco-industrial park a development must be more than:
>
> - A single by-product exchange or network of exchanges
> - A recycling business cluster
> - A collection of environmental technology companies
> - A collection of companies making 'green' products
> - An industrial park designed around a single environmental theme (i.e. a solar energy driven park)
> - A park with environmentally friendly infrastructure or construction
> - A mixed-use development (industrial, commercial and residential)
>
> Although many of these concepts may be included within an eco-industrial park, the vision for a fully developed EIP needs to be more comprehensive. The critical elements are the interactions among the park's member businesses and the community's relationship with its community and natural environment.

There has been some confusion about eco-industrial parks in the mistaken notion that they must resemble the Kalundborg industrial symbiosis, shown in Figure 1.1. In fact, Kalundborg is more of a network than a park. There is no common management group, and all the relations are bilateral. Further, they stretch across the region rather than residing on a single property. Paul Hawken (1993), musing on the Kalundborg example, wrote in *The Ecology of Commerce*: 'Imagine what a team of designers could come up with if they were to start from scratch, locating and specifying industries and factories that had potentially synergistic and symbiotic relationships.' Taking up his challenge, many

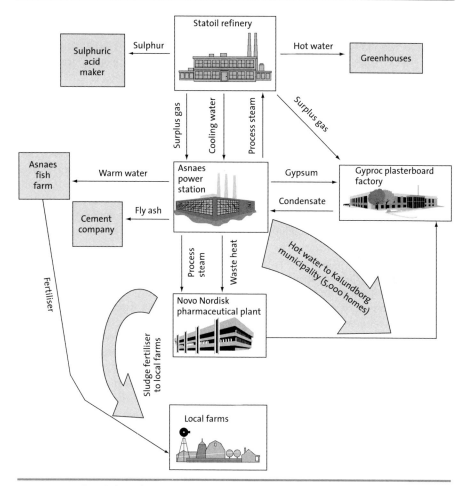

Figure 1.1 *Industrial symbiosis, Kalundborg, Denmark*

communities and entrepreneurs have tried to design a similar set of interconnections, with mixed success. The founders of the Kalundborg system will readily say that it is the relationships they formed that are key to its success, not necessarily the pipes.

Eco-industrial parks offer a discrete parcel of land where companies locate for maximum resource efficiency. The President's Council on Sustainable Development (PCSD 1996c) defines the eco-industrial park (EIP) as follows:

> [It is] a community of businesses that co-operate with each other and with the local community to efficiently share resources (information, materials, water, energy, infrastructure and natural habitat), leading to economic gains, gains in environmental quality and equitable enhancement of human resources for the business and local community.

Connection on contiguous property is not always possible or desirable. The network can take more visible and physical form when collocated. However, there are other options for eco-industrial strategies. In North Carolina, the Triangle J Council of Governments (TJCOG)[2] has created a successful network within a multi-county area, where mutually beneficial exchanges are identified, shown on a computer in data and GIS (geographical information system) format. Eco-industrial connections are possible within an existing industrial area where the businesses are not going to move to a new location. They can, however, seek out interconnections and preferred relationships that range from materials to marketing. If the goal remains the simultaneous achievement of the broadest possible business and environmental success, then they can be considered an eco-industrial network (EIN). This distinguishes an EIN from any association of businesses. I am much less concerned with maintaining the purity of the label than I am with encouraging businesses to look at their own practices and to connect with others to lead to sustainable businesses and communities. An EIN can also be connected to an EIP when it adds value to the park, including allowing for economies of scale for aggregating materials or to achieve other objectives.

These remain to be experimented with, but localities can carve out manageable geographic areas with diverse businesses that are encouraged to explore and self-organise eco-industrial opportunities. For rural areas this may incorporate a large territory, whereas in a dense city it may only be a few blocks. I would also argue that a larger EIN is desirable even with an EIP on environmental grounds. Any new development, no matter how 'green', will have an impact on the local eco-system. The only way to maintain or reduce impact is to work simultaneously with other businesses to improve their practices so that the overall impact on the water- or airshed is positive.

Figure 1.2 shows the advantages of thinking about eco-industrial development on multiple scales. The possibilities and permutations increase exponentially when a broader cluster is in the picture, yet it requires very clear communication, defined values, a mission, objectives and structures to be able to take full advantage of scale. A larger scale may increase the probability of mutually beneficial connections, but an eco-industrial framework moves beyond serendipity to consciously increasing the chances of positive outcomes.

1.4 Eco-industrial development as a framework

In essence, then, eco-industrial development is an overarching framework for the recreation of enterprises at the micro level, how communities are organised and how we live and work at larger scales. When first introduced in the 1990s and popularised by Paul Hawken, eco-industrial parks were seen as possible venues for waste exchange cleverly connected on a contiguous piece of property. It scented of interesting but improbable industrial ecology amusement parks. Seen in this manner, eco-industrial parks were indeed oxymoronic rarities—similar to a new particle generated in a high-speed physics cyclotron, requiring enormous energy, good fortune and being short-lived. It can also be

---

2   See the TJCOG website at www.tjcog.dst.nc.us/indeco.htm.

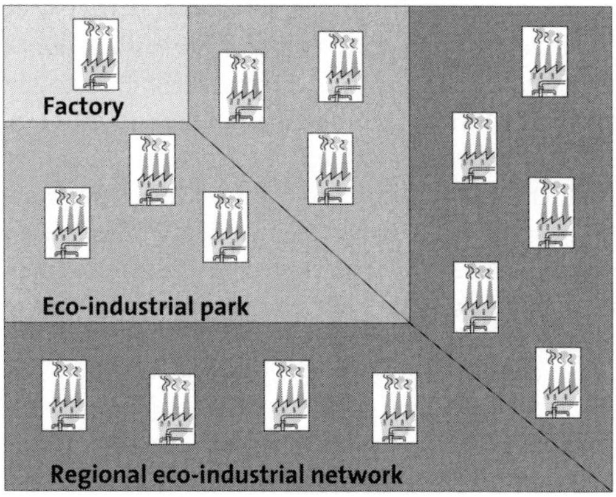

Figure 1.2 **The multi-scale approach to eco-industrial development**

seen as a jigsaw puzzle for misplaced resources. Success was possible but not probable, especially with small and medium-sized enterprises.

Instead, we need to understand that eco-industrial connections occur all of the time in all kinds and sizes of businesses and communities. This occurs in advanced industrial companies such as Hewlett Packard, which concentrates on only its top several competences and develops dynamic partnerships in areas in which it can neither excel nor wishes to excel. It also occurs at the traditional village commerce level in many developing countries where webs of relationships broaden capacity, redirect waste towards reuse and manage inventory, among other functions. But is this activity conscious and systematic? Do those involved know how to adjust to new markets, technologies and materials? Does it add maximum value to shareholders, stakeholders and the environment? I suspect not.

In eco-industrial development, the issues of scale are central. Holonic solutions,[3] which operate simultaneously at various scales, from the product to the workplace to the company to the region, provide different dimensions and call for different strategies. A framework of eco-industrialism helps us to systematically ask the questions of scale and strategy. We can make things better within a specific scale without affecting the larger framework. These are good acts, but the lack of impact on the larger scale tells us that good work does not necessarily sufficiency make. The stakes are too high and the opportunities too great to be piddled away. Scale is also relevant in understanding the design and architectural implications of eco-industrial applications. A design sense of scale will take expression in technical and architectural breadth and reach. Some of the work of William McDonough demonstrates the power and aesthetic of looking at the molecular level of benign materials to the design of buildings that follows the swallows of shape and function (McDonough and Braungart 1998).

3   Holon as a term was introduced by Arthur Koestler (1967) and expanded on by John Matthews (1995) to describe the way that whole systems are reflected and refracted at different levels of scale.

There are a large number of acronyms and programmes for sustainability and environmental improvement. Many, if not all, can be included in particular eco-industrial applications. This is an embracing concept that seeks systematic solutions to economy and community. Environmental management systems such as the Eco-management and Audit Scheme (EMAS), Green Productivity and the ISO 14000 series can all be incorporated at various system levels with eco-industrial development. It can subsume design for environment, by-product synergy, pollution prevention, cleaner production, green accounting and a host of other techniques. The subject of this book is far from the only answer to environmental challenges, but it represents an intellectually defensible framework from which profitable, equitable and healthy actions can emerge.

1.5 The 'eco' in eco-industrial

The use of the term 'eco-industrial' has been a matter of some discussion. Some who have no problem with the goals of eco-industrial applications avoid using the term because it ostensibly sends a message that raises the neck hairs of some business people who see environmental concern as only more cost and regulatory interference. 'Eco' echoes the twin mantras of eco-industrialism—simultaneous concern for economic success and commitment to environmental excellence. Let me first take on the issues of economic success. Eco-industrialism is a voluntary, market-driven approach that uses the discipline of internal and external markets to assure price, performance and quality. Eco-industrialism supports the end results of profit enhancement and frugal use of resources, but it asks us to rethink our relationships, the effect of our products on ecosystems and the impact of the processes of production on employees and affected communities. The elegant solution, the one we may need to dig deeper to find, is one that accomplishes both business and environmental improvement (Fig. 1.3).

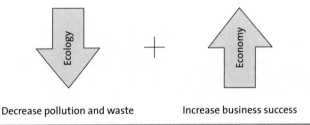

Figure 1.3 **Principles of eco-industrial development**

The goal of eco-industrial development is not to do the same with less. Its charge is to do far more with far less. The economic measure is return on assets (ROA). I have long suggested an internal benchmark of at least 30% higher ROA compared with performance of comparable traditionally organised companies. Current practice has not yet evolved far enough to be able to test this figure, but I suspect the number is low. This is dependent on a few variables. First, cash liquidity must be preserved based on shared investments in infrastructure, technologies and services. Although control over core technologies may be necessary, the uptime of equipment provides a significant return. Second, strong environmental management reduces liability, lowers energy costs while it increases reliability, lowers disposal costs and seeks to obtain value (or cash) from recovered materials. Third, eco-industrial companies are looking for a broad range of opportunities to partner: from transportation, to trade shows, to improving the value of employee benefits. Fourth, they are looking for revenue enhancement either by partnering to bid on certain jobs or by creating new products or services that serve definable unmet needs in the marketplace and reach companies with environmental supplier requirements. Hence, income goes up while costs go down. The flexibility of cash management provides a means to weather more difficult times and to take advantage of opportunities.

I promote the ROA focus not only as an industry advocate but also as a committed environmentalist. Wasteful use of resources should not be rewarded in the marketplace. The stock valuations of companies that aggressively look for new more resource-efficient strategies and practices ought to be reflected in their share price. This has already been shown in several models of 'green investing' (Foecke 2000). As green investing focuses less on not investing in the 'bad' companies with violations and more on supporting those that exhibit good practices, the more the differences will become apparent from complacent and wasteful companies.

I make no excuses for environmental companies that manage themselves so poorly that they cry into the kerchief of their environmental purpose. They should manage in a way that uses their assets wisely, meets customer needs and avoids overengineering of the product. Or they should go out of business and see if someone else can do it better. Propping up poor practices only enables more waste and hurts the reputation of the industry.

The 'green' consumer has represented about 3–4% of the market over the past 20–30 years, with little variation. Joel Makower (1993), editor of *The Green Business Letter*, has pointed out that surveys have shown that, given similar price and quality, 70–80% of consumers will choose the product with the environmental attribute. For eco-industrial companies the goal should be to significantly beat their competitors on cost and quality. For there to be a significant impact on the economy, green approaches have to come out of the shadow of micro-niches and take on or influence the behaviour of the larger market in their product or service segments. Today's approach of overpriced 'green products' and the appearance of poor performance makes any significant impact whimsical. Eco-industrialism can provide them with the difference to play in the big league. As we demonstrate the superior value of eco-industrial management, it will easily attract capital.

Strong management does not mean autocratic styles. Team-based management reflects the values of eco-industrial use, where skill flexibility, interdependence and full asset utilisation are valued. One can introduce new ways of working that reflect ecological principles rather than outmoded, mechanistic, heavy-handed management practices (Cohen-Rosenthal 1998c). Eco-industrial companies seek to increase their productivity through better indoor air quality and well-lit and attractive workplaces and motivate employees with fair family-wage jobs with benefits.

Let's turn to the 'eco'-logical face of eco-industrial. Eco-industrial networks have before them a range of environmental measurement, control and abatement strategies for transport of goods, production and maintenance processes. This is a baseline requirement for participation in these networks. This is not to balance the tough business approach described above but to complement it. Excellent managers know that unnecessary processes and waste need to be rooted out. Environmental management systems that focus on process *and* outcomes are integral to eco-industrial applications. By working at a system level, more accurate information is provided about environmental impacts across various environmental media. Monitoring can be more complete yet less expensive. Mutual assistance in solving environmental problems leads to better results and lower ecosystem impact. I am very much in favour of umbrella permitting systems that look at the cumulative impact on those in a neighbourhood (or affected by the broadcasts of smokestacks), where the real customer for pollution reduction is found. These can also be made more results-oriented than paperwork-focused, more concerned with the impact on seniors and children in a community than with sampling techniques, less expensive and time-consuming in terms of data collection and more attentive to finding solutions that work.

If the extent of the goal is to do less harm, it doesn't make a compelling case for the adoption of these techniques. From an environmental perspective, effort to restore habitat and clean up polluted land or waterways is high on the agenda. Green space taken out for construction of eco-industrial facilities should replace at least what was ripped up. Good land-use planning would use higher-density developments to minimise impact as well and would add more green space to compensate for the environmental load of operations as well as serving as an amenity for employees, the community and the birds and other wildlife.

There is also a different sense to the word *ecological* than environmental care. It has to do with the system level of analysis and action. Eco-industrial is not a strategy you do by yourself. It takes at least several co-operating parties. The by-product synergy process used by the World Business Council for Sustainable Development Gulf of Mexico looks to find that synergy in dyadic relationships between two companies. The Kalundborg industrial symbiosis is in reality a set of agreements between small sets of individual companies. In our construction of eco-industrial development, two parties working together around a few select issues is a good but suboptimised system. Seeking out the overlapping ringlets of subsystems or parallel functions in other organisations provides the venue for eco-industrial co-operation. This look at multiple organisations, multiple stakeholders and multiple functions shines a light on possible mutually beneficial action. One school of thought is the Charlie Atlas brigade, where each individual business or component demonstrates muscular excellence. The other takes a more communitarian route, where a common group and mutual support are sought. Some businesspeople prefer the sense of control they feel by flying solo. Then again, this is the least safe form of aviation! *Bonne chance!*

I prefer an ecological model that reflects natural systems. Although I recognise the importance of excellent engineering as part of the implementation of a process, I much prefer a more organic approach. In these systems, all duplication is not banished. Indeed, flexibility and multifunctionality are reflected in redundant systems that protect against glitches and simultaneously offer more opportunity. The equation of environmental awareness or organisational effectiveness with lean asceticism misses the mark widely. Natural systems are bountiful—so must be eco-industrial applications. Systems learn; they adjust to changing conditions. Rigidly welded eco-industrial proposals tumble apart when the system preconditions fall away or when the tolerances of each component are exceeded. A process-focused eco-industrial setting learns to address challenges along the way, growing and changing as they go.

1.6 Renewed urbanism and rural regrowth

Eco-industrialism looks to urban areas where in years past the kind of interconnection, live work and density of commerce was characteristic. It seeks to rebuild such communities not just with the 'backsplash' of new immigrant communities that have reclaimed abandoned areas. It serves as a magnet to bring back to the city working-class families of all backgrounds burdened with two car payments and the costs of moving out to the fringes of a city to be able to afford housing and safe schools. The key to reclaiming these areas is not 'yuppification' of old 'brown' houses but the creation of good jobs where now there are abandoned eyesores, and reclaiming affordable and desirable housing. Locating good jobs and industries near current neighbourhoods with full neighbourhood support and participation can make long bus rides history while building real opportunity and wealth where people are now.

The city represents the nexus of markets, transport, raw materials, capital and skilled labour (Jacobs 1984). A city reborn revels in that advantage. That advantage works best as a package where all are connected, seeking to highlight the particular character and competitive competences of each city. Cities provide a unique scale of market and materials. There was a golden rule of industrial location made popular by the consulting company McKinsey and Company that it is best to locate closer to your market or your raw materials, based on what is heavier. With imagination, entrepreneurs in cities can mine the material resources lying fallow in landfills and abandoned buildings and properties, service particularly significant local markets, take advantage of transportation and infrastructure investments and draw on a variety of education and research institutions to attract a skilled workforce. The availability of these assets, where the overhead for maintaining most of these is a shared cost, demonstrates one of the key eco-industrial operational principles: keep fixed overheads low while expanding the range of options.

Urban brownfields are not the only venues for eco-industrial approaches. Rural communities 'back on their haunches' also can serve as potential applications for eco-industrial approaches. Too much farmland is being eaten up by suburban sprawl; urban applications of eco-industrialism would take some of the pressure off this unfortunate land grab. But there are ways to help young people find good jobs in their rural hometowns or regions by rethinking the use of agricultural products and by-products. Research and investment can help bioproducts that are grown or are tossed away be turned into valuable raw materials used for value-added manufacturing or processing industries. For example, there is a way to use the whey from cheese processing to create various plastics. Society is moving from the processing of petrochemicals, a by-product of detritus from millions of years ago, to use of bioproducts that helps lower the level of global warming while increasing the level of global economic opportunity. These come not just from land-based farming but also from greater use of the wonders to be found in marine environments. The rural ethics of respect for the land and frugality dovetail with eco-industrial principles. Communal barn raising on the prairie and dale provides a model for co-operation in a community that can create new economic opportunity.

1.7 Eco-industrialism: an alternative to environmental elitism

Eco-industrialism strikes at the heart of environmental elitism and architectural pretension. Rather than focusing solely on pristine neighbourhoods within new urbanist walking distance to a Starbucks or trails for off-road vehicles, eco-industrial development seeks the full range of jobs. It does not reject the making of things but asks us to make the things we use with more care and more resource efficiency, thus making them better products. It asserts that all citizens, all workers, deserve workplaces and living spaces that are healthy.

The making of artefacts is not an ancient artefact itself nor an activity to be relegated to the lowest-paid and worst working conditions in exploitable parts of the globe. This is a dangerous mythology, especially in the United States. When Sweden and Germany, both with higher labour and social costs, can remain worldwide leaders in manufacturing, then it says more about the paucity of the US effort to provide good jobs than the fact that manufacturing is passé. Someone makes the mugs for the coffee salons, the chairs people sit in on the patio, the bricks that provide the veneer of the building, the cleaning fluid for washing the dishes, the flour that goes into the croissants and so on. There are metals and plastics fashioned to use in the computers that serve the Internet economy. Yet, if one listens to the insistent drone of academic and government pundits, these jobs have disappeared for good and the future is all about e-trade and virtual everything. Internet applications provide amazing opportunities as part of the new economy, but they represent only one aspect of the economy and are often linked to the purchase of goods, transport of these goods and can't escape consideration of end use. Behind all of these activities are valued jobs—jobs for a broader range of people than those touted in cyber-industries. All sectors require or produce products, which continue to draw on the material and energy storehouse of our planet.

When industry is viewed as poison, then it is segregated, if possible, outside the walls of the city or is cordoned off in the city with buffer zones. The segregation of industrial parks into areas that no one else wanted for residential or retail use assumes that the interaction of industrial and commercial workplaces with human habitation is to be avoided. Industrial estates become 'leper colonies' as part of the development process. A look at the master plan for most towns has considerable detail on where public facilities will be and where houses will be located. It rarely has much more detail other than hatch marks on a planning map for the industrial portion.

A victim of afterthought, industrial developers create a self-fulfilling prophecy because industry sees no other alternatives available. Now there is. Eco-industrialism challenges the outcast view of industrial development. Rather than accept the assumption that production facilities need to be shunned, eco-industrialists would prefer to see them as safe workplaces for community and employee health: old industries recreated in much better editions. These will be new industries so attentive to community and employee health that workers can proudly live near where they earn their livelihoods. This kind of development can be more profitable for developers than putting up a sign on willingness to build to suit, more convincing to new businesses since it is based on maximising business opportunity and more acceptable to communities as new development reflects community values. I hope to see 'green' housing being located within easy reach of 'green' industry where internal process, exterior architecture and land-use planning promotes welcoming environments for nature, neighbours and employees alike. Planned communities need not be offshore islands where the morning traffic jam is its residents driving off to work. They need to be affable and affordable to working people who deserve as much as anyone else the benefits of New Urbanism properly conceived.

All of this is possible now with today's technology and processes. Individual companies have demonstrated that being environmentally proactive pays off at the bottom line and with positive public attention. But for most small and medium-sized industry, environmental management and industrial ecology are foreign concepts where the arteries leading out of the city are home to industrial shantytowns. Locked in isolation by public policy and narrow management thinking, they often don't understand that it is the absence of systemic thinking and common action that leaves them as dishevelled as the fences around the property. When they band together for mutual gain, then they can realise the benefits of industrial networks. Corporate neighbours move from being eyesores to offering helping hands.

1.8 Eco-industrial equity

Eco-industrialism is also about equity. Eco-industrial development is an environmental justice strategy where placards are replaced by jobs, and toxics by transparent concern for worker and community health. In community after community documented in this book, local populations have made the decision that a trade-off between toxic workplaces and jobs need not be made. In the United States, most communities experimenting with this approach have large percentages of people of colour; they have demanded that the buck stops here when it comes to environmental racism. Communities are defining what they want—and are getting it. The same is true for working-class communities where people are recognising that many of their jobs are in unsustainable industries or factories and that if they aren't proactive they will be tossed out on the rubbish heap as well, leaving their families and communities to pick up the pieces. It is the goal of eco-industrialism to provide for family-wage sustainable jobs in companies that can profitably compete in the marketplace over time by constantly looking for new alliances and adjusting to new supply and market conditions.

Local equity can be affected by adopting an eco-industrial approach, but the approach has global implications as well. Unless whole systems of production make quantum-level jumps in resource productivity and functionality, then as the population rises so does the danger to us all. Current resource-intensive approaches to agriculture, so-called durable goods, energy production, building construction and so on paradoxically mean increasing hunger, homelessness, pollution and disease.

Proclamations of the end of work are wrong. The aphorism that there isn't enough work to go around ignores the billions of inhabitants on this planet that need transportation, communication, education, healthcare, housing, etc. Eco-industrialism is about how to produce those goods and deliver those services in ways that shore up communities and universalise opportunity.

My larger argument is that we must take on the primary way in which we allocate resources—price. Price is inherently about scarcity—raise prices until the numbers who can afford a product approximate supply. In a world where there is greater demand on the Earth's resources than can be supplied, profligacy increases price and concentrates power in ways that increasingly disenfranchise the poor. Use of price to allocate resources means that most of the world's population is locked out. The World Resources Institute estimates it would take four Earths to be able to make widely available a moderate income to all the current inhabitants on the planet, based on current levels of resources needed per unit of output.[4] Providing basic goods on a global scale is impossible with our present wasteful processes. Eco-industrial development draws people together in tighter, more efficient and more effective ways. Efforts such as Factor 4 or Factor 10 (von Weizsäcker *et al.* 1997) have demonstrated that it is technically feasible to achieve quantum levels of improvement. I am less impressed with the technical pyrotechnics of Lovins and von Weizsäcker than I am inspired by the goal as a means to the end of a more just and livable world.

I have always seen eco-industrialism as far more than a nifty eco-efficiency trick; I see it as an avenue to take to the next system level a shift in production systems that extract far more value from resources and create opportunities to lower cost so that basic goods become more affordable to a much much wider swathe of the population. It is not the only solution to broad-scale poverty but it is an important one. It allows us to draw on cultural practices of commercial co-operation from diverse farms, villages, towns and cities around the world and adapt them to new technologies and access to markets. It allows us to create a new measure—the resource intensity per unit of basic need—and to continuously strive to drive down what it takes to provide basic human rights as articulated by the United Nations. It is in these considerations of equity that eco-industrial strategies are revealed as key to any hope of sustainable development on a global scale. The growth of industry in developing countries has a choice to make: adopt old pathways that require continued transfer of wealth to the North and indenture to the oil economy, or realise local sources of energy and materials that employ a growing and impatient population. One leads to disaster and the other to proud prosperity, where meeting basic needs is a baseline from which quality of life emerges that celebrates human community and ingenuity.

1.9 Virtuous cycles

It is all about relationships. By seeking out rewarding interconnections, eco-industry pulls together three virtuous cycles connected to sustainability's triad. At the economic

4 Go to www.wri.org.

level, it circulates money within the local economy as much as possible by creating partnerships and networks locally. As such, it creates real wealth within local communities. Any community developer knows that the more times money circulates in the local community the better effect it has on the local economy. It may eventually wend its way elsewhere, but the local connections need to be maximised.

At the materials and energy levels, the environment is better off and the economy strengthened the more times that any given unit of material or energy is collected and recirculated in a local economy—there is less damage, dissipation or dumping the more this occurs. Imported materials that are buried or burned after single use close off the opportunity to reduce costly extraction, processing and imports later on. The more they cycle through various users and uses the better it is.

Yet, at its heart, eco-industrialism is about the recreation of community: first and foremost among businesses and those who work within them. Second, there is a rapprochement with communities that have spurned their industries and have cast off appreciating the power and magic of making new things in service to our larger society. In days gone by, chamber of commerce meetings were places of information exchange, bartering and bantering. Today, we live in a world where one neighbour doesn't know his or her neighbour in the next house or apartment, nor does one company know what the next does inside an industrial park. It is in this connection—company to company, company to community, person to person—that eco-industrial development's invisible magic works its wonders.

The eco-industrial agenda is broad, bold and sweeping. It encourages innovation and new relationships. Yet unless these take specific shape then this rhetoric and other hopes for a vibrant and sustainable future evaporate. Can we cross the boundaries that need to be crossed? I don't know. Can we be as innovative as we are challenged to be? I don't know. Will this spread wide enough and fast enough to make a real difference? Again, I don't know. But what I do know is that the pioneers highlighted in this book represent seeds for a century to come blooming with opportunity. And, reaching out, together we will try to make the difference we know we can make—with imagination as far and as fast as we can.

2
MAKING SENSE OF INDUSTRIAL ECOLOGY
A framework for analysis and action

Edward Cohen-Rosenthal
Work and Environment Initiative, Cornell University, USA

The end concern of industrial ecology is actually fairly easy to state. The goal, at the minimum, is to generate the least damage in industrial and ecological systems through the maximum circulation of materials and energy. Highest value use with the least dissipation of resources forms the core of systematic application of industrial ecology.

In this chapter I set out a hierarchy of analysis to frame future research activities on the technological side and on the economic development aspects of the equation. The core question that prompts this chapter is whether there is any change from energy or material abuse to re-use. In current industrial ecology literature the answer is at best unclear, if the question is asked at all. For many, finding any connection is wondrous proof of an industrial ecology. I set out two system conditions as guideposts

- The entropic effect of the transition is less than other possible choices.
- The next iteration of energy and materials can be transformed yet again into new and useful associations and cycles.

The simple waste-to-input exchange proposed in classical industrial ecology illustrates a possibility but not a probability. At the macro level, identifying that there is 'waste' of a considerable amount of materials in a process or larger system only flags its possible re-use. To come close to making the environmental and economic impact that industrial ecology implies, one cannot ignore the means for maximum resource and energy looping or re-use.

There must be a theory behind the desired flow or connections that extends beyond serendipity and information publication. That theory crosses many disciplinary boundaries. Built on the theory, strategies and technologies are required to turn industrial ecology from concept into reality. These include identifying ways to increase the value of the materials and energy recovered and a means for assuring transport, quality compatibility, sufficient volume and timeliness to make it a practical alternative. It is only when we assume that industrial ecology is the norm that choices become necessary.

2.1 Simply working with complexity

All things that we work with come bundled in a degree of complexity. This occurs at the physical, mechanical, biological and chemical levels. Large buildings are a complex interconnection of various building systems. Most machines or objects used by humans are an array of materials and parts. Biological systems are a complex arrangement of cells and subsystems. Managed materials are usually used in some combination to produce a desired result. We use various recipes of materials and energy to 'bake' the various things we use and consume. After many years I finally figured out what chemistry is all about: it's about bonds made and bonds broken. The rest is elaboration. The forms are infinitely complex; the principle is simple.

The admission of complexity and constituent parts does not necessarily lead to reductionism or an erector set approach to complex structures. Each combination and each variation results in uniqueness, with special properties and possibilities. The ability to break things apart, to separate segments or to create new combinations results in new properties. Each new set of relations offers renewed possibility. New configurations establish the field for new potential.[1]

Ivan Amato, in a wonderfully written book aptly called *Stuff*, notes:

> Just as a sociologist seeks to understand the dynamics of human interaction in lesser or greater collectivities under a variety of conditions, so the chemist tries to grasp the ways in which the elements of matter interact—aggregating, segregating, rearranging, mingling and repelling to emerge as multi-tiered structures, each with its own set of material traits. The hierarchical structure of a material is the result of the interaction of its ingredients. That is why a little more or less heat, a different proportion of alloying elements, a finer grade of a pulverised ingredient, this or that contaminate and any number of variations on a standard process can result in what seems like an entirely different material (Amato 1997).

How to work with 'stuff' forms a central focus of this chapter. How do we create it? How do we use it? How do we lose it? How do we re-use it? The distinction of products and by-products is often a transitory snapshot of a particular use. A more dynamic and systemic industrial ecology approach recognises that materials and energy flow is about *all* flows and not just those deemed desirable or undesirable within a particular process.

---

1 This argument does not exclude the creation of objects, tools and concepts that endure over large periods of time. At the extreme, we know that our Sun will eventually burn itself out like others in the universe. This ultimate fact needn't dissuade us from our acting over the next several billion years!

2.2 Entropy: creative dynamism

Entropy stands at the centre of the analysis and the solution. To some, entropy seems like an obtuse and academic entry. Our goal in industrial ecology is to assure the conversion of a product or material to another use when its initial use is completed, as a whole, as parts, as a material input or as an energy catalyst. On some occasions, it may serve multiple functions.

Entropy is about dispersion of mass and energy in the system, the homogenisation of different concentrations. Heat disperses, gases move to areas of lower intensity. Eventually, equilibrium is approached. The dynamics of the natural world occur because our world is far less uniformly diffuse. Ore for metals occurs in greater concentrations in some places than in others. Differential temperatures and topography contribute to wind movements. Water, especially potable water, is not equally available. In economic terms, wealth occurs at various levels of concentration. Population and industry tend to occur in particular areas. These patterns create a vibrant world.

When molecules disperse into the atmosphere, as with greenhouse gases, it is difficult to 'put the genie back in the bottle'. When materials are trucked away and tossed in a tumble at a landfill, it is harder to recover them for re-use. Although we can use some of the energy by burning wood or other biomass, pasting the materials burned back together is a real challenge. If one wants to create new chairs and the various parts are dispersed across a city, it takes lots of energy and will cost more to assemble them again than if the parts were closer together. If people are dispersed geographically, as in an urban sprawl, more energy and materials are required to link them together and meet their needs. The impacts of entropy are very practical.

Often, entropy is confusingly portrayed as the breakdown from order to chaos, as if entropy represents an ultimate anarchy. Gunter Pauli, promoter of zero emissions, asserts that entropy 'prescribes that all on earth will move from a state of order to a state of disorder, confusion and disorganisation . . . the law of degeneration, the evolution towards ever-more inefficient systems' (Pauli 1998: 32-33). He goes on to say that 'The law of entropy should be replaced by the law of regeneration. The present law does not make sense for the world we need to create' (Pauli 1998: 33). Pauli and others cannot wish away entropy because it is inconvenient for them.

Instead, the law of entropy answers our questions about diversity and change—not just on the Earth but universally. In the process of diffusion, new alliances are made. Many are tested, and some work. The second law of thermodynamics would have it no other way. It puts *dynamics* into thermodynamics, since perfect order and absolute chaos are abstract concepts, not realities.

Noted physicist Sir Arthur Eddington writes, 'Entropy is only found when the parts are viewed in association, and it is by viewing or hearing the parts in association that beauty and melody are discerned. All three [aesthetics, musicality and entropy] are features of arrangement' (Eddington 1958). All ecologies are reflections of relationships. Attempts to construct industrial ecologies from static descriptions of what is or ought to be connected are shredded by entropy's force. Entropy isn't the problem; it is a vital driver for solutions.

2.3 Living entropy

There is a curious relation between entropy and 'living' or biological systems. Ayres and Simonis (1994) note that 'A living organism, by virtue of its metabolism, can be regarded as an entropy generator.' As such, evolutionary biology is about the creation of systems to draw on entropy that display new functions through species adaptation and variation. These forces are not sucking the life out of existing systems but are challenging them to be creative in the future. The biosphere is a repository of time-bounded manifestations of an ever-changing materials and energy context.

Today's predominant practices of bury or burn lead to two kinds of distortion. At one level, collection of materials in one area too often raises levels of concentration of heavy metals and dangerous chemicals to a toxic level that would not pose such a threat if they were used in a different way or safely disbursed. Second, this practice races recklessly along the value chain from beginning to end, without exploring the possibilities that intermediate or transformative strategies might present for cycling materials or energy into productive and responsible re-use. Any extraction of a resource leads to some of it being used for its intended purpose, but often a large amount is lost through dissipation or waste. Redirecting this residual reduces the entropic outcome of the initial process, although never entirely.

The reason to extract far higher value and to conserve more falls along two lines of analysis. First, finite resources will last longer (i.e. be more sustainable). Avoidable waste makes no sense—on any timescale. As such, options for creating useful materials increase as a result of the greater availability. It is also true that excessive dissipation of a particular resource can have secondary negative impacts by generating later-order complications (e.g. the downstream effect of dioxins), creating dangerous levels of concentration and releasing energy at a disruptive degree.

Pollution can be defined as reaching an ending point where materials or energy are not re-used in productive ways or where the last configuration blocks positive alternatives for reconfiguration. The only possibilities available are dangerous to the system in which they are embedded.

The second reason is tied to equity. When prices and access to goods incorporate high dissipation costs, then needed goods and services are denied to those in developing countries or who have lower incomes. Classical economics confronts the issues of scarcity but denies scarcity based on monopoly or state power. Scarcity enforced by thoughtless or wanton waste is also a distortion of economic distribution. Theoretically, competing businesses will try to outdo their competitors on this dimension; in practice, they rarely do and then often only by tinkering at the margins. Getting away with pricing that makes charging acceptable for the costs of waste in production and the externalisation of unused resources and energy leads to higher levels of pollution. The poor can't escape air pollution from cars they don't own, nor buy bottled water to avoid contaminated water sources.

2.4 Confronting waste

Put very simply, the goal is to reintroduce materials and energy back into productive re-use with the minimum energy required and the least waste of material in the process. Why? Because we waste far too much. Von Weizsäcker et al. (1997) point out:

> Actually we are more than ten times better at wasting resources than at using them. A study for the US National Academy of Engineering found that 93% of the materials we buy and 'consume' never end up in saleable products at all.

Moreover, 80% of products are discarded after single use, and much of the rest is not as durable as it should be. Business reformer Paul Hawken estimates that 'over 90% of the original materials used in the production of, or contained within, goods made in the United States become waste within six weeks of sale' (quoted in von Weizsäcker et al. 1997). Hawken also observes

> we are far better at making waste than at making products. For every 100 pounds of product we manufacture in the United States, we create at least 3,200 pounds of waste. In a decade, we transform 500 trillion pounds of molecules into non-productive solids, liquids and gases.[2]

Energy is not any better. Von Weizsäcker et al. (1997) provide the following examples of massive energy haemorrhaging:

> the heat that leaks through attics of poorly insulated homes, the energy from a nuclear or coal-fired power station, only 3% of which is converted into light in an incandescent lamp (70% of the original fuel energy is wasted before it gets to the lamp, which in turn converts only 10% of the electricity into light) [and the] 80–85% of a car's petrol that is wasted in the engine and drive train before it gets to the wheels.

Only one BTU (British thermal unit) in twelve of world energy production is used to heat and cool the US building stock (Wellesley-Miller 1977).

At one level, this amount of room for improvement should be a golden opportunity for entrepreneurs. Ironically, even gold comes at a steep environmental price. Worldwatch notes that the overburden of two gold wedding rings is 'over six tons of waste at a mining site in Nevada or Kyrgyzstan' (Gardner and Sampat 1998). The major product of most human activity is prodigious amounts of 'waste'. If we consider the apple the product, then the same can be said about an apple tree. But an apple tree has other 'products' as well—CO_2 converted to oxygen, erosion prevention, provision of shade, home to insects and birds, and nutrients to the soil from its humus, to name a few. In natural systems, this happens naturally; in human systems we need to apply human intelligence to seek beneficial products from by-products. At times we do, or it happens out of good fortune; we need to make it systematic, and the rule rather than the exception.

2   See The Natural Step web page, at www.naturalstep.org.

2.5 Make-or-break combinations

All complex systems, when there is a certain level of energy applied to them, can be broken down into lower levels of system complexity—even if the initial outcome appears chaotic. Various levels of materials can then be recombined into new objects that serve new functions. It takes a specific amount of energy to create things and a specific amount to break them apart. This can be seen at the gross level in using a wrench to disassemble a bicycle. It occurs in the production and moulding of steel from iron ore and other ingredients. It is seen when complex hydrocarbon molecules are cracked in chemical production processes. It occurs in particle accelerators where atomic structure is broken down to the constituent parts. It occurs when we incinerate fossil fuels and convert them to energy. The question, then, is not whether we can alter the structure of the materials we use but to what purposes, with what technologies (broadly construed) and at what costs? This is a formidable design challenge.

Developing a structure for considering material and energy ecology is essential to prevent several prominent traps in environmental thinking. First is the mythology that a magic bullet can solve all problems. The receiving end of a system requires a diverse set of inputs with diverse characteristics; a boundary has multiple interfaces and multiple ways of entry (some easier than others). As such, a singular approach would be counterproductive. It would make the system more brittle and would raise the risk of an inflexible response to the impact of a particular material. Second, a reasonable framework helps the larger society explore value-added possibilities before lower levels of deconstruction of the materials makes it impossible or too expensive to recreate what could have been used. Third, an inchoate policy that jumps from one approach to another has a broader environmental impact by dissipating material and energy resources at a far greater rate than necessary, with potentially disastrous consequences.

2.6 What are the options?

The answer to waste and dissipation seems simple: use less. If we produce less widgets, then it usually takes less stuff to make those widgets. Asking what do we really need to produce and the degree by which it enhances the planet and its people is a reasonable starting point. Others are at work on the strategies for dematerialisation (Wernick et al. 1996): how to use much less material input to produce a unit of functionality. Notice I said 'functionality', not simply 'a unit'. For example, 'the minimum scale of electronic devices has decreased by a factor of 10^4 (to 0.5 microns) while the scale of machines has fallen by a factor of 100' (Rohatgi et al. 1998). Computer capabilities are witness to the fact that increased power can come in smaller sizes and weights.

The reduction of material input required to form a product or to perform a service reduces the rate of draw on the larger materials system. For example, if, instead of counting the number of square feet of space constructed, we measured the amount of time that space is used productively, then it would reduce the building of unnecessary spaces and spur innovative ways of combining or co-ordinating functions. Material intensity is one measure of dematerialisation, but the impact is just as dramatic on energy use where lighter, more temperature-adaptable materials require less energy to operate. The embedded energy saved in the fabrication of excess material also has a strong energy impact. Given the contribution of construction materials (Fig. 2.1) to materials flow in the larger economy, this could have a major impact.

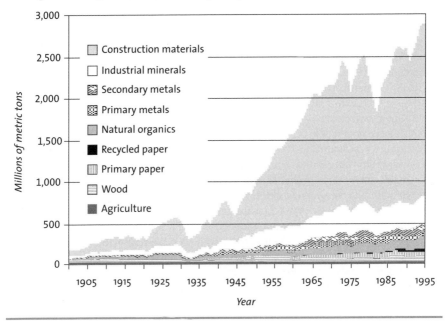

Figure 2.1 **US consumption of raw materials**

Hence, before we talk about materials flow there are two primary questions to ask: What are we using the materials for? How can we reduce the amount of energy and materials needed to obtain the desired result in all phases of the materials cycle from extraction, to refining, to transportation, to fabrication and then into the chain of product use and re-use? We need to examine whether there are significantly less-resource-consumptive alternatives for the service or function desired.

Attractive as this alternative is to deep ecologists who seek a far shallower imprint in the planet or to engineers seeking product efficiencies, there will be a continuing demand for the use of materials for products for human use. Energy demand will in all likelihood remain at a high level. Rising population alone will place pressure on the system, and the series of demands, many legitimate, for greater quality of life moves us beyond asceticism and extreme eco-efficiency. Hence, the hierarchy shown in Table 2.1 assumes that we will continue to fashion tools and products from the material world.

Each of the approaches described in this chapter provides economic opportunities. Depending on local scale, having the full range of approaches makes sense in an intensively populated area or region. The technologies associated with these various approaches provide market niches for profitable companies

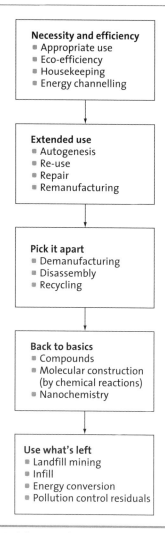

Figure 2.2 *Hierarchy of material use and re-use*

We can, with foresight and creativity, harness in positive and productive ways the cycles of construction and deconstruction of materials while linking to the flows of energy that ebb from one area to another. We can look at the broad approach that raises primary questions on what we are doing, seek design solutions that elegantly connect and contain resource use and deploy better housekeeping to plug leaks and tie up the loose ends that plague moving from abstract design to dissipative reality. Painting such a picture sets forth a new and vibrant vision of sustainability.

2.7 A heirarchy of resource strategies

Briefly described below is a range of approaches to meet the criteria of seeking decreased entropy through higher value and lower input of materials and energy in human activity and our associated environment. In most cases the order they appear represents an estimate of the degree of entropy involved—but that may not be true in all cases. Further, the criteria of highest use in particular situations may leapfrog one approach over another. In some cases there simply isn't the technical capability at the present moment to use this strategy—but that should be taken as a challenge rather than an absolute barrier. Accomplishing these goals draws on a broad range of techniques and processes. For different audiences this may entail ways of renewing products, creating new products, identifying components, introducing primary materials, generating energy or providing services.

The core dimension of value means that the entropic state such as solid, liquid or gas of particular materials is directly related to the potential use and the chemical or material conditions that they will be used under as well as the cost and safety of the process (including transportation and storage).

2.7.1 Energy channelling

The release of energy caused by changes in chemical state or composition has an impact on the larger environment. It is most often manifest in thermal energy but can be found in other forms as well. In metal refining and fabrication, large amounts of ambient energy are released as part of the process. This energy can be dismissed as dissipating into the larger atmosphere and of little consequence. Yet such emanations, whether in aquatic systems or in the atmosphere, can affect other ecosystems in those environments, including fish and birding patterns as well as microclimates that impact people and other organisms. This lost energy provides a valuable source of potentially capturable energy to help power other activities. So-called lost energy can be channelled back for commercial or domestic purposes. In Europe, the fuller use of district energy increases the efficiency of energy generation, with great impact on greenhouse gases and overall energy efficiency. All the energy that 'escapes' makes us prisoner to using more and more feedstock to compensate for inefficiency. For example, the Cornell Work and Environment Initiative examined the possibility of a metals eco-industrial park where energy loss from an open hearth steel process is turned into energy gain through co-generation processes in addition to potential connections of metal slag and excess materials.

Energy cascading is a dynamic engine of the natural world (Sternlicht 1979). All energy systems used by humankind are about the capture, channelling and use of energy released. The goal should be to harness as much of this energy as possible to reduce the amount of material consumed to produce the energy we need. This aspect is one of the most overlooked areas for improvement in materials and energy use. Energy is about 'the ability or capacity to do work'.[3] Lost joules reduce our capacity to get valued work done efficiently and effectively.

---

3 See *Encyclopedia Britannica*, www.eb.com:180/cgi-bin/g?DocF=index/ch/emi/10.html.

2.7.2 Autogenesis

The development of 'smart materials' adaptive to their environment is an important aspect to explore. At some level this also provides a metaphor for adaptive industrial systems and social resilience. 'A smart structure is one that monitors itself and/or its environment in order to respond to changes in its condition' (Culshaw 1997). It may be self-repairing or may respond to external stimuli such as temperature, voltage, pressure, pH, light, new chemicals and magnetic field and adjust its composition and functioning. Smart materials can respond to external sensors that signal the need for change or the material itself can be responsive to various stimuli. All smart materials work within a range of adaptability and thus obviate the need for new materials to survive or function when within those bounds. Today, smart materials tend to be made out of virgin stock; as the applications spread, they will get even smarter and use unique properties of recycled composite materials.

Amato (1997) predicts that:

> Smart materials that can respond to external conditions by changing their colour, shape, stiffness or permeability to air or liquids could become the stuff of future cities whose buildings are more comfortable, and better able to field sudden violent challenges from earthquakes to terrorist bombs. Smart materials also could lead to cities whose infrastructure can sense—and even automatically compensate for—the wounds of corrosion, metal fatigue, age and other slings and arrows of urban decay.

Materials that can 'learn' from their environment by building in a wider range of responsiveness and responses serve to expand their functionality and durability. Today, pastiches of materials are difficult if near impossible to disentangle, thus making re-use very difficult. Smart materials address this issue in two ways. First, smart composites reduce the overall need for materials by providing superior performance standards. Second, smart materials, through various labelling, sensing and keys for unlocking bonds, provide a greater ability to intelligently re-use the materials from which the product was created.

2.7.3 Design for durability

At some level, design for durability seems like a foolish formula for companies to follow. Vance Packard many years ago in *The Wastemakers* (1960) described the throwaway society where obsolescence is a planned strategy. In a finite market structure, this destructive process may make some kind of sense, but it loses its meaning in a global market where boundaries of delivery are disappearing and where customisation and innovation in value-added characteristics provide broad market opportunities. Design for durability leads to customer loyalty, whereas engineered breakdowns lead to dissatisfaction and generalised distrust of the company's products.

From an environmental perspective, design for durability has considerable advantages. It takes less energy to produce, discard and transport than more traditional practices. There is a downside to this strategy. Overengineering of features can lead to bulky and short-lived uses. For basic needs, durability makes sense. For emerging needs, adaptability of the product provides durability and flexible functionality for future demands. Multifunctionality can be as much a durable design feature as tensile strength.

There is a real reason why many antiquities are still standing when one doubts whether most modern artefacts will stand the test of time. When planning is done for the seventh generation and not the next fiscal quarter, then care is taken in the creation of useful objects. In ancient times, objects fashioned from the world carried a certain magic. This magic was in their ability to transcend time and place: to transform and be transformed. As impressive as the Pyramids are, the tents of the Bedouin, where 'temporary' structures are designed for long-term use, are deceptively simple and admirable. Objects are created not to discard but with pious regard for approximating eternity.

2.7.4 *Repair, re-use and remanufacture*

Stuff breaks down or falls apart. Design for repair means that the whole item does not need to be thrown out. For example, bicycle repair shops that fix old bicycles provide a way to extend product life. Design for easy maintenance and repair is an important part of a materials strategy. Disposable societies that do not provide ways to revive product function feed landfills with their thrown-away units. There is a risk that the repair of an item such as an old refrigerator could be more harmful than its replacement if it required CFCs (chlorofluorocarbons) and used considerably more energy. Repair or durability of inherently energy-intensive or material-consuming artefacts may not be desirable, and other strategies may need to be deployed to re-use their materials. However, the ability to replace what is needed rather than discard a whole object is a strategy worth promoting.

At the end of the useful life for a particular application the first question becomes 'Can the product or its constituent parts be re-used in some way?' At the gross level, this is an extension of the Salvation Army practice of redistributing items, where stuff no longer of value to one person can be of use to another. This is one form of re-use. Garage sales are another common manifestation of re-use. Although one can question the 'junk' sold and bought in yard sales, the lower price increases accessibility and little energy is required to transport an object to a new place where it serves a function other than cluttering the garage.

At the industrial level, 'waste exchanges' offer equipment or materials no longer usable in one place to others to purchase or take off the hands of the initial owner. These have become quite sophisticated information exchanges which go beyond newspaper 'want ads'. Re-use can lower prices and lessen the environmental footprint. Thrown-away stuff can contain real treasures.

A second level of re-use comes from re-using the parts that make up a structure. For example, deconstruction—the systematic taking apart of buildings—which takes usable mantles, doors, windows, bathtubs, fixtures and so on is another form of re-use. Bricks, beams and aggregate can all be reclaimed. In Minneapolis in a re-used building in the Phillips neighbourhood, the Green Institute's ReUse Center provides a haven and a retail outlet for such finds. In deconstruction, when the fixture cannot be sold, the materials embedded in a structure such as the metals, wood and glass are harvested and sold as commodities—often for re-use.

Remanufacturing is a term increasingly in favour for a process belonging to a subset of re-use. Remanufacturing 'restores durable products to serve their original function by replacing worn or damaged parts' (US EPA 1997a). Xerox is the premier example, based on its commitment to design for remanufacture.[4] Copying machines are returned to the factory, refurbished as new (often including service enhancements) and then re-enter the market either back to the original customer or to some other customer. This results in less waste and eliminates the need to remanufacture a whole product, thus saving on materials and energy.

German studies estimate that worldwide remanufacturing results in savings of 10.7 million barrels of oil a year and 155,000 railroad cars of raw materials. In addition, there are market advantages for manufacturers who maintain connections to customers and can provide products at lower costs. Remanufacturing works with a number of products, including automotive parts and engines, furniture, pallets, personal computers, photographic equipment, refrigerators, toner cartridges, steam turbines, manufacturing machinery and retreaded tyres. As of 1997, the US Environmental Protection Agency (EPA) estimated that there are over 73,000 firms involved in remanufacturing, with revenue over US$53 billion. This is a larger industry than extractive mining. In 1996, the remanufacturing industry in the United States employed ten times as many workers than did the metal-mining sector and earned US$53 billion—a sum greater than the sales of the entire consumer durable goods industry (Gardner and Sampat 1998).

The concept of remanufacturing is not about the product being second hand or second rate but that it has a second life with full functionality and value for the end-user. Hence the quality of the product, including appropriate aesthetic repair, is part of an effective remanufacturing process. Remanufacturing is a process used in many instances to provide what is needed without starting from scratch.

2.7.5 *Disassembly, demanufacturing and recycling*

When objects and other assorted contraptions cannot otherwise be re-used there is a possibility for component re-use. Visiting a computer disassembly plant in central New York, I watched as factory-line disassemblers took apart old computers and separated the re-usable parts for return to productive functions. Dismantling taught not only ways to channel materials but revealed design lessons for manufacture. Design for disassembly to take advantage of products after they have been used requires forethought during the product design phase. Product parts can be designed for re-use. In production, it requires process and product simplification to ease disassembly. When the parts can be inserted or repaired in other machines, it is great. When that is not possible, disassembly parcels out components according to the kind of materials from which they are made. To ease the sorting process, parts should be made from the minimum required types of materials.

If there is one strategy that has caught the public eye it has been recycling. Curbside paper, glass, aluminum and plastic collection has become a regular part of many communities. The natural world recycles all the time through the carbon, nitrogen and hydrologic cycles and by taking other similar journeys. Human systems must consciously ask what constitutes the maximum possible safe and economical looping of material flow. The answer has to be broader than the paper, glass and metals that find their way to recycling bins.

4 www.xerox.com.

My supposition is that upstream and downstream connections are more likely when there is a common materials domain or chemical compatibility serving as the basis for looping materials back into productive uses. For example, synergy between metals companies may be more possible than synergies between metals companies and companies in other sectors. Because of the variations involved this is far from a guarantee, but industrial ecologies linked to material domains makes sense. Organic materials represent one flow, and other materials, such as metals, inert materials and water, represent other flows. Other families of materials can be imagined, from various polymers to ceramics.

Water may seem like a strange item to include in this list but the limited amount of usable water on the planet requires strong conservation and re-use strategies, especially in drier climates. Heavy users of water can link together to create a common stream of water use and re-use. This includes closed-loop water-cooling systems, which make use of evaporative cooling towers, air heater exchangers, reclaiming treated water and use of grey water (Ross and Sasser 1998).

Recycling can be deceptively attractive. According to Ayres and Ayres (1996), 'Materials recycling can help (indeed, it must) but recycling is energy intensive and imperfect, so it cannot fully compensate.' The viability of recycling is dependent on two variables: (1) aggregation and (2) transportation. While others point to the commodity markets as a key driver, *a priori* it requires the ability to economically collect a sufficient amount of a particular material or substance. Second, the energy and other costs associated with transporting recycled materials need to be reasonable or preferably lower than for other options for disposal.

There are significant economic opportunities for those who identify ways to collect streams of different valued material, for those who can sort effectively and for closer links between supply and re-use. The trick is to move beyond recycled steel, aluminium and glass and broaden the range of recyclable streams returned to industrial feedstock. The enterprise of recycling is hardly a new idea. Our ancestors knew it intuitively; we have had a lapse in judgement (Walker and Desrochers 1999):

> By the early nineteenth century, slaughterhouses' by-products were commonly turned into glue, soap, fertiliser, hog and chicken feed, combs, knife-handles, even a glass substitute suitable for lanterns. As chemistry advanced, Chicago's stockyards evolved into what might be called an 'eco-industrial park,' as small satellite industries producing everything from cosmetics to explosives appeared around the giant cattle-killing plants . . . All this without the benefit of a curbside recycling programme . . . People *paid* the plants for their wastes.

There is another form of recycling that escapes the public eye: in-process industrial recycling. The volume is far greater than all kerbside or municipal solid waste combined. These apply to organic and inorganic by-products of an initial manufacturing phase. If processes were not designed to recover this material and energy, then the volume to be handled externally would be magnified, as would the cost of products.

There are many processes used for industrial recycling, including distillation, decanting, solvent extraction, use as aggregate, chemical precipitation, re-extrusion, energy conversion, metallurgical reprocessing, bio-reduction, chemolysis and pyrolysis, in addition to storage of by-products and reintroduction into the same or other process (Smith *et al.* 1995). In some ways, the broader strategies outlined here are reflected in the choices of in-process recycling. The chemical industry has become very adept at the re-use of process wastes to reduce the amount discharged to the larger environment. The general questions of quality, concentration, availability, transport and costs all come into play in this arena as well. Lessons learned for in-process by-product recovery can serve as valuable insights for communities dealing with the same issues. The viability of recovery strategies often depends on scale. Large users of materials have often worked out ways to use industrial leftovers. Communities or company networks can serve as similar levels of aggregation providing opportunities for revenue-producing use of under-utilised assets at current industrial sites or for the creation of new business opportunities.

2.7.6 Detrivore technologies

The changed composition of materials and objects doesn't just happen that way. Moss on trees and stones serve a function of changing the tree bark or the rock into smaller particles that make up soil. Parasitic bacteria break down ingested food into a level that can be absorbed into the bloodstream. These are all examples of detrivore technologies. The earliest proponents of industrial ecology saw a possible trash-to-cash connection linking waste streams and input streams. It is possible to find veins and nuggets of gold; but most gold comes from sifting the soil to find the precious ingredient. The application of energy over time transforms them from one condition to another. Whether it is the scavenger population that thrives on the rubbish heaps of Cairo, to anaerobic bacteria that sponge up oil spills, to composting strategies, to pyrolysis approaches that change states of materials, all of these are frameworks for decomposition and reclaiming subsequent components.

2.7.7 Compounds

By definition,

> compounds are any substance composed of identical molecules consisting of atoms of two or more elements. There are millions of known compounds, each of which is unique. They range in complexity from water, which has two hydrogen atoms bonded to one oxygen atom, to the nucleic acids, which contain thousands of atoms. Most common materials are mixtures of different chemical compounds. Pure compounds can usually be obtained from them by physical methods, such as filtration and distillation, which do not alter the way in which the atoms are combined within the compound. Compounds can be broken down into their constituent elements or transformed into new compounds by chemical changes (reactions) (Considine and Considine 1984).

It is this property that creates a distinct category of looping material back into productive use. In the wizard's world of different compounds, differing quantities of this and that can lead to a diffcrent potion. Separating out various compounds that emerge from a primary process may yield valuable mixtures that can be used essentially in their current form in some other process.

Petroleum refining and pharmaceutical manufacturing is largely about the cracking of complex molecules and reformulating them into a new product. Many natural biological systems rearrange compounds to meet other functions (e.g. bulk food may be broken down into glucose). When complex substances, especially organic materials, are being handled, chemical and thermal manipulation can convert large molecules into a variety of new formulations.

Various kinds of renewable compounds can be used in various industrial applications, such as those shown in Figure 2.3. These primary compounds may be found intact or may be the by-products of other physical, chemical or biological processes. Awareness of the ability to remix the ingredients of various compounds provides virtually an infinite number of alternative uses with different and unique properties. Breakdown of more complex substances into other usable products can also occur through biological processes such as composting and phytoremediation.

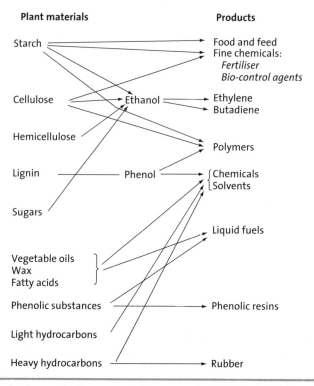

Figure 2.3 **Production paths from renewable raw materials in industrial products**

2.7.8 Molecular reconstruction and nanochemistry

As if the previous levels weren't difficult and confusing enough, the world ahead promises amazing ways to rearrange or rethink materials at the molecular and atomic level. Amato (1997) describes

> tools for seeing, moving and understanding individual atoms and the ever larger atomic collections that become the materials we use. The emerging

ability to micromanage materials even at the atomic level is giving researchers unprecedented access to the mostly untapped materials wonders of the periodic table of the chemical elements.

This enters the brave new world of nanochemistry. Ralph Merkle of the Xerox Palo Alto Research Center writes:

> The properties of materials depend on how their atoms are arranged. Rearrange the atoms in coal and you get diamonds. Rearrange the atoms in soil, water and air and you have grass. And since humans first made stone tools and flint knives, we have been manipulating atoms in great thundering statistical herds by casting, milling, grinding and chipping materials . . . Even in our most precise work, we move atoms around in massive heaps and untidy piles—millions or billions of them at a time. Theoretical analyses make it clear, however, that we should be able to rearrange atoms and molecules one by one—with every atom in just the right place . . . This technology, often called nanotechnology or molecular manufacturing, will allow us to make most products lighter, stronger, smarter, cheaper, cleaner and more precise.[5]

These new processes not only have major importance for the artefacts humans use and for manufacturing but also can have a significant effect on the environment.

> With processes based on molecular manufacturing, industries will produce superior goods, and by virtue of the same advance in controls, will have no need of burning, oiling, washing with solvents and acids and flushing noxious chemicals down their drains . . . Molecular manufacturing processes will rearrange atoms in controlled ways, and can neatly package any unwanted atoms for recycling or return to their source. This intrinsic cleanliness inspired environmentalist Terence McKenna, writing in the *Whole Earth Review*, to call nanotechnology 'the most radical of the green visions' (Drexler and Petersen 1991).

The future development of atomic architecture, the ability to manipulate materials at the submicroscopic level, rivals the potential of genetic engineering. Once again, it requires judgement and an ethical context to avoid frightening versions in either realm. The question is not whether it can be done, but under what value system and with what consequences. Used wisely, it could be an extraordinary tool to alter what was thought to be unalterable and render that material even better than harmless—it can be restorative and healthy.

2.7.9 Energy conversion

Waste-to-energy plants have become expensive fixtures in many urban centres. Biomass is burned to convert cast-offs into energy. This can be done through connections to steam or linked to turbines directly into the grid. In 1996 over 100,000 tons of refuse was consumed each day in the USA in the 114 waste-to-energy plants nationwide (Zannes 1997). About 15% of US waste is consumed at these facilities, and they generate 2,600 MW of electricity. Many of these facilities are attached to recycling and other materials recovery approaches, with more than 850,000 tons of common recyclables and 740,000

---

5   See R.C. Merkle, 'It's a Small, Small, Small, Small World', at www.techreview.com/articles.

tons of metals. Overwhelmingly, what is burned is organic waste, with the residue a fraction of the initial volume. With adequate pollution control and aggressive recycling diversion methods, waste to energy could be a contributor to materials re-use.

Waste-to-energy is not without its critics. My problem is not that such conversion mechanisms exist but that we resort to them too quickly, burning inefficiently large amounts of material that could have had higher value further up the hierarchy. Von Weizsäcker et al. (1997) point out that 'one of the biggest enemies of materials efficiency ... is waste incineration, among the most dissipating waste disposal technologies'. Putting up a big box that converts waste to electricity seems like a great way to use what we didn't want. Giant plants for energy conversion are hungry vultures that need to be constantly fed feedstock, and flow control agreements needed to maintain expensive investments can lock in resources that could be better used in another way. Peter Grogan in *Biocycle* notes that 'the reality is that waste to energy and recycling composting options are in competition' (1996). The urge to combust not only precludes more useful product uses it is at the core of the dangerous growth of volume of greenhouse gases, which threaten planetary survival.

Another type of waste-to-energy system is found in smaller-scale operations where combustible industrial residue is used on-site. Smaller-scale technologies that convert biomass to refuse-derived fuels or directly link to a generator are increasingly appearing on the scene. Many of the technologies cycle process gases back through the system as part of the technology. When waste-to-energy facilities increase the thermal efficiency and utility of their output, then they can be significant contributors to an improved environment. Much of this has to do with more efficient turbines, better co-generation options and tight transmission connections.

2.7.10 Rubbish: landfill mining and infill

Alas, even with the best of efforts, some materials will have to be stockpiled into rubbish dumps and landfills. Yale industrial ecologist Marian Chertow (1998) points out that the actual volume of municipal solid waste in the United States went down in absolute terms over the ten-year period from 1985–95, from landfilling 83% of 158 million tons (131 million tons) to only 57% of 208 million tons (119 million tons). The notion of mining landfills for their resources has been discussed in various ways. In 1974, Cappello proposed a trash-to-cash process as a way of resolving the dilemma of the wide separation between points of scrap generation and processing facilities. Such facilities could be placed at landfills, such as in the effort to consciously convert resources to economic opportunities in Tucson, AZ, and Dallas, TX. It also can be found in particular processes such as flue gas desulphurisation (FGD), where scrubber waste is used to replace natural gypsum in plasterboard. In describing 'waste mining', Rohatgi et al. (1998) assert that 'this strategy simultaneously reduces (a) the environmental damage due to the primary waste stream, (b) the rate of exhaustion of the second resource, and (c) the environmental damage due to mining the second resource'.

The systematic cycling of metals, minerals and other materials makes sense. New mines of the future need to be opened in inner cities where residual resources lie as eyesores and dangers. According to Robins (1991), 'every year, the nine Northeast [US] states bury 40 million tons of valuable materials and cover them with 20 million cubic yards of soil. An additional 11 million tons of valuable material is burned.' Can we not work out a way to effectively reclaim resources from its urban ore and re-use them to rebuild cities and neighbourhoods? It is not just in cities that resources labelled as 'waste' lie fallow. Why can't we examine the possibilities in agricultural and aquacultural waste and use these discards in new ways that generate economic opportunity?

According to the industry magazine *Waste Age*:

> Large landfills are a storehouse of valuable steel and aluminum scrap. For example, a 50 acre site might contain as much as 240,000 tons of steel and 20,000 tons of aluminum depending on the site's depth . . . There are estimates of more than 400 million tons of steel in these closed (or soon to be closed) sites. Considering that the US steel industry consumes about 100 million tons of scrap each year, the scrap recovered from landfills represents a significant supply source (quoted in Fisher and Findlay 1995).

These statistics are amazing enough. When you consider that, in copper mining, ore contained less than 3% of the metal in 1900 and is now being extracted at a 0.5% level, concentrations in landfills may be much higher than those in traditional mines (Gardner and Sampat 1998). This will require new technologies and new entrepreneurs to obtain the highest returns.

The process of filling in currently concentrated areas and building on top of landfills is a major factor in many areas of urban development and infill (literally). As such, new urban or industrial growth builds on the mounds of detritus of past development in places such as London's Docklands and Battery Park in New York. In ancient cities, the layers uncovered by urban archaeology demonstrate that this is not a new phenomenon. In natural systems, siltation provides a way in which discarded dirt over time provides a base for further land development. Construction on urban landfill is a more rapid expression of this phenomenon. As such, the 'rubbish' serves as a useful platform and reduces the need for use of greenfield sites. In some cases, the sites have been toxic pits that endanger human health.

The approach to handling rubbish or pollution at the 'end of the pipe' is to contain or make 'inert' the residue so that it does little if any harm. However, according to Ayres (1998):

> it is ultimately limited in its effectiveness by the fact that wastes can never be completely inert as long as they differ chemically or physically from the composition of the environmental medium into which they are discarded.

2.7.11 Pollution

The other bin that residual resources go into is the world around us. Carried through the air, laden in the soil and slurried by water, pollutants despoil our planet. Traditional approaches to pollution have looked separately at each avenue for delivery of pollution's poisons. Pollutants do not occur in neat carrying cases. They slosh together, and impact rarely comes by a single method or from a single substance alone. The systemic nature of pollution requires that responses be systemic as well, not simply a shuffling of pollutants. They need to be eliminated or turned to constructive uses.

Not all dissipation to the environment is bad. Water transfer can be beneficial as can the introduction of certain metals and nutrients into a natural system. But many human-made substances do not easily or quickly decompose or contribute to other systems. In many cases, they block the natural connections necessary for re-use to occur. As such, they become pollution that, either because of the nature of the substance or the quantity or concentration in which it appears, disrupts the cycle of healthy natural or human re-use. Our natural system is resilient. It can absorb a certain level of stress and exhibits an adaptability to carve new pathways to accommodate or cleanse materials released into the system. However, in many cases, we have dangerously overloaded this capacity in ways that have led to the extinction of many species and that threaten our own survival. This is why all the strategies listed above are so important to pursue. Armed with alternative approaches, we can address many if not most of the challenges that humankind's activities have generated.

2.8 Desperately seeking solutions

What to do about this range of possibilities is an arena for spirited debate. I would posit the need to work at a system level significantly beyond the management of particular chemical elements or compounds or individual pieces of technology. Hardin Tibbs (1998) illustrates this point:

> In systems terminology, structure implies more than physical features, it includes the pattern of relationships in the system, the way the set of stocks, flows, loops and delays are connected together. The relationships between beavers, their dams and the landscape are to a large extent programmed into their genes. Similarly, we need to embed appropriate conceptual structures for human eco-structure design into organisational planning and government policy.

Taking aim at particular chemicals or compounds is an 'up the down-escalator' approach to the management of potential dangers. Chemical inventories do serve a useful function for internal awareness of toxic materials and for community awareness of particularly nasty 'potions', but for each chemical to which years are devoted to investigation and information-gathering (often through rear-view mirror epidemiological studies) we miss many more and often overlook the cumulative or combined affects of various chemicals with significant effects on human and ecosystem health. For years, Greenpeace has targeted chlorine as a rogue chemical. The issue is not diabolical chlorine atoms but the uses, situations and combinations in which chlorine finds itself in relation to other systems. It is distribution, concentration and combinations in the system that matters.

At the enterprise level, we have tried to deal with each company as if it were a separate entity, but no company exists outside of its larger ecology of market, materials, geography, community and so on. Our attempts at organisational improvement are often short-sighted attempts to try to find boxes for problems that then will have tidy solutions. However, the results of these approaches are limited, and too often the problems will not fit tidily into the boxes we find for them. We attempt to create contractual understandings through laws and written agreements that will order the world around us. Paradoxically, we spend more time managing and untangling what seemed to be simple. The solutions to our frustrations are not necessarily found in new gadgetry of the technological or in social variety but in ways of dealing with ecologies of organisations and markets that respond in much different ways. Rigid approaches that seek a single right answer are folly to pursue.

According to Robert Ayres (1998):

> The solution must surely be a combination of radical dematerialisation of goods (including capital goods) together with dramatically increased recycling of non-renewables driven by solar energy. The details of this combination need to be addressed seriously by economists, engineers and environmentalists.

There are considerable policy, technological and economic opportunities embedded in the levels described above.

Rather than play roulette with materials transformation and collection, new technologies that systematically address the re-use of materials offer new vistas for technological innovation. There are major areas for scientific and engineering research. Managing and connecting these processes opens economic opportunities for those daring enough to see the niches and fill them. Policy-makers can provide incentives and regulation to encourage wise and healthy material and energy transformations that respect the principles of entropy and value. The possibilities are literally infinite. We have barely scratched the surface of potential ways to harness materials and energy to benefit humankind and the planet.

2.9 Not one loop but many

The analogy of natural cycles has been a powerful metaphor in rethinking material and energy productivity. Yet nature exhibits not one cycle but many. These cycles may intersect with each other, or they may not. As a whole, this circumlocution process is characteristic of our known world. Not only can machines be cycled back into re-use but so can materials—and those materials can be cycled back into the parts that are used for remanufacturing. Altered compounds and molecules can become the stuff that makes up those materials that make up parts that make up the remanufactured machine that is shared among others to minimise the amount of times those machines need to be fashioned. The complexities of the possibilities are mind-boggling. But underlying the infinite number of combinations and recombination are simple principles of seeking the most positive associations again and again and again.

In some ways, I see the application of the principles of this chapter as analogous to a 'genome project' where scientists and technologists with far more specific knowledge of these processes than I possess map out our current use patterns and current technologies along the hierarchy I've described. We need to learn where are the gaps and where is the connective tissue that enables us to move from one strategy to the next in the most cost-efficient and simplest way possible. Knowledge of what one doesn't know is one form of wisdom. I know we need to do far better than we have in shepherding resources on our Earth; others need to join in the hard work it will take to conceive and implement how to make it better grounded in the real world.

Walter Stahel (1998) of the Product-Life Institute in Geneva provides a means for conceptualising these multiple loops (Stahel 1998; see Fig. 2.4). Each species, each element of our universe, is engaged in the transformation of a range of resources. If humans are made of 'smart material', then we will exercise the wisdom of systematically expanding the value, range and quality of what we transform. Each of the approaches described above seeks to seriously address the viability of the planet through stewardship. No one approach can meet the magnitude of the challenge. No one person can make the change. By systematically engaging the challenge, we can live on a more supportable, sensible and meaningful planet—one that constantly changes and creates new possibilities.

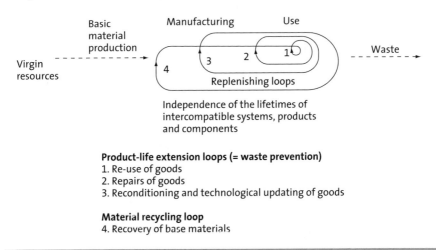

Figure 2.4 **The self-replenishing system of product-life extension**

Source: Stahel 1998

3
A WALK ON THE HUMAN SIDE OF INDUSTRIAL ECOLOGY

Edward Cohen-Rosenthal
Work and Environment Initiative, Cornell University, USA

Knowledge of kinds of waste-streams can provide a means to determine potential linkages. But this doesn't link the waste-streams; decisions by people do. Turning wasted materials into potential new enterprises is an elegant idea, but enterprises are started by entrepreneurs, not printouts. Computerised linkages can provide information, and computer models can illustrate possibilities, but the best they can do is stimulate imagination, not bring about the connections. What role do people and social processes play with regard to industrial ecology and sustainable economic development?

In a World Bank report on sustainability, Michael Cernea (1994), a sociologist, points in the right direction:

> The case for environmentally sustainable development is usually argued in economic and technical–ecological terms. As has happened in other areas, many are tempted to think that if they can 'get the economics right', everything else will fall into place. Soothing as this econo-mythical invocation may be, it is nonetheless one-sided. The social components of sustainability are no less important . . . The environment is at risk not from extraterrestrial enemies, but from human beings, including both local and distant resource users. Thus, the call for 'putting people first'.

Industrial ecology is all about connections: material connections, energy connections, organisational connections and human connections. It is the application in the world of commerce and industry of more cyclic and sustainable flows between businesses (and their communities), especially with regard to energy and materials. This should lead at the same time to less pollution and higher profitability, but, fundamentally, industrial ecology is a social construction. It deploys human intention to create more integration, more cycling of resources and better use. It draws on the natural world and impacts the natural world, but is inherently a form of 'biomimicry' (Benyus 1997).

The Interagency Working Group on Industrial Ecology, Material and Energy Flows of the US federal government defines 'industrial ecology as an emerging science which provides the conceptual tools to analyse and optimise the flow of energy and materials in our production systems' (IWGIEMEF 1998). One school in industrial ecology (Allen and Behamanesh 1994; Graedel and Allenby 1995; Socolow et al. 1994) reflects a notion that if only there were the right plotting of where materials reside or hide then we could find a proper place for each type of waste. Hence the essential requirements are chemical and engineering know-how, proper technologies for aggregation and separation, and market-based incentives for industrial ecology connections. Would it were that simple. In Brownsville, TX, an input–output model used for a large petroleum refining complex has been adapted by Bechtel Corporation (1997) to a regional materials flow to simulate a virtual eco-industrial system. Elaborate lines connect the reported by-product from one company to those materials needed by other companies. The US Environmental Protection Agency (EPA) has devised a similar product. Although the intentions of these models are excellent, the adequacy, reliability, timeliness and quality of the data are highly suspect. They use static models to describe a complex dynamic ecology.

In this chapter I put forward the proposition that this kind of information on materials flow is helpful as a pathway to creative interconnection, but it isn't sufficient and may very well be a secondary consideration. The human dimension can be broached at a conceptual level on the nature of systems and change processes; it can also be enjoined at the practical level in describing what it takes to create effective industrial ecology connections.

Industrial ecology rightly challenges assumptions that our production systems are the best possible. Since wasteful practices or undervalued use of resources are considered economic sub-optimisation, it is too often assumed that companies and institutions will seek the highest and best use. Theoretically, those with better and more efficient resource use will win in the marketplace and thus move industry towards continual improvements in resource and energy productivity. One way to address the issue is to make costs transparent and account for the real costs of artificially cheap resources. Classical economic and industrial ecology theory also asserts that better information will lead to optimisation, as sure as the sun rises in the morning.

Yet it is the actual ecology of business that can get in the way. The ruts of established pathways, entanglements with others, willingness to accept waste as a normal cost of doing business, technological funnels and organisational inertia all combine to make this process cumbersome at best. Replenishment and restoration of nature's capital stock is rarely factored into the equation. Largely, the gaps between resource use and possibilities are ignored because it doesn't hurt enough, others are doing the same thing anyway and it takes too much effort to bring about significant change. Industrial ecologists Robert and Leslie Ayres (1996) caution:

> It is not safe to assume (as many macroeconomists tend to do) that every technology in place is the optimal choice, as if it had been the winner in a Darwinian competition against all possible candidates. In fact, the process of technological evolution is much messier than that. As often as not there are superior 'off the shelf' alternatives that simply never had a chance to get established. If the initial hurdle could be surmounted, the superior alternative(s) might quickly take over.

3.1 Self-organising compared with engineered systems

'As strange as it seems, industrial ecology will conduct business the way a sun-soaked hickory forest recycles its leaves', according to Janine Benyus in Biomimicry (1997). All branches of industrial ecology pay obeisance to the notion that the more cyclic patterns in natural settings should serve as a lesson for deliberate interventions. 'The ideal anthropogenic use of the materials and resources available for industrial processes (broadly defined to include agriculture, the urban infrastructure, etc.) would be one that is similar to the ensemble biological model' (Jelinski et al. 1992), where waste equals food, with strong linkages in and around an organism. There is a significant difference, however, in how that should be accomplished. One school emphasises detailed scientific analysis of particular materials concentrations and flows to identify optimal realignment for improved cycling of resources—and especially to deal with priority chemicals based on toxicity or volume. The National Science Foundation, under its industrial ecology solicitations, has funded for several years a number of these engineering-oriented studies through a joint grant programme with the Lucent Corporation.

Hardin Tibbs, an early and insightful proponent of industrial ecology, describes the principal elements of industrial ecology as industrial ecosystems, industrial metabolism, biosphere interface, dematerialisation, energy systems and policy innovation. Tibbs (1992) says, 'our challenge now is to engineer industrial infrastructures that are good ecological citizens'. By this he means that at the least we need to achieve a balance between industrial systems and ecosystem carrying capacity and renewability. This is a noble goal. The means for accomplishing this would be to calculate carrying capacity effectively, identify the load provided by industry through the life-cycle of its activities and products and adjust industrial activity accordingly (Adriaanse et al. 1997). By identifying the stray use of resources and redirecting them to the right targets, economic opportunity can expand. That's it. Problem solved. Yet in Tibbs's list human citizens are not central to industrial ecology.

In an actual ecosystem, achieving a sufficient level of certainty about these metrics is highly uncertain and dynamic. Humans, as partial actors within the larger ecosystem, by definition lack the perspective and knowledge that would make that level of engineering possible. My issue is not with the desired outcomes but its epistemology. Applied in an industrial setting it is also problematical. Experience has shown that different-sized companies have different data-collection capabilities, that different measures are used and that processes change at a differential rate while issues of concentration, quality, timeliness and volume throw wrenches into neatly assumed connections.

I believe that the natural system as applied to industrial ecology should operate in a different way, where self-organising behaviour leads to the maximum amount of connections and broader system adoption (Cohen-Rosenthal 1998a; Lowe and Warren 1995; Wallner 1998). There is less attention to the specific exchange mechanisms and an emphasis on the process of fostering interconnections. Although this is arguably true within a particular organisation or organism, it is when these are seen in relation—in a larger industrial ecology—that the notion of living systems gathers more power. Industrial ecology is a relational set of observations and behaviour. As Stamps (1998) describes organisational ecology:

> This organic, ecological model of a company as a set of relationships and interdependencies is one that doesn't square well with the traditional corporate model of management and control—a model that still favours technological and mechanical interventions.

Margaret Wheatley and Myron Kellner-Rogers (1996) provide a different notion of system, though they share the same objectives that Tibbs, Allen and others have for harmony between natural and industrial systems: In *A Simpler Way* they offer the following:

> We live in a universe which seeks organisation. When simple relationships are created, patterns of organisation emerge. Networks, living or not, have the capacity to self-organise. Global order arises from local connections. It was these co-operative structures that first created life. Life linked with other life and discovered how to continue discovering itself. Life learned how to self-organise (Wheatley and Kellner-Rogers 1996).

Hence industrial ecology can occur as synaptic connections based on particular ecological conditions or it can be constructed by building dedicated 'bridges' from the locus of one material locus to another. An industrial ecology should be an adaptive and regenerative living system. It should enliven the people within it and other forms of life on the planet. Biologist Paul Weiss, as part of extraordinary symposium organised by Arthur Koestler in 1968 in the Austrian Alps, says:

> Life is process, not substance. A living system is no more adequately characterised by an inventory of its material constituents, such as molecules, than the life of a city is described by the list of names and numbers in a telephone book. Only by virtue of their ordered interactions do molecules become partners in the living process; in other words, through their behaviour.

If an industrial ecology is truly to be an ecology, then the number of permutations of interface and interconnection are of such a magnitude that in real time they cannot be managed as simple exchanges. Consider what goes on in our bodies:

> The average cell in your body has about 80% water and for the rest contains about 10^5 macromolecules. Your brain alone contains 10^{10} cells, hence about 10^{15} (1,000,000,000,000,000) macromolecules . . . Could you actually believe that such an astronomic number of elements, shuffled around . . . could ever guarantee your sense of identity and constancy in life without this constancy being insured by a superordinate principle of integration? (Weiss 1968).

Natural systems can manage this challenge; mechanical systems go berserk. Hence there is a fundamental and qualitative difference in the rules and process of natural decision-making when the order of magnitude of choices goes above a certain level. In the lifespan of a single brain cell it connects with other macromolecules 10^9 times. When considering lots of people working in lots of industries, if industrial ecology is truly to draw on natural systems, then it cannot use an abacus to calculate what to do. The way a network learns is very different from what plodding linear traditional models of learning have led us to believe.

Peter Senge (in de Geus 1997) offers an excellent 'riff' on visualising the difference between views of organisational change as mechanical or biological; machine or living:

Seeing a company as a machine implies that it is fixed, static. It can change only if someone changes it. Seeing a company as a living being means that it evolves naturally.

Seeing a company as a machine implies that its only sense of identity is that given to it by its builders. Seeing a company as a living being means that it has its own sense of identity, its own personhood.

Seeing a company as a machine implies that its actions are actually reactions to goals and decisions made by management. Seeing a company as a living being means that it has its own goals and its own capacity for autonomous action.

Seeing a company as a machine implies that it will run down, unless it is rebuilt by management. Seeing a company as a living being means that it is capable of regenerating itself, of continuity as an identifiable entity beyond its present members.

Seeing a company as a machine implies that its members are employees or, worse, 'human resources', humans standing in reserve waiting to be used. Seeing a company as a living being leads to seeing its members as human work communities.

Finally, seeing a company as a machine implies that it learns only as the sum of the learning of its individual employees. Seeing a company as a living being means it can learn as an entity, just as the theatre group, jazz ensemble, or championship sports team can actually learn as an entity.

3.2 The role of social process in the development of industrial ecologies

It is not human intention as reflected in engineered solutions that is at issue. Rather, it concerns what degree of prediction is possible or desirable. Bounding of possibility by narrow analysis may send us down the wrong paths or limit the invention of new approaches. Interactions occur between chemicals, species, people and so on all the time. The connections industrial ecology speaks of happen—indeed frequently. The power of industrial ecology as a conceptual framework is not the description of mass flows and their consequences but in applying human intention to explore potential connections so that we create interactions with more value and less waste. Industrial ecology is an intervention at the organisational and social level. Natural ecology and chemical processes are not necessarily predicated on people, but people can bring their processes to the 'on-ramp' of natural processes for nature to work its magic.

I have limited faith in future technological salvation as an independent process. Too many questions keep me from bowing at the alter of ever-improving technology. Why is it that tribal communities can use every last shred of a fallen tree or dead animal, but modern society treats its leftovers as offal? How can it be that American pioneers used and re-used everything they could but modern America is the most wasteful nation on Earth? How can it be that with incredible technologies that can do more faster we do not take the time to use what we have better and more wisely? It is not in the technologies that the answer lies, but in the ways humans make choices, their willingness to seek out new connections, to invent new combinations, to explore the possibilities of the world around them. Technology application is a function of human intent. Hence technological determinism or optimism alone is insufficient.

In studies based on US Toxics Release Inventory data regarding the effectiveness of different source reduction methods, the startling results showed triple the decrease in toxic releases when there was significant internal employee participation compared with relying on external or purely technical solutions (Bunge et al. 1996). The results quadrupled when technical expertise was coupled with internal engagement. When technology is viewed as tool—as an extension of human capability—then it has a powerful influence on the use of resources. When technology is analysed as an object, as an independent driver, it becomes a hollow instrument. A society sitting on the sideline waiting for technology to solve its problems is waiting for Godot.

The Industrial Transformation project of the International Human Dimensions Programme on Global Environmental Change has laid out a broad agenda for altering the trajectory of industrial practice. Its research includes examination of the role of organisations, management and network theories that I explore in this chapter. It notes, 'surprisingly little scientific effort has been devoted to the role that organisational research can play in "greening of industry" ' (IHDP 1999).

The first step in industrial ecology is to imagine the possibility of connection, of linkage for mutual support. Data remains the humus from which the seeds of connection are nourished and take flower. An understanding of what occurs is important. An understanding of what should occur is even more important. In actual applications of industrial ecology, vision is the starting point. The birthing of industrial ecology linkages requires interconnection of those engaged in its pursuit. In many situations where sustainable communities or eco-industrial activities are being designed, a key component is the use of group processes as a way to build a vision, generate buy-in and to move along planning. For example, in both Cape Charles, VA, and in Minneapolis, MN, architecturally based design charrettes were an important historical part of the development of their ideas. Architectural charrettes place a premium on developing images and plotting relationships on the map to articulate what the configuration or site might be. In both cases, the end results of the conferences were not exactly how the local applications turned out. At one level, this doesn't matter because it broadened the community of commitment to the idea and moved the process along in a direction that worked out. On the way, there was also a sense of frustration from an energised community about visioning processes when the unfolding reality looked more like the shadows in Plato's cave than staring into the light of a bright new day.

We have used future search conferences as a method for gaining stakeholder involvement. Search conferences are excellent vehicles for creating a notion of a desired future (Weisbord 1992). In Baltimore, MD, for example, over 200 participants from every segment of the community over three days worked intensively on developing an idea of what an eco-industrial park would be in their area. The first day was an afternoon briefing providing background information on the area and on ecological thinking. The next two days led the group through a process that started with understanding future trends that would affect them at a global and local level. It ended with specific plans for creating an eco-industrial strategy. The goal of such processes is to 'put the system in the room' and challenge its constituents to create options that find consensus among diverse participants.

Social aspects of industrial ecology stretch beyond the interorganisational relationships within a symbiotic connection or eco-industrial cluster. The environment in which industrial ecology operates includes the larger community and social context. These factors can be enabling or inhibiting in terms of achieving broad goals for industrial ecology. In the permeable distinctions between workplaces and their surroundings, each can add value to the other. Transport systems, access to municipal services and links to educational institutions are all ways that can enhance effective eco-industrial activity. The workplaces also need to have a positive impact on the local area by buying materials locally, sharing facilities, diverting potentially polluting materials from the environment and creating quality jobs. Part of what drives this kind of action to improve industrial performance is a broad concern for the health of the global commons; hence profit-maximising strategies are linked to strategies that improve public welfare. The use of social processes such as those described above to engage the broader community is essential for effective strategy development and implementation.

3.3 The role of business systems in industrial ecology

Shifting by-products from one place to another is not hard; making them safe, valuable and economical is the real trick. Before a by-product has value it has to have a market. Even if it has theoretical value, to be useful it has to have perceived value in the marketplace. Furthermore, the resource has to be handled through manageable transactions—it has to be obtained, transformed, transported, sold and then reintroduced back into productive re-use. Hence, there need to be strategies to assure the most economical approaches to the collection and distribution of potential products and by-products. The more poorly that is managed, the lower the likelihood of beneficial connections. Assumptions of connections cannot stand outside the realm of economic viability. I have placed a great deal of emphasis on business systems in applied industrial ecology (Cohen-Rosenthal 1998b). The same material can be transferred from company A to company B as is shipped from company C to company D. In each case, the outcomes of the exchange are affected by how well the collection, shipment, receipt and use are handled. Poor management systems can mask potentially viable linkages, forcing us to think that the attempt is flawed rather than its implementation.

Strategies for encouraging industrial ecology of materials while remaining rooted in chemical and engineering parameters are equally dependent on assuring that management systems obtain customers, conduct the process efficiently and manage with quality, integrity and reliability.

Dexter Dunphy and Andrew Griffiths in *The Sustainable Corporation* (1998) make a persuasive argument that the next round of organisational change will require a shift in thinking:

> Fundamentally rethinking the basis of products and services around the principles of sustainability is already producing major breakthroughs in efficiency and effectiveness, and these breakthroughs are becoming the nuclei of the new industries of the future. These industries will transform, bypass or supplant our wasteful and dangerous smokestack industries and modify our impact on the environment. In this process some corporations will play a major transformative role; those who seize the opportunities of the major breakpoints between old and new industries will survive and grow.

As we have moved from managing environmental issues at the end of the pipe and looked upstream to change product processes and determine life-cycle impacts, a systemic analysis addresses the total structure of industrial operations to identify areas of embedded waste of time, materials and energy. Efforts to improve environmental performance have evolved towards life-cycle analysis and design for environment. These efforts at a single unit level are often helpful and necessary. To move to the next level of environmental and business performance requires going to the next level of organisational configuration by examining interorganisational and intersectoral issues. This is why you see an emphasis in leading businesses on building strategic partnerships and creating alliances around core competences. It makes sense, and it makes money.

No company can do everything well. Any company is tail-tied to its ecology of market, industry, resource access, labour availability, weather, geography, capital costs and so on. Rather than ask how all those factors can be controlled, the better question is: why control them? Why not use those variations as assets and opportunities? Building on the available resources and finding ways to mutually sustain each other requires less energy than it takes to enforce rigid orbits. As in physical science, 'strong' attractions between elementary particles are what gives each iteration identity and defines function. I find the 'strong attraction' among organisational systems helps define their possibilities as well.

I have long held that for industrial ecology to be successful it has to demonstrate superior business as well as environmental results. Clever connections between specific material flows can have a significant business impact in a limited number of cases. Applying a broader business-based industrial ecology concept helps participants to discover more connections and roots out upstream barriers to reduced waste generation or possibilities for integration with other processes or facilities. As such, it drives up the economic impact of industrial ecology and increases interest in its application. There are impressive illustrations of the impact of industrial ecology's connections such as in Kalundborg, Denmark (see Fig. 1.1 on page 19), where energy and by-product exchange have demonstrable and significant economic and environmental benefits. (Ehrenfeld and Gertler 1997). Most observers, while impressed with this effect, have difficulty in seeing how to replicate this icon of industrial ecology in their own situation. That is because the focus is on the wrong end: it is not just about waste heat and materials, it is about co-operation among firms to support symbiosis and integration into their own operations. It changes the way they do business.

I have used the model illustrated in Figure 3.1 as a way to think about how business systems both within and between firms can seek more synergy, greater opportunity and resource stewardship. In my opinion, integration along these drivers of industrial success can be just as important (or more so) in terms of overall environmental, business and social impact as can be particular material flows. Unless co-operating facilities can see a significant impact on these core business drivers, industrial ecology will become, of necessity, marginalised. Environmentalists, too, frustrated with many companies patronising their concerns, can broaden appeal by expanding the range of strategies for improvement beyond single-facility control strategies by looking at business systems.

In some organisations, materials and energy form a large part of their budget and time. Some of those companies have not sufficiently addressed the issue of waste. These are the companies that traditional industrial ecology can easily address because there are real improvements to be made and savings to be accrued. However, this is not the situation

Materials
- Common buying
- Consumer–supplier relations
- By-product connections
- Creation of new material markets

Energy
- Green buildings
- Energy auditing
- Co-generation
- Spin-off energy firms
- Alternative fuels

Transportation
- Shared commuting
- Shared shipping
- Common vehicle maintenance
- Alternative packaging
- Intra-park transportation
- Integrated logistics

Marketing
- Green labelling
- Accessing green markets
- Joint promotions (e.g. advertising, trade shows)
- Joint ventures
- Recruiting new value-added companies

Human resources
- Human resource recruiting
- Joint benefit packages
- Wellness programmes
- Common needs (payroll, maintenance, security)
- Training
- Flexible employee assignment

Information and communication systems
- Internal communication systems
- External information exchange
- Monitoring systems
- Computer compatibility
- Joint MIS system for park management

Environment, health and safety
- Accident prevention
- Emergency response
- Waste minimisation
- Multimedia planning
- Design for environment
- Shared environmental information systems
- Joint regulatory permitting

Production processes
- Pollution prevention
- Scrap reduction and re-use
- Production design
- Common subcontractors
- Common equipment
- Technology sharing and integration

Quality of life and community connections
- Integrating work and recreation
- Co-operative education opportunities
- Volunteer and community programmes
- Involvement in regional planning

MIS = management information system

Figure 3.1 **Possible areas of eco-industrial networking**

in most companies—and, if industrial ecology is to have broader application, then it must affect more than this subset of energy-intensive and material-intensive companies.

My rule is that, if industrial ecology is to be accepted by corporations, it must have a significant and measurable impact on the top three to five drivers for organisational success. If market image is the major concern, then the industrial ecology strategy must address that issue. The same holds true for cost leadership, product innovation, risk management, quality, market access, reliability of supply chain, customer core requirements and so on. Although a value-based commitment to environmental stewardship is welcome, even in those situations the particular industrial ecology strategies must demonstrate a difference in the success of the organisation. If it doesn't, commitment will wither quickly and, more importantly, opportunities to make a real difference will be overlooked or dismissed.

3.4 The role of environmental management systems

One particular management strategy used frequently is to deploy environmental management systems that go beyond regulatory compliance. Although specifically addressing environmental issues, these systems deliberately link to quality, production, service and managerial systems. In a number of applications for eco-industrial activities a strong emphasis is being put on environmental management systems. At the eco-park in Londonderry, NH, complying with the ISO 14000 series standards of environmental management systems is a basic requirement of participating companies as a way to demonstrate environmental good faith. In the Cape Charles Sustainable Technology Park, more favourable treatment in leasing comes to companies who demonstrate sound environmental management approaches. In a proposed eco-industrial park in Plattsburgh, NY, a key element of the design was to certify the whole park to the ISO 14000 series standards and then assist individual companies in their compliance. The ISO 14000 series and similar systems include as part of the certification process ways to integrate environmental management with employee awareness, involvement and training. I have placed a strong emphasis on continuous environmental improvement as baseline criteria for involvement in an eco-industrial network.

At several levels environmental management systems become a critical component of industrial ecosystems. First, it is hard to argue ethically that industrial ecology should strive to maintain waste-streams so that it can divert some or all of it to other purposes. This is especially true if real accounting of the cost of the waste-stream is considered. Hence, sound environmental management is a prerequisite for effective and valued industrial ecology connections.

Second, even under the most optimistic of circumstances, emergency management plans and containment alternatives need to be available in case of a failure in the system. This is regardless of the elegance of a zero-emissions design. Such precautions include transport of goods into and out of a zone with industrial symbiosis and storage capacities.

Third, organisations and contexts change. New materials emerge, new customer requirements prompt changes to processes, and scientific understanding about the dangers and value of differing ingredients increases. As such, a fluid system is better able to respond to these contingencies than systems based on rigid recipes. Whereas environmental management has been a stopgap against fines for non-compliance with laws and regulations, under industrial ecology it becomes an integral part of all the operations of a facility or organisation.

3.5 The role of industrial network theory in industrial ecology

When industrial ecology goes beyond the internal circulation of materials or energy and is more than a set of transactions between two companies, then it is a network. There is a qualitative difference in network relationships compared with transactional relationships. A broad literature on networked organisations provides valuable insights for industrial ecology (Chisholm 1998). For the most part, this literature does not deal directly with environmental management or outcomes, but it does go into great detail on resource productivity and market responsiveness of networked industry (Porter 1998). Most of the descriptive work in industrial ecology focuses on dyadic exchanges without reflecting network approaches to economic development. Chisholm draws on the work of organisational change pioneer Eric Trist, whose own thinking evolved from a sociotechnical to a socio-ecological approach over the years (Emery and Trist 1975). Describing network principles, Chisholm (1998) writes:

> First, interorganisational networks operate largely as abstract conceptual systems that enable members to perceive and understand in new ways. Developing shared understanding makes it possible for members to create ways of organising to deal with these complex problems.
>
> Second, networks differ from mere interorganisational relationships. Networks improve the ability of organisations to deal with ill-defined, complex problems or issues that individual members cannot handle alone. Network activity is oriented to the shared vision, purpose and goals that bind members together . . .
>
> Third, loose coupling of members is another feature of these systems. Members represent diverse organisations that are physically dispersed and meet to conduct activities required to carry out the higher-level system purpose . . .
>
> Fourth, network organisations are self-regulating. Members, not a centralised source of power, are responsible for developing a vision, mission and goals and for initiating and managing work activities. Members share their understanding of issues and devise ways to relate to each other in carrying out the work necessary to bring about a shared vision of the future.

In economic development, the use of cluster models for economic opportunities has had wide appeal (IDeA 1997; Mathews 1996). Whether in Italy, Denmark, Australia or parts of North America, industrial networks have proved successful as a model of resilient economic development that leads to greater market impact and versatility, lower costs for doing business and more effective resource use. In my visits to successful areas such as Emilia-Romagna in Italy there is little overt connection to industrial ecology. Michael Porter, competitiveness doyen of the Harvard Business School, makes the point clearly:

> Clusters affect competition in three broad ways: first, by increasing the productivity of companies based in the area; second, by driving the direction and pace of innovation, which underpins future productivity growth; and third by stimulating the formation of new businesses, which expands and strengthens the cluster itself. A cluster allows each member to benefit *as if* it had joined with others formally—without requiring it to sacrifice its flexibility (Porter 1998).

Implied in the descriptions of these economic organisational models is that they form 'industrial ecologies' more adaptive to their environment and thus result in resource efficiencies. Dan Esty and Michael Porter (1998) point to that same conclusion: 'By forcing attention to the interdependence of various parties in the production process and identifying the potential for synergies among these companies, industrial ecology can help overcome a variety of obstacles to more efficient resource use.' At times, that connection has a materials or technology basis, but it also affects the organisations at a meta-level that leads to systemic improvements in resource utilisation and responsiveness. In our work at Cornell, we have tried to make an explicit interconnection among organisational entities as mutually supportive, seeking synergy at all possible levels. Industrial network and industrial ecology theory have missed each other like two ships in the night, yet they sail the same waters. Traditional industrial clusters would benefit from a more refined and explicit materials–energy connection *and* industrial ecologies can learn from the particular characteristics of networked organisational structures.

Peter Wallner (1999), working with eco-industrial clusters in Austria, has observed:

> The most critical innovation of the industrial ecology concept is the level of inter-enterprise co-operation. It is not the single element of the production system, the company, that is the subject of analysis, but the network of region-wide settled enterprises for which a spatial proximity can be defined. A broad variety of networks with primarily ecological, economic and partly socio-cultural spheres of action and furthermore the framework to develop a culture of co-operation have become the focus of action. It is the complexity of the industrial system and the new responsibilities in the social context that make the new challenge an interdisciplinary approach.

There is often confusion that creation of networks saves because duplications are eliminated. A successful network, however, reduces draw on current asset use while increasing flexibility through redundancy of function. In other words, the network offers more—not less. Tighter interconnections characteristic of industrial ecology contracts what would have been lost spaces, materials and time, leading to higher productivity and customer responsiveness. Industrial ecology is more than reducing the costs of disposal and perhaps creating a revenue stream through waste exchanges. Tapping into the power of networks alters the sense the organisation has of itself and how it operates. It is at this more profound level, which is often difficult for those in successful networks to articulate, where networks generate their most power. Those who have studied networks point out that

Having a basis for new ways of thinking and operating is crucial. The socio-ecological approach has provided this conceptual foundation . . . Building on this perspective, network development work emphasises the importance of members having new ways of perceiving and experiencing their local organisations within the larger network system (Chisholm 1998).

3.6 Eco-industrial networks as learning organisations

Many organisations today are recasting themselves as learning organisations, where individuals and groups constantly comprehend what is going on around them, share insights and test ways to continuously find new and better ways to work. Implementation of industrial ecology is a learning activity at a number of levels. It requires discovery of materials, energy and business system synergies. Those involved must learn new ways of working together that builds trust and generates shared value. Information between interacting systems provides a currency for collaboration. As various approaches are tested, those involved must learn from what works and from what doesn't. Shifting contexts and players may change old assumptions. As industrial ecology unfolds, it calls for new policies, new technologies, new roles and new behaviour.

This learning is embedded in the process of forming a vital ecology. As de Geus (1997) points out, 'ecology, after all, is itself a process of Piaget's learning through accommodation. Learning in ecosystems takes place constantly, as entities adapt themselves to new understandings, based on changing conditions in their environment.'

Adaptive behaviour is a core characteristic of industrial ecology. This requirement is why I so adamantly argue against time-frozen analysis and static solutions. As industrial ecology theory and practice advances, the requirements for speed and flexibility of adaptive behaviour will increase as well. Systems that can make connections quickly will be much more effective than slogging analysis, persuasion on each possible connection and patient implementation.

According to John Seely Brown, chief scientist at Xerox's Palo Alto Research Center:

> Relationships are crucial because they not only determine how work gets done, they play a role in how meaning gets constructed . . . Corporate America doesn't like to even admit there is a social and emotional component to learning; they want to believe that this is just information-sharing. But that doesn't recognise the richly textured social fabric you need in order to have a learning environment. Without this rich complexity, learning just doesn't happen (quoted in Stamps 1998).

David Rejeski (1999) of the White House Council on Environmental Quality has written:

> In an effective industrial ecology relationship there is constant feedback between various systems that determines needs, timing, material composition and quality as ways to assure the proper integration between two systems. They need to learn from each other.

This connection to learning is made explicitly by Tachi Kiuchi (1999), Managing Director of Mitsubishi Electric and Chairman of the Future 500, a high-level corporate group exploring industrial ecology:

> When I visited the rainforest, I realised that it was a model of a perfect learning organisation. A place that excels by learning to adapt to what it doesn't have. A rainforest has almost no resources. The soil is thin. There are few nutrients. It consumes almost nothing. Wastes are food. Design is capital. Yet rainforests are incredibly productive. They are home to millions of types of plants and animals, more than two-thirds of all bio-diversity, so perfectly mixed that the system is more efficient and more creative than any business in the world. Imagine how creative, how productive, how ecologically benign we could be if we could run our companies like the rainforest.

The system of management of the members of an industrial ecology as a whole reflects the concept of double-looped and triple-looped learning put forward by Chris Argyris of Harvard. Argyris (1992) explains:

> The term is borrowed from electrical engineering or cybernetics where, for example, a thermostat is defined as a single-loop learner. The thermostat is programmed to detect states of 'too hot' or 'too cold' and to correct the situation by turning the heat on or off. The thermostat is able to perform this task because it can receive information (the temperature of the room) and therefore take corrective action. If the thermostat asked itself such questions as why it was set at 68 degrees, why it was programmed as it was, then it would be a double-loop learner . . . Single-loop learning occurs when matches are created, or when mismatches are corrected by changing actions. Double-loop learning occurs when mismatches are corrected by first examining and altering the governing variables and then the actions.

One can liken simple reporting of materials flow to single-loop learning, and this represents one level of industrial ecology. Questioning the underlying logic of production represents another level of industrial ecological inquiry. Robin Snell and Almaz Mankuen Chak (1998) provide concise definitions of the types of looped learning:

> It is **single-loop** when they [group members] make simple adaptive responses. It is **double-loop** when members begin to see things in totally new ways, e.g. changing their views of their roles, of the business or the business environment. **Triple-loop** individual learning entails members developing new processes or methodologies for arriving at such reframing.

It seems plausible to assert that the best forms of industrial ecology must move beyond single-loop learning (the acquisition of information) to double-loop learning, where assumptions are tested openly and new approaches tried, to triple-loop learning, where processes shift to engage in regular adaptation. Industrial ecology is formally about seeking cycles in industrial systems, where materials do not pass once through a system and then dissipate to who knows where. In fact, what are called 'closed-loop' systems are truly not about simple circularity but loops connecting to loops connecting to loops; not repetition but reinvigoration, a renewal of purpose and use. The organisational system that seeks cycles has most success when it moves towards triple-looped or continuous learning, where new insights and possibilities are a natural and fluid part of organisational functioning and reinvention.

3.7 The role of human resources in industrial ecologies

Imagining technologies and processes that loop waste and other materials back into the system and are vigilant about environmental problems is easier to put in a goals statement than to do in practice. People enact those policies. One cannot institute a regime of respect for natural materials and energy resources without congruent respect for human resources as well (Hopfenbeck 1993; Wehrmeyer 1996). This requires management leadership and incorporation into the core activities of the enterprise while engaging the whole workforce, customers and suppliers in a common cause. It involves the whole value chain and set of stakeholders. Their interconnection is vital. Management practices that absolve any responsibility once out of immediate contact are as ethical as believing that no one will see where you dumped your rubbish, or rationalising that one pipe with toxic effluent dissolved in a stream isn't so bad. Protestations of 'it's not my problem' leads to lost customers and despoiled environments. Once we acknowledge that all aspects of a business affect resource and energy use, then we come to recognise that all employees are 'environmental managers'.

Environmental integrity and avoidance of waste are, in my experience, frequently highly valued in a workforce. Seeing unnecessary waste is disheartening. Watching systematic waste at all levels, jaded workers wonder whether they should care. Counter-pressures and dysfunctional hierarchies often rob those at the front lines of their opportunities to be workplace conservationists (APO 1997). Values explicit in the larger society do not disappear on the shop floor or in the office—but our workplaces sublimate these desires.

In an effective strategy, all employees at all levels will have an awareness of the systemic nature of environmental stewardship and its link to business performance, along with the organisation's and eco-industrial cluster's strong commitment. The Natural Step process developed in Sweden by Karl-Henrik Robèrt around four system conditions for sustainability has been used in this fashion in a number of companies around the world (Nattrass and Altomare 1998). Whether with this particular regimen or not, to operate systematically it works best if its stakeholders can see the broader picture (Cohen-Rosenthal 1998a). Peter Dobers and Rolf Wolff (1996) define this kind of 'ecological competence' as:

1. Making use of an interdisciplinary scientific knowledge base
2. Managing inter-organisational relationships
3. Integrating ecological technology into multi-technological operations
4. Creating and maintaining a value-based discourse

Unless those who carry out these practices have the commitment and skills needed to understand and troubleshoot the processes and to operate effectively and efficiently, the technologies involved then it is of little avail (Vale 1996). You cannot have a flexible industrial ecology in a company if the skills and human resource patterns are rigid and narrow. A corollary to industrial ecology is broad-based skills development to enable workers to adjust to their environment and variations affecting their work. Further, rigid classifications that do not allow for multifunctionality block use of the capabilities and redundancies essential to meeting industrial ecology's aspirations. The team becomes a responsive element of an industrial ecology approach that seeks the highest and best performance. Expending unnecessary energy or potentially disrupting lines of communication threaten survival. The approach of making decisions closest to the point where situations are manifest is both more efficient and more effective. Work teams represent a primary organism in the organisation.

An effective strategy for industrial ecology cannot ignore human commitment, skills and social organisation. These enable not only potential physical connections to be accomplished well but also the engagement of the membership of the organisation in a common search for greater connections that increase performance, reduce resource use and add to the vitality of the workplace. In an ecology, different parts may play different roles at different times, but all of the parts are part of the whole ecology all the time. We may be moving materials or cascading energy, but the people who inhabit that system are integral all the time as well. That connection can be a vital and contributing connection through involvement of all who work with a system. Or it can lay dormant. When people are left as rocks on a flood plain, even though they are not active they channel where the water can and cannot go. Our question is: what channels do we choose?

3.8 The role of imagination and creativity

Scientific studies of materials flow and the recording of differences in mass balance, the tracking of particular chemicals and analyses of waste-streams all have value, but nothing surpasses the possibilities of imagination. Common permutations of possibilities can be programmed and can be useful to stimulate ideas, but actively exploring what could be done or what would be the best that could occur is as critical. It nurtures the potential discoveries that randomness and ecologies provide; what we may not have thought of as possible may emerge as the most desirable or splendid possibility. The essence of invention is recognising that connections and combinations that were not perceived as valuable suddenly come together (Stacey 1996).

My belief is that networks of businesses brought together to seek common benefit will uncover possibilities that preordained pathways will miss. These 'Ahahs!' create a commitment that recipe books cannot approximate. Nor can the nay-saying of what is supposed to go wrong a sufficient answer to creative problem-solving about how things *can* be done. Ecological decision-making is inherently inventive and opportunistic. To assert that people are an essential part of industrial ecology is not an anthropocentric argument. In pond ecology, people may or may not have an impact. In industrial ecology, they are central. Celebrating the role of people in industrial ecology leads to better use of materials and energy, but such a strategy makes a material difference in the quality of our lives and the life-giving energy that sustains us.

Part 2
ECO-INDUSTRIAL DEVELOPMENT FOR WHOM? THE ROLE OF STAKEHOLDERS

4
THE ROLE OF GOVERNMENT IN ECO-INDUSTRIAL DEVELOPMENT*

Bracken Hendricks
Institute for America's Future, USA

Suzanne Giannini-Spohn
US Environmental Protection Agency

4.1 Introduction

Eco-industrial development (EID) is defined more fully and eloquently elsewhere in this book, but to understand the role of government in advancing eco-industrial strategies it is helpful to review a few core principles. At its root, EID is about advancing multiple objectives simultaneously through more carefully designed development. By optimising the inputs of capital and labour, energy and raw materials, EID hypothesises that it is possible to greatly increase the benefits that we derive as individuals and as a society from our investments in production.

By integrating public purposes and private-sector tools it is possible to allow the natural functioning of the economy to produce prosperity that is more equitable and more sustainable. By harmonising our development investments within environmental constraints while meeting the social needs of citizens it is possible to reduce the need for regulation while still increasing the wellbeing of the general public. EID proposes not only a more efficient use of resources, both private and public, but also a more effective means of meeting our needs as a society.

EID grounds our understanding of economic productivity within a larger context of the basic systems that sustain and drive productive activities. By looking more closely at material and energy flows, the assimilative capacity of natural systems and the flow of capital investments to communities, individuals, facilities, regions and clusters of industries, EID is able to optimise the production of benefits while reducing waste and its attendant harms.

> * This chapter reflects the views of the authors; it does not necessarily represent the official position of the US EPA, which has no formal policy on eco-industrial development at the time of writing. Discussions of the role of government and specific examples refer to the US federal government, unless noted otherwise.

EID is compelling for many reasons. For businesses and industry it presents new opportunities to increase profitability, differentiate products, avoid government interference and be good corporate citizens. For governments it also holds promise. EID is a practical tool for meeting public challenges efficiently with concrete solutions that improve people's lives. Governments should care strongly about this emerging industrial strategy, and EID practitioners must understand the strong role of public incentives in shaping patterns of development.

4.1.1 The influence of government on development

EID uses ecological models to understand how social, economic and environmental systems influence production. Clearly, government policies are a critical component of this larger network of forces influencing economic and industrial processes. Policy decisions affect the investment, design, production and hiring practices of industry, the consumption patterns of individuals and the spatial patterns of community development. Wherever public funds play a role in development activities, or where regulations create impacts on industry, policy has a profound effect. Understanding the role of public policies and institutions is therefore essential to designing eco-industrial systems.

Unlike many industrialised nations, the US federal government does not have a department or agency charged with setting an integrated, long-term national industrial policy. Even without a national industrial policy, however, federal agencies also have important direct and indirect impacts on the nature of industrial development. The combined impacts of policies on energy, environment, transportation, labour, banking, trade and economic and community development amount to a de facto industrial policy strongly influenced by public agencies and resources at all levels of government, with wide variation nationally. In addition, numerous states and local governments do develop such policies and channel substantial resources into attracting and retaining industrial development.

The evolution of eco-industrial development in the USA, like other industrial approaches, has taken place within this complex policy context. The system of incentives and regulations in some cases encourages, and in others has a negative impact on, the use of eco-industrial strategies to invest in public goods, internalise externalities and increase resource efficiency. If it can be assumed that government policy is a relevant concern in promoting industrial development generally, and eco-industrial development specifically, then two key questions emerge:

- What type of development will result from the current pattern of industrial policies (or non-policies)?
- What policies are needed to encourage long-term stewardship of people, places and the environment through EID and similar systems-based approaches?

In the rest of this chapter we examine how eco-industrial activities may be helped or hindered by ongoing traditional activities of government.

4.1.2 Government as a stakeholder

In addition to understanding the role of government in shaping eco-industrial practice, it is also useful for governments themselves to examine the model presented by EID. EID not only promotes concrete, market-driven approaches to meeting the needs of communities—grounded in the activity of individuals—but also it forges real and lasting partnerships among stakeholders.

Government is increasingly challenged to co-ordinate multiple, complex and competing interests and priorities within a dynamic management setting. Governments are being asked to do more with less and to make more efficient, streamlined and highly targeted interventions into private-sector activities. By keenly analysing the larger systems that extend beyond the boundaries of specific firms, and by building reciprocal relationships among public and private partners, EID introduces a very valuable model for improving the economic productivity of industry and meeting the demands of investors and citizens.

This could be an especially useful strategy in the USA given the current political climate of devolving federal powers and the increased reliance on incentives and market mechanisms to achieve public aims. Within this changing landscape of government roles, EID offers a mechanism for building public–private partnerships that are rooted in performance standards, yielding public as well as private benefits.

At this time, our global population is presented with an enormous challenge: to provide for economic necessities while adapting to ecological constraints. EID offers a real-world solution for reducing the environmental footprint of development while increasing the gains to society. It is also a model that promotes transparency and increases public voice. Governments and policy-makers from many disciplines and jurisdictional levels can benefit from encouraging such 'industrial symbiosis'.

4.1.3 A history of public involvement within eco-industrial development

EID in the USA exists as a voluntary activity of individual firms seeking to enhance resource efficiency and save costs rather than as a legislatively mandated, government-funded programme to reduce pollution or rebuild the industrial base. Eco-industrial projects in the USA have largely originated from the local community rather than in federal or state government and are financed through similar channels to traditional industrial development, but with individual firms also reducing their costs through enhanced efficiencies.

The earliest EIDs in the USA were closely identified with key public supporters. Projects driven in part by local governments attracted the attention of federal agencies through the recognition of the President's Council on Sustainable Development (PCSD) and organisations such as the National Association of Counties (PCSD 1996b). These innovative experiments became laboratories for interagency co-operation. In many cases federal agencies provided critical seed money to local authorities to initiate the projects and, over time, funds from many agencies have provided partial support for several EIDs across the nation. The motivations for agency involvement have come from a range of public mandates: to support sustainable development (US Department of Energy [US DOE]), to minimise the environmental impacts of development near sensitive coastal habitats (National

Oceanic and Atmospheric Administration [NOAA]), to promote brownfield redevelopment and economic revitalisation (Economic Development Administration [EDA] of the US Department of Commerce [US DOC]), and to encourage pollution prevention practices (US Environmental Protection Agency [US EPA]).

Governments are interested because EID can benefit local, state and national economies and the environment. In the USA it is significant that strong public-sector interest has come from local governments, where policy is played out on the ground. Because it integrates multiple objectives within the context of a site-specific installation, EID is an important public and private strategy, because too often government programmes can become overly 'stovepiped' and difficult to navigate for community practitioners. EID can channel these various federal programmes and funding streams toward beneficial outcomes in particular places, to meet the goals set forth by firms and communities.

Federal agencies have for several years been struggling with the question of how to channel economic (re)development money toward more sustainable development projects that leave smaller environmental footprints. Traditional economic engines for neighbourhoods such as industrial parks and shopping centres are very land-intensive and are designed to operate with use of large and often wasteful flows of materials and energy. Historically, agencies have not evaluated projects in terms of their efficiency or environmental performance; nor have they monitored the contribution of federal development investments to urban sprawl and the disinvestment from existing communities. The federal government has supported EID as one tool for tying public support of development projects more closely to outcomes.

Several other nations explicitly promote EID as part of their industrial strategy. For example, Japan's 'central government provides both technical and financial support to local governments to establish an area (eco-towns) where zero waste (mostly solid waste reduction) is promoted regionally through various recycling and industrial symbiosis efforts' (Morikawa 2000). The Industrial Estate Authority of Thailand has launched a pilot project to develop five of its industrial estates into eco-industrial estates (Chavanich 2001). In the Philippines, the Board of Investments/Department of Trade and Industry Priorities Plan promotes the conversion of existing industrial estates into eco-industrial developments.[1] In the USA, although numerous agencies have used their programmes to support eco-industrial projects it is uncommon for policy guidance to explicitly mention EID as a preferred policy tool. One notable exception is the US DOC EDA, which awards preferential consideration to grant and loan applications that employ eco-industrial strategies in industrial development projects.

4.1.4 The role of government in a market economy

According to economic theory, a properly working competitive market results in the most efficient allocation of resources, without the interference of governments. However, in practice, there are numerous market failures that interfere with realising these efficiencies. These market failures frequently justify government actions to protect public interests and the functioning of the markets themselves. Information failures can prevent prices from fully accounting for the impacts or benefits of economic activities. Distorting

---

1 For more information, see the Private Sector Participation in Managing the Environment (PRIME) Project Industrial Ecology Module website at www.iephil.com.

subsidies can also lead to price signals that do not reflect the true costs of production. Barriers to entry can prevent new firms from developing innovative technologies and production methods and can even discourage entirely new markets from developing. Some costs are not borne either by the producers or by the consumers of products, and some benefits accrue to society as a whole outside of market transactions. These 'externalities' further skew production away from what would be efficient or optimal to society as a whole. Patterns of inequality can further hurt the functioning of markets, as lack of access to capital, prejudice, inadequate education or decaying infrastructure disadvantage communities and regions unfairly.

Together, these market failures can result in the perpetuation of inefficient or destructive practices and the development of a mix of products in the economy that does not provide for the greatest benefit to consumers, firms, workers or the environment. It is the unique role of the public sector to establish the framework within which market activities unfold. Without appropriate regulations, the enforcement of rules and interventions to correct for distortions, markets cannot be free, and the playing field will not automatically be level. In industrial development, policy decisions directly shape whether companies bear the cost of pollution and whether firms develop programmes for investing in their communities and the skills of their workers. These decisions then have large secondary impacts on the adoption of clean or dirty technologies for production and on the long-term vibrancy of the regional economy.

Economics says much about improving the efficiency of production, but much less about establishing equitable and socially desirable priorities. Determining goals for just outcomes and establishing mechanisms for economic redistribution also fall outside the realm of the private sector. In short, governments and markets are closely linked. Business is the engine of production, but government is essential to protect and maintain the systems that allow for orderly commerce to unfold. Policies establish incentives, create codes of conduct and direct private activities, for better or worse. Governments harmonise trade, establish currencies and ensure that the vulnerable do not bear the costs of development in unfair proportion. When markets fail, it is the obligation of government to intervene. Some of these traditional public activities are shown in Box 4.1.

4.2 Correcting market failures

4.2.1 Internalising environmental externalities through regulation and enforcement

A universal precept of most countries' environmental protection laws, the 'polluter-pays' principle, makes polluters liable for the costs of waste they generate. Landfill tipping fees, smokestack scrubbers, permit fees, Superfund liability (CERCLA 1980) and administrative costs of compliance—all attach a charge to those activities that result in pollution of the environment. Such a foundation of environmental law must support any effort to promote EID.

Environmental laws, regulations and enforcement establish a baseline (floor) of environmental performance that is needed to protect the environment. Uniformly

Correcting market failures

Market failures are addressed by:

- Correcting price signals to account for externalities such as pollution
- Ensuring competition through regulation of market power and monopoly
- Overcoming collective action problems

Providing public goods

Public goods concern:

- Access to education
- Public data and information systems
- Research and development
- Common infrastructure (roads, bridges, publicly owned treatment works)
- Public safety, regulation, national defence and other police powers

Achieving socially desirable objectives

Socially desirable objectives are achieved by means of:

- Taxes, subsidies and other means (to achieve social equity)
- Equal opportunities and economic development
- Civic engagement, democratic access and participation

Box 4.1 **Traditional activities for government intervention**

enforced laws 'level the playing field' because all companies will bear similar costs for the waste they generate and dispose of. These standards internalise social costs into the market prices consumers pay for using the products and services of those companies. Only then do markets reward cleaner and more efficient producers who can eliminate waste entirely either through redesigning production processes or through creating secondary markets for their waste-streams. Public policy choices are directly tied to production costs and thus to companies' motivation to minimise waste through recycling and re-use, environmental management systems and other pillars of EID.

Many environmental laws also arise from public mandates to protect public health. Without a foundation of human health and safety standards, governments have not been able to build successful 'beyond compliance' voluntary EID programmes. The failure to regulate effectively creates incentives to produce pollution and waste. A regulatory framework that rewards firms for cleaner production and that forces polluters to internalise the costs they currently transfer to the public at large will go a long way to promoting a level playing field for EID.

However, reliance on 'command-and-control' environmental regulations that specify the type of technology to be used rather than allowing firms the flexibility to meet set emission or effluent standards can also impede innovation. Where regulations specify safe pollution levels but grant producers the freedom to meet those thresholds through a variety of cost-effective means, development and diffusion of new technologies is greatly encouraged and the economic burden associated with regulation is minimised.

Within command-and-control systems there is no incentive for business to improve on the best available technology at the time the regulations are promulgated. The long lag-time in legislative and regulatory processes in most countries results in the use of technologies even after they become obsolete, often encouraging outdated technology even before the laws and regulations specifying their use can go into effect. This type of regulation can lock industry into poor performance and discourage adoption of more innovative solutions such as EID, preventing the full realisation of their potential benefits.

Like emission standards, the regulated treatment of industrial waste is highly relevant to EID. Restrictions on the duration that waste may be stored before recycling, restrictions on the transfer of certain waste between different companies and the high administrative costs of shipping have greatly increased the cost of hazardous waste generation. Internalising these costs through regulation provides a strong incentive for source reduction and changes to use of less toxic process chemicals. However, some hazardous waste disposal regulations also may have unintended negative consequences for eco-industrial networking, by making it more difficult to 'close the loop' for certain waste-streams, which end up being sequestered in hazardous waste landfills. Innovative regulatory tools, such as bubble permits, may provide a solution for measuring the net output of pollutants from industrial systems, allowing more creative approaches to reducing hazardous materials within those larger 'industrial ecosystems' while still protecting public health.

The fragmented nature of environmental protection laws in many countries, including the USA, does not integrate well with this 'systems approach', which characterises EID. For this reason, the comprehensive yet adaptable performance-based management strategies advocated by EID not only can improve industrial performance but also can offer insights into improving the process and performance of regulation.

4.2.2 Creating and expanding markets

Often, recycling and re-use of materials are impeded because there is no efficient way for buyers and sellers of recyclables and waste to find each other; in other words, there are very limited markets for many types of waste.

Government procurement policies have created markets for some recyclables. In an often-cited example, a US presidential executive order mandated post-consumer recycled content in paper to be bought under contract for federal government use. Although small quantities of such paper were being manufactured at the time of the executive order, it was very expensive, retarding market expansion. The huge market demand created by federal government preferential procurement policies allowed new economies of scale to be realised, lowering prices and making recycled paper more competitive to private-sector purchasers (see US EPA 2001b: 111-42). In a similar fashion, the first US commodities market in recyclable paper and plastic by the Chicago Board of Trade (CBOT) was funded by the US EPA in 1996 (Field 1996). Though CBOT no longer trades recyclables, this market void is now filled by several electronic marketplaces; for example, the Global Recycling Network lists 13 recycled commodities for purchase, sale or exchange (www.grn.com).

A second type of market can also be created—one that does not facilitate trading in the waste product itself as a raw material but one that allows the sale of a portion of reductions in pollution levels. In the USA, the 1990 Amendments to the Clean Air Act (CAA 1990) provide market mechanisms for trading air emission credits for sulphur dioxide and nitrogen oxides, among other air pollutants. Firms that reduce their air pollution to levels *below* the permitted level gain air emission credits, which may be sold or traded to other companies. Thus the enhanced resource efficiency of EIDs can become a source of income, whereas less-efficient producers must pay additional costs for their 'right' to pollute.

A second approach to ensure that the prices of goods and services include the full costs of production, in order to motivate eco-efficiency, is the imposition of charges. Pollution charges are particularly useful in reducing pollution that is not subject to regulation. In the USA, for example, some local governments charge households for kerbside rubbish collection, with fees reflecting volume (rather than a flat rate). A recently released study sponsored by US EPA indicates that such programmes result in decreased volumes of rubbish going to landfill and increased recycling and composting (US EPA 2001b).

A similar programme in the Netherlands prohibited unlicensed discharges into surface waters and imposed charges on polluting emissions. The Netherlands programme has been one of the most extensive and successful experiences with charges for water pollution in OECD countries (World Bank 2000). Eco-industrial networks may reduce their disposal charges, saving costs and leading to lower prices for goods and services than non-networked companies. Without such policies, firms and individuals will see no reward for changing behaviour, and the costs will be distributed inequitably, regardless of performance.

Even small pollution charges can motivate companies to decrease pollution if the resultant reduction in levies reduces their operating costs. In the People's Republic of China, companies that exceed their permit limits are fined proportionately to the volume of pollution released. Although in some cases fees have been set too low to substantially alter production processes, many small to medium-sized enterprises have been motivated to implement low-cost or no-cost behavioural or production changes in China, even as a result of these small fees.

4.3 Providing public goods

Public goods are 'necessary economic goods and public services that are available to everyone without restriction or qualification' (Manki 1997). These products and services benefit a large segment of the public and often require large upfront investments. These goods include such 'commodities' as highway construction, public education, information systems and public health. When markets leave the provision of public goods to the actions of individuals, they will be supplied in insufficient quantity. Because any consumer receives only a small fraction of the total benefit, he or she will not choose to invest in these goods at a level consistent with their true social value. This gives rise to 'collective action problems', where groups of independent actors will not maximise their collective wellbeing.

It is just this sort of systemic breakdown that EID was designed to address. Like business improvement districts, condominium associations or other co-operative, voluntary, quasi-public partnerships, EIDs create opportunities for private entities to come together to create shared solutions to common problems on matters of collective interest. In Kalundborg, Denmark, the first eco-industrial exchanges were born out of a collaborative investment in shared water treatment infrastructure among several firms.[2] By devising shared institutional arrangements, collective investments in mutual gains became possible.

Public–private partnerships are becoming increasingly important tools for governments, developers and industry as they try to harmonise their activities to make development both faster and more responsive. Public goods are a central driver of partnerships in EID, and it is the ability to maximise overall benefits for communities and firms that makes EID compelling and that drives a continuing government interest in fostering the approach.

4.3.1 Community development: encouraging positive externalities

So far we have primarily discussed environmental systems, but each of the distortions mentioned above also plays out within communities and in the valuation of 'social capital' and workforce investment as well.

Just as overproduction of harmful by-products is encouraged by policies that allow costs to be externalised, it is possible, too, for firms to under-produce beneficial goods because these benefits accrue to the general public rather than being fully captured within the price that firms can charge. Investment in higher wages and safety standards, employee training programmes or choosing to site new facilities in distressed communities on urban brownfields can all produce social benefits beyond those that are realised by firms.

Improved community stability, quality of life and more secure social networks all result from investing in the people and places that support businesses, but these benefits are not captured in a company ledger. As a result, private activities alone will consistently under-invest in these public goods. This trend can be seen clearly in the decisions of manufacturing industry over the past half-century, as firms have increasingly competed by depressing wages, eliminating investments in career ladders for worker training and moving production offshore.

Investing in public improvements to local communities, both people and places, frequently is a central consideration in EID. EID tries to optimise not only the material flows through industrial systems but also recognises the importance of human capital and investment in social networks as primary inputs to the industrial system. An excellent example of using EID as a strategy for community reinvestment is found in the work of the Minneapolis Green Institute and its extensive economic development and job training activities, profiled elsewhere in this book.[3]

Distortionary subsidies and weak regulation can have a damaging effect on workforce investment and labour markets. The heated competition among states to recruit and retain new industry has led some areas into 'smokestack chasing', based on weakening labour and environmental standards, driving down wages and escalating tax incentives. These industrial recruitment strategies have often had unintended consequences, depressing standards of living, locking in low levels of public services and committing

---

2 See Chapter 16 and www.unepie.org/pc/ind-estates/casestudies/kalundborg.htm.
3 See Chapter 17 and www.greeninstitute.org.

localities to large-scale construction of new infrastructure without proportional rises in revenue (LeRoy 1994). Inadvertently, this trend has also encouraged abandonment of older industrial communities, sprawl development patterns, disinvestment from existing skill development programmes for low-wage workers and increased public expenditure on duplicative infrastructure while existing infrastructure is not maintained.

EID instead offers a strategy where communities compete for development not based on reducing the costs of the inputs to production but by maximising the overall value of production. Through location near existing networks of resources and improved relationships with the suppliers of their primary production inputs producers can gain cost advantages, and by investing in local capacity and reducing social costs they can create long-term benefits that improve their competitive position. These advantages include proximity to a network of related suppliers and opportunities for exchange of materials and energy, access to a skilled workforce and improved access to information and knowledge of production processes. Producers also benefit from shared regional investments in transportation networks and infrastructure that can sustain a clean and innovative manufacturing base.

Development approaches that rely on high standards of quality and increased value added to production are often termed 'high-road' strategies (Bosworth and Rogers 1997). These approaches have been proven to be viable in a number of communities where public–private partnerships have demonstrated that economic competitiveness is compatible with public protection. The Corporation for Enterprise Development (CFED) has articulated a set of principles for effective economic development that outlines the proposition of high-road economic development and closely mirrors the public and private collaborative partnerships that can drive EID (AFL–CIO 1998). These principles state that:

- Strong economies compete on the basis of high value, not low cost.
- Investments in development capacity provide the basis of future economic growth.
- Government is an indispensable partner in the process.
- Government services must meet quality standards for accountability, effectiveness and accessibility.
- Economic development is for everyone, not just business persons.
- Competitiveness and equity are two sides of the same coin.
- Quality leadership can turn economies around.

Empirical data supports the long-term viability of this approach. Although many industry sectors have increasingly moved production to less-regulated and rural locations for cheap land and low tax rates, this growth has been strongest among low-skilled, low-value-added and low-capital-intensity manufacturing. In contrast, urban areas continue to be highly attractive to high-value-added and highly capital-intensive industries such as printing, electronics, transportation equipment and instruments. These producers benefit from their proximity to markets, suppliers and transportation hubs. They are also highly dependent on access to skilled workers and thus are willing to accept higher local taxes and land prices in order to receive the benefit of higher expenditure on education and infrastructure (Nelson 1993). For these industry sectors, the value proposition driving their location decisions is more complex than 'low-road' economic development models suggest.

In the high-road scenario and in the low-road scenario policy choices are at the root of business decisions. In both cases an efficient and profitable market equilibrium can be achieved, the biggest difference between these two development paths being their relative creation of harm and benefit to the shared public realm—felt by citizens and communities and managed by public institutions. For EID these observations are critical. If it is to become more widely adopted, EID will compete on the value of networks, the benefits of increased capacity to address collective action problems and the use of improved efficiency to improve value creation. In an incomplete system of private accounting that externalises costs and diminishes community values, EID will operate at a competitive disadvantage. It is this public-good component of private EID that demands partnership models—including collaboration with public and non-profit entities.

4.3.2 *Education, training and research*

Education and training are public goods because their value to society as a whole far exceeds the benefit that is captured by individuals through improved earning potential. Industrial development at one time provided reliable career ladders into long-term stable skilled and semi-skilled employment. With the increasing pressure of global competition, the costs of providing the tools and mechanisms for skill development and career advancement through employer-driven training programmes has been increasingly squeezed out of corporate budgets as reflecting non-productive overhead costs. This disinvestment from workforce capacity can be sustained over the short term, but in the long run it represents a drain on the capacity and competitiveness of the industrial base. There is, as a result, a pressing national interest in ensuring that less-skilled workers have direct access to meaningful training and opportunities for career advancement. This realisation is a primary justification for many economic development investments. EID, however, is uniquely positioned to actually build the connective institutional capacity with community-based educational and research institutions and to support the creation of a real track toward sustainable living-wage jobs in growing industries.

Similarly, investments in research that expands our collective understanding of the world around us but that does not directly yield profitable, patentable products will not be undertaken at sufficient levels without some public investment. Much basic (as opposed to applied) research is government-funded. Some areas relevant to EID that have been funded by the US government include:

- US DOC, EDA: feasibility studies for site-specific EID
- US DOE: combined heat and power technology
- US EPA: computer models such as the Facility Synergy Tool (FaST, www. smartgrowth.org/library/typelist.asp) and the *Handbook of Codes, Covenants and Restrictions for Eco-Industrial Development* (WEI 2000)

Applied research and commercialisation of experimental technologies are further areas for potential public–private collaboration through EID.

4.3.3 Data and information

EID is a knowledge-based production model that leverages information on regional inventories, production processes and material and energy flows to add new value to the overall industrial system. This can require costly collection and analysis of large amounts of data to identify opportunities for material exchanges and energy efficiency. Thus it may be an appropriate government activity to provide access to this 'public good'.[4]

Access to high-quality local data and information management systems are also an increasingly critical strategic resource, as businesses and others must manage production and distribution information in real time and examine the broader environmental and community impacts of their activities. GIS (geographical information system) tools and rapid growth in demand for place-based data will be an increasing factor in public policy debates, community development strategies and industrial management techniques. EIDs can serve both as useful anchors for local data collection and maintenance and as important users of information products.

More than raw data, though, the public needs access to complex information to make better development decisions. The management of industrial environmental performance has been revolutionised in the past half-century through increasing requirements for public reporting and disclosure of information. Pollutant Release and Transfer Registry/Toxics Release Inventory programmes in the USA, authorised by the 1986 Emergency Planning and Community Right-to-Know Act, allow public access to plant-specific information about the release of toxic chemicals, without ranking performance; these programmes often lead to public pressure on plants to reduce their use of toxic chemicals (US EPA 2001b: 153-72).

Similarly, 'red-lining' of minority neighbourhoods by lenders has been blocked and community finance has been greatly improved through increased transparency required under the 1977 Community Reinvestment Act (CRA 1977). Similarly, reporting laws on public subsidies have been used effectively at the state level to demand accountability for public outcomes in subsidised economic development activities. In international development, the Programme for Pollution Control, Evaluation and Rating (PROPER) is credited with improving compliance with environmental laws in Indonesia through a government programme that publishes ratings of industry compliance levels (World Bank 2000).

Recognition of superior 'beyond compliance' performance also provides good publicity. The PCSD in 1994 named four US EIDs—Cape Charles, VA; Chattanooga, TN; Brownsville, TX; and Baltimore, MD—as pilot projects that helped the communities attract funding through several government programmes as well as promoting recruitment of new tenants to the eco-industrial parks (EIPs). When the Cape Charles EIP opened, it was fully rented with a waiting list of potential tenants to locate in a rural Virginia community that had for many years struggled with recruiting new economic development opportunities.

Inadequate knowledge of available technology, low levels of environmental awareness and poor enterprise management are often linked to waste and inefficiency. The creation of national information centres and the diffusion of new technology via workshops have been two strategies used by many governments. A good resource for information on

---

4 For examples of government-funded resources that industries can use, see www.oit.doe.gov/bestpractices/pubs.shtml for energy-management 'best practices'

sustainable community development is available through the US DOE's Centre of Excellence in Sustainable Development website.[5] This resource provides links to computer tools for community design and decision-making, land-use planning, green building, transportation, economics, industry and disaster planning. EIDs have provided proving grounds for testing many of these innovative models and best practices, operating in an integrated and comprehensive manner.

4.4 Achieving socially desirable goals: income redistribution and an end to incentives to create waste

This chapter so far has discussed numerous examples of how policy drives development incentives. Improving incentives can have far-reaching impacts—extending the influence of policy well beyond the areas directly affected by public investment and regulation. For eco-efficient strategies to become a widely embraced approach to industrial and community development, public policies must be refined, with greater attention paid to the unintended consequences and long-term objectives of public actions. Demanding more return on public investments by linking subsidies to high performance levels is an effective way to help ensure that industrial subsidies lead to the desired changes in job creation, income and tax revenues. Public investments and regulations are directly linked to social equity by reducing harm and providing broad access to benefits.

Taxes and subsidies are the most basic tools for achieving equity and are the primary mechanisms for encouraging industrial and economic development. As previously mentioned, these market interventions can have wide-ranging effects. Some of these impacts will enhance eco-industrial activities. If not carefully designed, however, taxes and subsidies can hurt incentives for EID and improved efficiency of operations. Politically popular subsidies for extraction of non-renewable energy sources and virgin materials can promote consumption and waste rather than encouraging conservation. Highway subsidies promote freight shipments by diesel trucks and increased commuting; further, this subsidised ease of mobility can impede the local trade that supports an eco-industrial network.

Subsidies to the photovoltaic sector in contrast have been used to favour the commercialisation of renewable energy (McVeigh *et al.* 1999). In a local government programme in California, for example, customers of the Los Angeles Department of Water and Power (LADWP) who take part in the Department's Solar Rooftop Incentive programme (the Solar Rooftop Incentive for Residential and Commercial Customers) will be subsidised in purchasing and installing a grid-connected photovoltaic solar system for their home or business. The incentive amounts to US$3 per watt for systems manufactured outside the city and US$5 per watt for systems manufactured inside Los Angeles city lines. LADWP will pay up to US$50,000 for each residential system and up to US$1 million for a commercial installation.[6]

5 www.sustainable.doe.gov/toolkit/toolkit.shtml
6 See the LADWP website, at www.ladwp.com/ whatnew/solaroof/solaroof.htm.

Similarly, subsidies for rail travel under the 1991 Intermodal Surface Transportation Efficiency Act (ISTEA) and the 1998 Transportation Equity Act for the Twenty-first Century (TEA-21) have reduced the vehicle emissions of workers commuting to jobs that exceed walking distance. TEA-21, which expires September 2003, is currently undergoing reauthorisation under the acronym TEA-3. The US government also subsidises travel costs of federal government employees that use car pools or public transport. Private companies seeking to reduce parking-lot use and traffic congestion caused by their employees have adopted similar programmes. Opportunities for greater cost savings and improved benefits can be increased when many firms establish co-operative arrangements under an EID approach, to share fixed costs and achieve economies of scale.

At a systemic level, government may eventually choose to reform the incentives built into taxation and regulation. Environmental tax shifting has been proposed as a means of moving government revenue streams from dependence on taxing positive activities (such as productive labour through an income tax) to taxing those causing harmful outcomes (such as pollution and waste). Such a system of taxation could be made equally progressive, while radically shifting incentives for the movement of private capital toward clean production and activities that reinvest in communities. Numerous individual policies have already moved in this direction. Tradable emissions credits, brownfield legislation and Smart Growth policies all make important improvements to regulatory and spending priorities that work with markets to improve public outcomes.[7] Reducing subsidies for the use of virgin raw materials will further strengthen the ability of efficient industrial practices to compete on a level playing field.

4.5 An eco-industrial policy prescription

The policies that will best support the promotion and expansion of EID are in many ways the same as policies advocated more generally for industrial efficiency, urban reinvestment, regional planning and the development of liveable communities. Because it takes a systems approach, however, EID places an additional focus on relationships among firms, the nature of work and the flow of materials within economic and industrial systems.

Because of synergies accruing to the networked sharing of resources, EIDs will outperform financially and environmentally in a policy climate that also improves environmental performance of conventional industry. Such policies: (1) establish a baseline 'floor' of environmental performance that protects human health and safety and the environment and that enforce it uniformly; (2) price resource use and waste disposal realistically to include the costs of adverse ecological and health impacts; and (3) promote access to information about voluntary strategies to improve resource efficiency.

EID will flourish in a policy climate that encourages greater density in industrial establishments in order to facilitate material exchange and to promote proximity to sources of labour. Policies that facilitate reintegration of industrial practices into the urban fabric

---

7 For a comprehensive review of US market-based mechanisms for environmental protection, see US EPA 2001b.

through cleaner production and use of less toxic processes, intermodal transportation access and modern infrastructure networks will all go a long way toward promoting EID. Policy can support the management of products for their entire life-cycle through re-use and reclamation of embedded energy and materials, and it must be made a priority to reinvest in existing social networks, human resources and built infrastructure. In short, the eco-industrial policy agenda is fully compatible with the agenda of a broad coalition of forward-looking economic development advocates.

The seeds of this policy framework can easily be found within existing policy. Policy-makers, developers and community advocates alike must integrate EID strategies and systems analysis into their economic development and planning activities. In this way, government policies can begin to promote more efficient use of all resources: capital, human, social and natural.

4.6 Summary and conclusions

Government has an invaluable role to play in every phase of EID. Public institutions at all levels of government function as important stakeholders with a responsibility to advocate for maximising the public value from investments in production. Clean production and pollution prevention are often a centrepiece of EID, but the policy issues involved are clearly much broader. Regulatory enforcement, appropriate incentives and public access to information all play critical roles in EID and in helping to prevent the negative social and environmental impacts that result from market failures.

To truly foster and promote EID, government must be ready to lead, support and follow in equal measure. It must **lead** in laying the policy foundation that will permit systems-based approaches to regional economic development to flourish. It must **support** EID activities where appropriate through programme funding, provision of technical support, information resources and planning assistance and through responsive and flexible regulatory policies based on real outcomes in communities. Last, government must **follow** the lead of innovators, advocates and leaders both in local communities and in industry. Government must be ready to respond as true public servants to ensure that development serves public needs.

Government is clearly a relevant partner in community economic development and in advancing the co-ordination of industrial systems through EID. Government programmes can provide critical pieces of the financing puzzle for EID as it does for development activities generally. But it is only one contributor to a much broader set of relationships. Public initiatives or interventions will never form the core of programme development. EID must be a partnership led by private firms, supported by public policies and potentially assisted with targeted public financial support. There is a clear public interest in ensuring that this cleaner, more strategic and more efficient approach to economic development receive a fair chance of success.

Appendix

Detailed policy descriptions

Policy shifts to support eco-industrial development

Eco-industrial development (EID) will be supported by several key policy shifts that are emerging in a wide array of public programmes. These new approaches include the following.

- Aligning government policies and programmes to promote efficient use of material and human resources: energy, environmental, fiscal, health, housing, industrial and transportation. This includes:
    - Establishing a baseline 'floor' of environmental performance and uniform enforcement, to protect human health and environmental integrity
    - Use of flexible performance-based regulatory tools that encourage innovation and continuous pollution reductions
    - Use of compliance-assistance programmes to build co-operative partnerships with industry

- Improving support for efficient and equitable investment in human resources, social capital and education. This includes:
    - Protecting workers from the consequences of production changes toward more efficient methods and from environmental enforcement actions on firms
    - Discouraging 'abandonment' strategies by firms toward existing workers, communities and physical infrastructure
    - Supporting re-use of current investments in people and places, including improved worker training and Smart Growth strategies

- Improving access to information of all types in production design and decision-making. This includes:
    - Promoting development of core information infrastructure, and the use of decision-support tools that account for community and environmental values
    - Improving pricing signals on resource use and waste disposal to include the full costs of adverse ecological and health impacts
    - Ensuring that markets provide stronger incentives for the use of clean technologies and resource efficiency
    - Reduction of subsidies rewarding over-consumption of limited resources

These changes are reflected in many ways within the following policy areas.

Energy efficiency and renewables

EID opens tremendous opportunities for improving energy efficiency and developing new generation capacity. Utility companies have shown great interest in EID as a technique for demand-side management, as a tool for revitalising a manufacturing customer base and as a potential source of co-generating production capacity. The Burlington, VT, eco-industrial facility is designed around a wood-burning power plant; that at Londonderry, CT, is anchored by a high-efficiency natural gas facility; the Red Hills EcoPlex, MS, is being built around material and energy exchanges, with lignite coal mining and a power generation facility.

Energy systems play a large role in EID from the perspective of both generation and consumption. By cascading energy flows through multiple productive uses one can greatly improve the productivity of each unit of energy consumed and create marketable uses for thermal energy that is currently lost as waste heat. Design solutions can play an important role in improving the efficiency of EID with strategies such as green building and day-lighting, providing cost savings and improved working conditions for employees. Policy decisions regarding energy supply, regulation and pricing will strongly shape the incentives and direction of energy conservation and therefore will affect the shape of EID over the coming years.

Federal programmes
The US Department of Energy (DOE) and Environmental Protection Agency (EPA) have a number of programmes to support such work, including Energy Star, Green Lights, and Brightfields; in addition, numerous research and development (R&D) programmes exist within the National Laboratories and the US Department of Commerce (US DOC) that can support this area.

Pollution prevention, environmental protection and sensitive ecosystems
Increasing emphasis is being placed on voluntary programmes in pollution prevention (P2) and environmental management systems (EMSs) by environmental agencies in the USA and other countries. It is cost-effective and a less adversarial strategy than 'command-and-control', and can yield additional benefits in terms of quality of life, worker safety and performance enhancement. Both P2 and EMSs are at the centre of eco-industrial exchanges. Additional environmental policy changes that will support EID include the imposition of taxes on pollution and the development of tradable permits such as the SO_2 programme. Policies to address global climate change and to reduce greenhouse gas emissions are likely to be a positive driver for EID, either through encouraging greater energy efficiency or even potentially as a source of revenue through the sale of carbon emission reduction credits.

Programmes for the protection of sensitive ecosystems can also be opportunities for partnership or investment funding for EID. The 1972 Coastal Zone Management Act of 1972 (CZMA), which is administered by the National Oceanic and Atmospheric Administration (NOAA), has been used to fund more environmentally sound economic development in sensitive estuarine environments. Some of the early seed money for the eco-industrial park at Cape Charles, VA, came from CZMA. Because EID manages the full flow of materials and energy through the 'industrial metabolism' of a firm, it is an ideal strategy for promoting economic development near areas where subtle pollution impacts can have significant impacts on the ecology.

Federal programmes
US EPA offers a wide range of programmes in support of P2 and EMS for improving businesses operations and in support of community-based environmental protection. The International Organisation for Standardisation ISO 14000 series of standards address environmental management and can provide useful voluntary guidelines for improving environmental performance. NOAA, the US Department of Agriculture (USDA) and the US Department of Interior (DOI) also administer numerous programmes that support more sensitive development and land use in critical habitat such as wetlands and coastal estuaries, either through programme funds or access to information.

Transportation systems and public transport
Transportation investment is one of the largest federal funding streams to state and local development. The money is generally awarded in block grants to state departments of transportation, which have latitude in how the funds are directed. Transportation systems have tremendous impact on the movement of goods to and from markets and on the planning of infrastructure and other public investment. The siting of industrial facilities is currently closely linked to highway access. Intermodal transportation access in older industrial and urban areas is often quite good. Transportation networks can be used as a source of competitive advantage for regional eco-industrial business clusters and as an engine for redevelopment by EID facilities interested in revitalising urban communities and reversing patterns of disinvestment.

Transportation patterns, social policies and labour markets are also very closely linked:

> While jobs once were concentrated in cities, two thirds of new jobs today are created in the suburbs. More than half of these new jobs are not accessible by public transportation. And 94% of welfare recipients do not have cars' (Center for Community Change, www.communitychange.org).

Transportation funds often include resources that can be designated for improving equity, job training and upgrades to public space and pedestrian access. These resources are likely to involve engaging state-level departments of transportation.

Federal programmes
The main source of all transportation funding is the 1998 Transportation Equity Act for the Twenty-first Century (TEA-21), which revised the 1991 Intermodal Surface Transportation Efficiency Act (ISTEA). The lead agency for transit development at the Department of Transportation is the National Transit Administration.

Land-use and infrastructure development

Land-use decisions are primarily controlled at the state and local levels of government and are often used as a powerful tool for industrial recruitment tied to tax incentives and other favourable treatment. Land use in many ways ties together all the other policy areas mentioned in this section, as it is the policy framework that organises activities in particular places where development activity takes place. Zoning can create difficulties for siting new industrial facilities, but innovative zoning approaches can also be used to favour cleaner industrial development near residential areas or sensitive environments. Brownfield redevelopment involves the reclamation of abandoned and possibly environmentally contaminated land to encourage infill development. This is fundamentally an eco-industrial and resource-efficient concept applied to land-use and spatial planning.

Maryland has pioneered land-use and infrastructure planning at the state level to encourage greater reinvestment in communities and 'Smart Growth'. Maryland has established a system of priorities that gives funding preference to maintenance of existing physical infrastructure, such as roads and sewer lines, over the creation of new infrastructure in greenfield locations. This is an important example of regulatory and budget incentives driving development patterns. Like transportation, public infrastructure investments can heavily subsidise development, and public investments play a major role in determining industrial location and community development patterns. Infrastructure development is often a key component of economic development and industrial recruitment strategies.

Federal programmes
The Economic Development Administration (EDA) in the US Department of Commerce (US DOC), funds planning, infrastructure and economic redevelopment projects in distressed communities, including military base closures and disruptions in the local economy as a result of flight or collapse of the main local industry. For projects involving industrial parks, which make up almost 40% of its portfolio, the EDA has specific guidance that gives preferential treatment to EIDs. The brownfields programme is administered jointly by US EPA and the Department of Housing and Urban Development (HUD) and currently enjoys strong bipartisan support. Many other agencies invest in some way in physical infrastructure development and planning. This area more than any other, however, requires true partnership and co-ordination at the local level.

Workforce investment, living-wage jobs and lifelong learning

Industrial development means jobs, and cleaner production is a better neighbour. EID is often seen as a strategy for returning a base of industrial jobs to the inner-city poor and as a potential engine for community revitalisation. The work of Michael Porter* and others has shown that urban labour and consumer markets are an undervalued and under-utilised resource in the USA. The erosion of the industrial base has also undermined the development of career ladders into 'middle-class' professions and skilled crafts for the US working-class population. These problems can especially be felt in inner cities and among many minority populations. Investing in the skills of employees and the system of workforce development are critical elements of eco-industrial strategies. Improving quality and standards for jobs, wages, skill development and workplace safety are key elements of designing industrial systems that maximise the real economic vitality of communities.

Building effective training programmes to reinvest in the skills of people and the quality of the labour market can be an effective means of attracting new resources to development efforts, broadening community partnerships in support of projects, promoting social goals and accessing new federal support for EID. Numerous federal policies address workforce development as a social programme, within education policy, or as a tool for industrial recruitment. Federal money in these areas is generally devolved to states, where it is made available to community-based projects. The guiding legislation for worker training is the 1998 Workforce Investment Act (WIA), which includes a specific role for workforce development within economic development strategies and incentives for job creation.

Federal programmes
The Welfare Reform Law of 1996 created the Temporary Assistance to Needy Families (TANF) programme, which provides cash assistance to low-income families with children while they strive to become self-sufficient. The Workforce Investment Act (WIA) authorised the Welfare to Work (W2W) tax credit for employers hiring TANF recipients, while the 1994 School to Work Act funded development of vocational programmes until it expired in 1999. The US Departments of Education, Labour, and Health and Human Services are important federal partners in supporting workforce development, adult literacy and skill-development programmes. The community college system is also a very important community partner for vocational training in economic development.

Housing, community development and finance

There are many agencies and programmes providing access to financing and other resources for community development. HUD is the lead federal agency for co-ordinating federal community preservation and development activities. It is active in promoting equity in access to economic opportunities for the nation's urban poor. As a result, its influence extends well beyond housing issues to include many forms of real estate finance, statistical research on issues relevant to regional economic development and business recruitment for low-income areas, including tax incentive programmes for business development and training programmes. In the Community Builders programme, for each major region or metropolitan area a HUD employee is designated as a point of contact to facilitate the development of local partnerships and to assist in the co-ordination of public resources toward reinvesting in communities.

In addition, the Small Business Administration (SBA) offers a wide range of financial instruments to support small businesses and economic development. The Department of Treasury runs the Community Development Finance Institution (CDFI), which funds private and non-profit financial institutions investing in neighbourhood revitalisation. The 1977 Community Reinvestment Act (CRA) uses the government's regulatory authority over lending institutions to create incentives for increased lending in economically distressed communities. Other tax and financing innovations, such as tax increment financing (TIF), which locks in low real estate tax rates, have been used by many levels of government to establish a more favourable climate for commercial investment in socially desirable locations, and these approaches have been used both to finance socially innovative development and as a corporate subsidy. EID is a good complement to many public financing approaches, as it offers the promise of a higher return in terms of public benefits than many development strategies.

* www.isc.hbs.edu/index.html

Federal programmes
Important HUD programmes for industrial development include community development block grants (CDBGs) awarded to states and local governments according to a formula; this money directly funds urban revitalisation and economic development activities, including commercial investment and job creation. The HUD Section 108 Loan Guarantee programme allows communities to borrow up to five times their CDBG award in community development loans, secured by the federal government. Enterprise zones, empowerment communities and renewal communities are designated urban and rural areas where special federal tax incentives are offered to all new commercial investments.

Research and development, and public–private partnerships

Public investment in R&D is made to reduce the risks of innovation and to ensure that the public benefits from new technologies and business strategies. Such investment can also be a source of capital for economic development. Technology-led economic development is a strategy that combines investments in state-of-the-art infrastructure with establishment of research-oriented industrial parks to establish the building blocks of a more diversified network of high-technology industries. This strategy has been followed by many regions in their efforts to attract new investment and new tax revenue. Manufacturing partnerships take this linkage a step further, to ensure that the best available scientific and engineering information is translated rapidly and effectively into improved industrial production practices. This focus on improving the productivity, quality and innovation of private industry through public partnerships is highly compatible with EID.

This regional development approach promotes the development of networks of highly interconnected small businesses and suppliers, and it encourages innovation and the development of shared infrastructure and information systems. Because the linkage between public subsidies, private investment and anticipated improvements to the regional economy is explicit in this development strategy, it encourages a basis of public–private partnership that can be very supportive to EID approaches. Research Triangle Park, NC, is a region that has benefited greatly from technology-led investment and has leveraged the improved economic development and public capacity to promote a strong network of eco-industrial exchange.

Federal programmes
The US DOC National Institute of Standards and Technology (NIST) runs a Manufacturing Extension Project (MEP) which provides excellent training resources and customised consulting to businesses for upgrading their industrial production processes.† The EDA and the National Telecommunications and Infrastructure Administration (NTIA) both operate grant and loan programmes specifically to support development of high-technology infrastructure and to reduce the digital divide. The Minority Business Development Agency (MBDA), also at the US DOC, provides access to resources and expert advice for the development of minority-owned firms. The National Aeronautics and Space Administration (NASA) and the National Science Foundation (NSF) have both played active roles in supporting the development of industrial research parks.

Information infrastructure, community process and civic engagement

EID is a strategy that is grounded in information. It is a data-intensive and knowledge-driven approach to manufacturing that uses improved design and more creatively adaptive management to find efficiencies and resources that had previously been overlooked. The benefits of this approach are clear, but to make it cost-effective it is necessary to develop a deeper capacity to collect, manage and assimilate public and private data and information on a regional basis—to track and understand the flow of resources within the regional economy. Investments in core information infrastructure and regional information management capacity will benefit private businesses with more

† www.nist.gov/public_affairs/guide/mep.htm

targeted industrial recruitment and resource management strategies. Local government and other stakeholders will also benefit through better use of location-specific information in public management, and improved opportunities for citizens to make real choices about the direction in which economic development will lead their community and their region.

Examples of this synergy exist in many places. Some economic development agencies, for example, perform resource audits, in which they actively identify, assemble and promote redevelopment of unused brownfield sites. This regional database helps overcome the information failures and increased risk that generally work against the more complicated task of re-using existing real estate and infrastructure. Enhancing the public capacity to inventory, track and access all sorts of information—from labour-market data, through regulatory constraints on development, public programmes, to industrial waste-streams within industry clusters—empowers meaningful choices and better decisions. In the rapidly accelerating and globalising information-based economy, knowledge is a vital resource that must be managed if communities are to preserve their individual character while benefiting from increased opportunities. Information infrastructure forms a critical link between education, physical infrastructure and institutional capacity. EID can form a bridge in channelling private and public investment into a co-ordinated and broadly usable resource.

The infrastructure for improving information flows in communities includes technical tools such as GISs, which map databases to reveal patterns of spatial distribution. This infrastructure, however, also includes a subtler capacity rooted in dynamic working partnerships among the many parties that drive development, with opportunities for real influence by the populations that development must serve. Developing eco-industrial systems has often served as the initiation of a dialogue that builds much wider trust and civic engagement. Cape Charles and the Green Institute, for example have dedicated facilities and meeting areas for use by local groups for public meetings and projects. The physical establishments of these eco-industrial parks, then, provides a home for expanding these partnerships, restoring the workplace to the centre of civic and community life and offering common ground for communities, businesses and government to come together and advance their collective interest in building real prosperity.

5
THE ROLE OF LOCAL GOVERNMENT IN ECO-INDUSTRIAL PARK DEVELOPMENT

Maile Takahashi
Harvard University, USA

5.1 Sustainable economic development: the case for eco-industrial development

In spite of the economic prosperity of the 1990s, many urban and rural communities are still suffering from economic distress. It is crucial that planners and economic developers create new strategies that address the unique concerns of these places. The issues, such as brownfields, high unemployment, high levels of crime, poor transportation systems, foreclosed farms and poor education systems, require tools geared to addressing economic development in the context of larger social issues.

There is growing interest in sustainable development (development that focuses on balancing environmental, community and business interests) in the USA and in other parts of the world. These strategies offer the hope of a win–win approach to economic development. Of course, there are no easy answers, but there are approaches that hold promise. Up to this point, sustainable development approaches have focused on housing (New Urbanism) and growth management (Smart Growth). These programmes address economic development on a nominal basis and have had a difficult time incorporating 'jobs' other than retail and service-sector jobs as part of their development strategies. Eco-industrial development is an emerging sustainable economic development strategy that expands the industry sectors that can be sustainably developed.

The overall concept of eco-industrial development integrates business, environmental excellence and community connections to create economic opportunities and improved ecosystems. This is not a 'cookie-cutter' approach to development. Eco-industrial development uniquely manifests itself as a result of the partnerships that form between government agencies, community members, businesses and industrial developers. The goals are to foster practical connections between waste and resources and to promote a networked approach to doing business and interacting with communities.

As environmental regulations and community pressures increase, businesses will become increasingly interested in proactive strategies that address those concerns, including eco-industrial development. The benefits to communities and businesses for adopting eco-industrial development strategies are numerous, depending on the local conditions.

A few of the potential benefits of eco-industrial development are listed in Box 5.1.

Some of the benefits of eco-industrial development are:
- Continuous environmental improvement
- Community pride
- Better use of resources
- Reduced waste
- Enhanced market image
- High-performance workplaces (i.e. increased employee productivity)
- Reduction of operating costs (for energy, materials, waste disposal and water)
- Innovative environmental solutions
- Improved environmental efficiency
- Increased protection of natural ecosystems
- Recruitment of higher-quality companies
- Higher-value development for developers
- Minimised impact on infrastructure
- Income from sale of by-products

Box 5.1 **Examples of the benefits of eco-industrial development**

5.2 Local economic development, environmental sustainability and social equity: achieving it all

Traditional economic development strategies focus on bringing new businesses to a community or on expanding existing businesses. The 'role of the public economic developer is to foster the growth and retention of business activity and, through a healthy local economy, provide employment opportunities and a strong tax base' (So and Getzels 1988: 287). This is often done through tax credits, infrastructure upgrades, job training, small business loans and support and can go as far as providing capital for businesses.

Planners and local decision-makers often focus their efforts on creating economic development programmes to bring jobs to their communities. In fact, the creation of jobs motivates most politicians and therefore most city economic development programmes focus on job creation. Most people assume that economic development must be at the expense of the environment and require the sacrifice of quality of life for residents of a community. In reality, industries are not indifferent to quality-of-life issues. Most if not all employees care about where they live and work. Happier employees increase productivity, and all industries are interested in higher-productivity workers. Positive work environments, healthy communities, shorter commuting times and stable work all contribute to more satisfied employees. Knowing this, local economic developers and local government planners have the ability to propose and pursue economic development strategies that will still attract industries to their communities without sacrificing the features valuable to the members of the community.

Industries do contribute to the economic base of a community and should be looked to as valuable members of a community; however, they should not be allowed unilaterally to define what the health and wellbeing should be of that community, especially if receiving public subsidies or support in establishing themselves. Some communities create requirements for industries that receive public assistance in meeting environmental sustainability and social responsibility criteria. Some of the case studies described in later chapters discuss these requirements in detail (e.g. Cape Charles Sustainable Technology Park has developed guidelines for admission to the park and for ongoing evaluation; see Chapter 18). Attempting to broaden the role that economic development programmes have in a community can also include incentives toward increasing environmental and social responsibility. These incentives range from expedited permitting to financial incentives. One specific example is the use of density bonuses, given to developers that include certain thresholds of affordable housing in their developments.

5.3 Eco-industrial development and existing economic development strategies

Eco-industrial development can be incorporated into the existing economic development strategies being used by many communities. In this section I will give a brief review of a few of these strategies and address how eco-industrial development can be incorporated into them.

5.3.1 Smart Growth

Smart Growth, initiated by the US Environmental Protection Agency (EPA), is gaining increasing support from a broad range of interests, including local governments. The goal is to protect existing open space and to concentrate development in urban areas. The Smart Growth initiative is based on the idea that all development should find a balance between community, business and the environment. The eco-industrial concept embraces many strategies that accomplish this goal (see Box 5.2).

The Smart Growth concept has gained acceptance as more and more people move into the 'downtowns' of communities. A survey conducted for the Rouse Forum in 1998 shows that US downtowns are experiencing a resurgence (Katz *et al.* 1999: 2):

> The preliminary survey—conducted by The Brookings Institution Center on Urban and Metropolitan Policy and the Fannie Mae Foundation—looked at 24 cities around the nation and found that all of them expect the number of their

> • Smart growth recognizes connections between development and quality of life. It leverages new growth to improve the community. The features that distinguish Smart Growth in a community vary from place to place. In general, Smart Growth invests time, attention and resources in restoring community and vitality to center cities and older suburbs. New Smart Growth is more town-centered, is transit and pedestrian oriented and has a greater mix of housing, commercial and retail uses. It also preserves open space and many other environmental amenities. But there is no 'one-size-fits-all' solution. Successful communities do tend to have one thing in common—a vision of where they want to go and of what things they value in their community—and their plans for development reflect these values. •

Box 5.2 **The primary goals of Smart Growth**
Source: ICMA with Anderson 1998: Executive Summary

downtown residents to grow by 2010. For example, the city of Houston expects its downtown population to quadruple; Memphis and Seattle anticipate twice as many downtown residents in the next 12 years, while overall population loss continues in many cities. The downtown trend holds for northeastern and midwestern cities with well-established downtown residential districts and for Sunbelt central business districts that have not traditionally supported much housing. Anecdotal evidence suggests that people are living downtown because they want to be near their work places and cultural amenities, and because they enjoy a bustling urban environment.

One of the attractions of urban development is the ability to have one's home and work in proximity. Eco-industrial development is considered a significant piece of the economic development strategy under a Smart Growth plan. Sensitive development of industries are essential in maintaining proximity to work while maintaining pleasant places to live.

5.3.2 New Urbanism

New urbanism is an exciting development strategy that has been gaining popularity. New Urbanism promises development that encourages the formation of 'community' and environmental conservation through design. A stated goal is often to increase efficiency by creating work–live situations and providing transportation hubs to support various transit options instead of traditional commuting patterns. One reason this is an exciting strategy is that a community is created from scratch. Many of the initial 'new towns' that were developed touting New Urbanism, such as Seaside, FL, and Celebration, FL, were largely made up of aesthetically pleasing neighbourhoods, with no opportunity to work (Henton and Walesh 1998: 2 [emphasis original]):

> **Discussion of the economy has been largely absent from discussions about New Urbanism and Livable Communities**. The much-heralded examples of New Urbanism—Celebration, Seaside, Kentlands—make little mention of where and how residents would earn their living. An outsider's impression of these new communities could easily be that they are so attractive precisely because no one seems to need to work! Apparently, residents who are

employed must be employed in the region surrounding the new community or by the few retail businesses and civic institutions in each community's center.

As the New Urbanist concept is maturing, attempts are being made to include economic development issues into the planning for these new developments

Eco-industrial development offers an opportunity for planners to create economic development that complements the goals of New Urbanism.

5.3.3 Industrial clusters

The concept of industrial clusters (see Box 5.3) has received much attention from the US Economic Development Administration (EDA) as a thoughtful approach to economic development based on the existing strengths of a community and the strengthening of local businesses. These principles are compatible with eco-industrial development. Eco-industrial development strategies are based on each unique place, which requires identifying the strengths of the locality to identify the industries that are best suited in that locality. One part of the strategy is to fill in the upstream and downstream connections for an industry (i.e. suppliers, customers). The eco-industrial approach takes this concept one step further by encouraging development of industries that have been traditionally unrelated but that could exchange materials, by-products and services.

Industrial clusters can be considered a risky public strategy in some ways because of the vulnerabilities of particular sectors or technologies, which could leave one-sector communities open to downturns in a particular industry. Eco-industrial development can

---

*Industry clusters are concentrations of competing, complementary and interdependent firms and industries that create wealth in regions through export to other regions. Clusters typically spill over multiple political jurisdictions—including multiple cities and even counties. Clusters are important for the communities in a region because they drive the vitality of support- and local-serving industries (for example, construction, retail and restaurants).

Economists such as Michael Porter and Paul Krugman point to several benefits to firms of participating in a cluster:

- **Access to specialised workforce**: companies can draw on large markets of people with specialised skills and experience working for related firms.
- **Access to specialised suppliers**: companies in clusters have access to concentrations of specialised suppliers for inputs and services.
- **Access to networks**: companies in clusters have access to information flows and technological spillovers that speed innovation.

Geography is important to clusters because firms and people gain from being in the same place. The ease and speed of sharing a specialized workforce, suppliers and networks are enhanced by close proximity. This proximity helps reduce the 'transactions costs' that are critical to the success of fast-moving firms. To benefit from the region, companies and people need fast access to resources in the region, including those based in other neighborhood, town, or regional centers.*

Box 5.3 **Industry clusters**

Source: Henton and Walesh 1998: 12-13

create buffers in front of fluctuations in one market by expanding the notion of 'industry sector' and creating the opportunity of cross-sector exchange. It is foreseeable that when non-traditionally linked industries are linked creative processes could evolve.

5.3.4 Brownfield redevelopment

According to the US EPA (1997b), brownfields are 'abandoned, idled or underused industrial and commercial facilities where expansion or redevelopment is complicated by real or perceived environmental contamination'. Brownfield redevelopment begins to identify, assess and determine how to clean up and re-use a brownfield site safely and sustainably. Brownfields sites are often associated with social and legal issues that discourage redevelopment. It has become clear at all levels of government—local, state and federal—that measures must be developed that will overcome these hurdles.

To date, much energy has been focused on brownfield remediation in order to upgrade brownfield sites for the 'market'. Other programmes focus on the development of legal tools to respond to concerns about liability, and the growing interest in redevelopment of the sites. Many brownfield sites, however, are located in neighbourhoods that have community concerns about the former pollution and potential redevelopment plans. Local governments would like to see brownfield sites contribute to the economic development potential of their area and to contribute to the local tax base.

Eco-industrial development is ideal in these conditions because it provides a forum for economic redevelopment planning that addresses the neighbourhood and environmental concerns regarding the site. Eco-industrial development strategies have been used for brownfield sites in Cape Charles, VA, Minneapolis, MN, and Dallas, TX.

5.4 Achieving environmental goals through eco-industrial development

Environmental goals are often achieved through regulation. Environmental policy focuses on meeting minimum standards established by the federal government, the states (e.g. California, Massachusetts) and occasionally local government agencies. The regulations focus on achieving 'safe' pollution levels. Brownfields are what can occur when there is a lack of environmental regulation and enforcement. Cities are responsible for ensuring that this kind of contamination does not impact future generations. Cities can take a proactive role in championing solutions to these issues that will lead to economic stability in the future by minimising waste, increasing energy conservation, reducing energy costs, using alternative energy sources and re-using materials.

The heart of eco-industrial development is the concept of 'continuous environmental improvement'. Once industries within an eco-industrial development meet the baseline standards, they are further committed to reduce their environmental impacts.[1] In Chapter 18, the Cape Charles Sustainable Technology Park entrance standards are discussed and

1 For more examples of standards and incentives to encourage continual environmental improvement, see WEI 2000.

serve as an example of minimum baseline criteria for environmental and social requirements and options for incentives to encourage continual environmental and social improvement.

Effective eco-industrial developments are encouraged by providing incentives for environmental improvement that exceeds baseline standards. This can be accomplished through a network exchange as industries identify common issues and are able to solve problems together, share ideas across industries and perhaps even exchange their waste products to make useful products from these. Common environmental areas that local government planners consider are waste and recycling, transportation and air quality, and open space.

5.4.1 Waste and recycling

In 1995 the USA generated 208 million tons of waste, which is an increase of nearly 120 million tons per year of waste compared with the 88.1 million tons per year generated in 1960 (US EPA 1997b). In some parts of the country, landfill space is at a premium. This is particularly true in the eastern states. Although the western states have fewer land constraints, they are all concerned about the volume of waste heading into landfills. Most communities have a recycling programme with the goal of reducing or eliminating waste. Sometimes these focus exclusively on residential and retail waste, but a large part of the municipal solid waste-stream is from businesses, which participate only as long as it is convenient and economical for them to do so.

One of the ways that eco-industrial development is focusing on improving business profitability is through minimising waste and reducing the cost of raw materials (or inputs). A city-wide recycling programme can serve as a 'broker' for products that come through the recycling programme and provide a logical link between buyers and suppliers. Local governments can help improve the costs associated with finding inputs or outputs by serving as a co-ordinating hub for goods and information. Triangle J Council of Governments acts as a broker for businesses in its service area. By doing this, Triangle J serves to reduce waste, increase business success and extend the life of landfill.[2]

5.4 2 Transportation and air quality

For many communities their road systems are at or exceeding capacity. This leads to people spending increasing amounts of time in their vehicles and to air pollution problems that threaten the health and wellbeing of residents. Further impacting localities is the threat of losing federal highway funding if air quality standards are not met. Many local governments are attempting to develop regional and local solutions to traffic problems. According to a Surface Transportation Policy Project report, 'though road building has been outpacing population growth, congestion has grown worse' (STPP 2001: 3). The study explores the reasons why this is occurring, such as the increase in the population of drivers, and new roads encouraging additional driving. The same report suggests, 'the best route to providing commuters with congestion relief is to provide more choices, not more roads. The burden that traffic congestion places on commuters

---

2 For a discussion of the efforts of the Triangle J Council of Governments to develop a waste exchange programme, see Chapter 21.

is considerably less when those commuters can choose to ride a bus or train, or walk or bicycle' (STPP 2001: 9).

In addition to impacts on commuters, transportation is a key factor in business location. It is crucial for the manufacturing, warehousing and services sectors to deliver and receive their products and inputs in a timely way. As part of eco-industrial development planning, identification of industries that are appropriate for the transportation network of a particular area should be one of the criteria considered. Eco-industrial development engages the private sector in innovative approaches to traditional transportation dilemmas by asking questions such as 'How can each business work with other industries to minimise truck trips?' 'Can I be located near my suppliers or my customers?' 'Can a facility co-ordinate with other businesses in the industrial park to provide transportation for employees and goods?' Local government can enhance the attraction of industries and allow existing businesses to continue to improve by addressing transportation issues in the recruiting of industry.

Among some groups, transportation networks are being approached much like watersheds, only these groups are calling them 'traffic sheds'. Planning with the entire traffic shed in mind is one of the techniques available to minimise the impact on traffic and air quality (Kendig 1999). Eco-industrial development considers the regional connections that businesses have to the larger regional economy. As more cities and counties begin to use regional transportation planning, consideration of industries that 'fit' in the regional transportation network will allow development that works with the existing network and burdens it less.

5.4.3 *Open space*

Localities often struggle to provide adequately maintained open space for passive and active recreational use. Eco-industrial development provides an opportunity to create non-traditional open spaces for a community. For example, the Cape Charles Sustainable Technology Park incorporates a boardwalk and wetlands on the property. This part of the project has been done in partnership with state agencies and non-profit organisations. The boardwalk is open to the public as an amenity to the community. In exchange, there is storm-water detention and treatment occurring naturally in the wetland. This is one of the ways that industrial development can create workplaces that augment a community's amenities.

With the goal of exchange and network formation—an essential for effective eco-industrial parks—parks need to be designed to encourage interactions between individuals and companies, much like the New Urbanism concept for housing. Designs can included shared parking and other facilities, which can increase available open space. Other features can include active recreation to further encourage interactions between firms within an eco-industrial park. These facilities, as in Cape Charles, can be made available to the public.

5.5 Achieving other city-wide goals through eco-industrial development

Eco-industrial development offers a tremendous amount of flexibility as to the focus of projects. This type of development not only provides ways to achieve continuous environmental improvement but also can address a locality's social concerns. The Cape Charles Sustainable Technology Park (see Chapter 18) provides criteria for standards for business participation in the community. This can run the gamut of job training, benefits, healthy workplaces and community use of common spaces such as conference rooms and athletics fields.

5.6 Influencing the decision-making process: inform, guide and create

Local government planners are in a unique profession where they serve as a link between the larger goals of creating liveable communities, developer interest and politics. Planners can choose to integrate eco-industrial principles into many areas of what they do. For a long time, planners were perceived as having the role of implementing the zoning code and creating city policy directed by elected officials. Over time that role has evolved into that of an 'advocate' planner. Planners have three key roles that can influence city policy: informing decision-makers, networking and guiding citizens through city processes, and creating policy.

As a provider of information, the information that planners put in front of decision-makers and the emphasis put on the information gathered can be influential in the decision-making process. Planners often set agendas for meetings, have the 'ear' of the city decision-makers and can provide validation for 'new' ideas in a politically insecure setting for decision-makers. In the end, however, planners do not make the final decisions about projects; the elected officials do. Planners can identify projects that will resonate with a community and help make a project more attractive to elected officials.

A planner is often the first city staff person that citizens encounter when a project with an eco-industrial focus is developed. Planners are usually well versed in the economic development and planning resources available and can guide citizen groups to the right places. Taking it a step further, the planner can also serve as an educator and networker for citizen and business groups. Local government planners can participate in this by providing in-house staff services for organisations, by reviewing projects and making connections with appropriate city services and community members. They can also provide information on resources that could assist in the success of a project.

The third role of a planner is in the creation of policy that supports development that achieves particular goals of the community. The creation of these policies helps to educate decision-makers and the residents of the community. Many cities undergo a visioning process as part of their general planning process. They have also started creating goals for 'sustainability' for their communities. The creation of these types of goals can seem of little practical relevance or application, but it does arm community groups and future development projects with guidelines and policy that they can point to later. As greater acceptance occurs, more-developed and specific policies can also follow. With these policies in place, it becomes easier to review all city policy and programmes from a filtered lens of economic development, environmental sustainability and social justice. For example, programmes such as solid waste management, education, transportation (i.e. bus, light-rail, shuttle and tram services), building codes, permit requirements, air emission limits, Welfare-to-Work programmes, preferred purchasing and links to other public services can all reflect these principles. Eco-industrial development should eventually be the standard for all development in a community.

5.7 Suggestions for incorporating eco-industrial development into local government planning

Networking is perhaps one of the most significant resources that the planner brings to an eco-industrial project. There are some unique issues associated with eco-industrial development that the planner can support, the most prominent concerning by-product exchanges. Imagine easements crossing property lines containing pipes delivering one firm's by-products (waste) to another firm that can use these as inputs to their production process. Planners should position themselves as facilitators for this type of exchange. By asking questions during initial contacts with developers and companies seeking to locate in the community, planners can ascertain the potential for these resource exchanges and make the connections, which add value to the community.

Peter Lowitt, former Planning Director of Londonderry, NH, and currently Director of the Devens Enterprise Commission, provides the following vision of the role that can be played by the 21st-century planner. An e-commerce grocery store seeking to build a 300,000 ft^2 distribution and food preparation facility is located just down the street from a corrugated paper distributor that receives one railroad car of recycled corrugated paper a day. The railroad cars return to the corrugated paper plant empty. The opportunity exists to address one company's waste disposal problem (the grocery store) and address the other company's expense by filling the empty railroad car heading to a corrugated paper plant with raw materials for recycled corrugated paper (cardboard delivery boxes from the grocery store).

Planners and code enforcement officials need to educate themselves about green building design and design for environment strategies, and play a role in creating policy that encourages and provides incentives for these types of development. This can help newer technologies and approaches gain an acceptance among developers, which creates a positive feedback loop to further allow them to make inroads into the marketplace.

Alternative economic development strategies need to go beyond the goals of job creation and new investment in a community to address other issues such as environmental quality, community impacts and 'good' job creation that provide better wages and greater economic multiplier effects. Eco-industrial development is a practical approach for developing a sustainable economic development strategy. It can bring the various stakeholder groups together to identify common goals and discuss how they might be able to find common ground. This approach requires upfront planning, but in the end it can create new partnerships in economic development activities within a community.

Some immediate ways that planners can incorporate eco-industrial development into their communities are to:

- Review solid waste management techniques
- Develop sustainable development policies and guidelines
- Review brownfield redevelopment projects for eco-industrial development potential
- Consider eco-industrial development strategies in the job-creation component in New Urbanism projects
- Review and evaluate economic development strategies to explore how eco-industrial development approaches could be applied.

Eco-industrial development offers a practical approach to implementing sustainable economic development. Planners can act as agents for change at the 'front line' within communities. Eco-industrial parks are located at the intersection of environment and economic development. Eco-industrial development represents a positive outcome to the question of what type of economic development should be promoted by a community concerned with environmental protection. It is also an important market niche that a community can use to position itself as progressive and sustainable.[3]

---

3 This final paragraph draws on a paper that was presented to the National American Planning Conference in New York in April 2000 (Takahashi 2000). Discussion was provided by Peter Lowitt (personal communication), Director and Land-Use Administrator for the Devens Enterprise Commission, the agency charged with permitting the redevelopment of the former Fort Devens in Massachusetts.

6
COMMUNITY ENGAGEMENT IN ECO-INDUSTRIAL DEVELOPMENT

Mary Schlarb and Judy Musnikow
Cornell University, USA

When President Clinton's President's Council on Sustainable Development (PCSD) met in 1996 it struggled with the definition of eco-industrial development (EID). Essentially trying to put teeth into the concept of 'sustainable development', the PCSD sought to integrate the three main principles the term evoked. The business community often talked about economic sustainability, whereas environmentalists were concerned about ecological sustainability; still others focused on social equity. Taking a broader definition than conventional industrial ecology, PCSD (1996a) defined EID as:

> A community of businesses that co-operate with each other and with the local community to efficiently share resources (information, materials, water, energy, infrastructure and natural habitat), leading to economic gains, gains in environmental quality and equitable enhancement of human resources for the business and local community.

Industrial ecologists have devoted considerable attention to the business and environmental benefits that by-product strategies can provide. An equally important yet less explored aspect of EID is the enhancement of social equity and environmental justice for community members through the application of eco-industrial approaches to the redevelopment of local economies. Beyond the technological, economic and policy dimensions of forging business-to-business connections, those who are 'on the ground' attempting to design and develop eco-industrial projects have incorporated strong social justice and anti-poverty measures into their EID planning.

What distinguishes EID from more traditional economic approaches is the emphasis on the relationship between industry and the surrounding community: 'Eco-industrial development is built on three overlapping strategies: applying ecological principles to work and workplace design, promoting business networks and building strong linkages to local community' (Schlarb and Cohen-Rosenthal 2000: 106). Côté and Cohen-Rosenthal (1998: 182) emphasise this notion: 'Since industry is a human creation and humans are social animals, we need an approach which brings industry and environment together with a social or community perspective'. Lowe (1998: 26) further suggests the valuable input of community collaboration to EID in its ability to help 'capture the benefits of this innovative concept'.

Each of the eco-industrial projects described in this book have integrated in some measure 'the three Es' of sustainable development: strong local economy, quality environmental health, and social equity and justice. In each case, community participation has shaped and supported the incorporation of these three goals through activism, joint needs identification, participatory design and/or community-based governance. These EID projects have relied heavily on the local community's contributions, acceptance, satisfaction and continued support. In this chapter we explore how communities can both benefit from and support EID to further their sustainable development goals.

6.1 A trend towards greater public involvement in decision-making

In the USA and internationally, greater community participation has transformed the way in which localities form policies and build programmes (e.g. Thomas 1995: 3; Grimble and Chan 1995: 113; Zialcita et al. 1995: 2). Citizens, no longer content with public hearings about decisions that have already been made by government, demand a greater voice in issues that affect their lives. Government agencies, too, have demonstrated a growing interest in creating processes by which the public can shape policy-making and development strategies (Box 6.1), and many grant programmes require multi-stakeholder engagement.

6.2 Community benefits of eco-industrial development

EID offers a number of potential benefits to communities and businesses. The approach seeks to bridge the perceived gap between the interests of businesses and communities by building partnerships. Public involvement processes allow businesses and community stakeholders to express interests and concerns and to co-operate in generating development alternatives. The rationale for this collaboration is that many of the interests of industry and citizens overlap, and so mutually beneficial strategies for sustainable development exist. Below we discuss the interlinked benefits of building linkages within and between communities and businesses, under the headings economic efficiency and profitability, job retention and growth, community development, environmental stewardship, building community capacity and pride, building social capital, and community support and buy-in.

> • The purposes of the policy are to:
>
>   - Strengthen EPA's commitment to early and meaningful public involvement
>   - Ensure that environmental decisions are made with an understanding of the interest and concerns of affected people and entities
>   - Promote the use of a wide variety of techniques to create opportunities for public involvement in Agency decisions
>   - Establish clear and effective procedures for conducting public involvement activities in EPA's decision-making processes.
>
> When final, the policy will apply to all EPA programs, including such activities as rulemaking for significant regulations, permit issuance or modification, selection of plans for cleanup of hazardous waste sites and other significant policy decisions. The policy will not replace public participation requirements established by existing laws or regulations but will supplement those requirements and enable EPA to implement them in the most effective ways.
>
> EPA's Brownfields Economic Redevelopment Initiative is designed to empower states, communities and other stakeholders in economic redevelopment to work together in a timely manner to prevent, assess, safely clean up and sustainably re-use brownfields . . . EPA is funding: assessment demonstration pilot programs (each funded up to $200,000 over two years), to assess brownfields sites and to test cleanup and redevelopment models; job training pilot programs (each funded up to $200,000 over two years), to provide training for residents of communities affected by brownfields to facilitate cleanup of brownfields sites and prepare trainees for future employment in the environmental field; and cleanup revolving loan fund programs (each funded up to $500,000 over five years) to capitalize loan funds to make loans for the environmental cleanup of brownfields. These pilot programs are intended to provide EPA, states, tribes, municipalities and communities with useful information and strategies as they continue to seek new methods to promote a unified approach to site assessment, environmental cleanup and redevelopment. Brownfields National Partnership—The Clinton Administration expanded the Brownfields Initiative to build partnerships between public and private organizations to link environmental protection with economic development and community revitalization.•

Box 6.1 *US Environmental Protection Agency, Draft 2000 Public Involvement Policy (FRL-6923-9)*

Source: US EPA 2000

6.2.1 *Economic efficiency and profitability*

The appeal of eco-industrial parks for tenant businesses and industrial development is the increased profitability and cost savings brought through economies of scale and added value to outputs (Gertler 1995). Value is added to by-products as they enter back into the production cycle as raw materials for another firm or process. With EID, companies can find opportunities to improve efficiency of energy and material use through waste exchange, recycling and innovative technology and production processes. Regulatory penalties for harmful practices may also be eliminated or reduced. Increased economic efficiency within and between tenant companies will most probably increase the value of real estate for private and public developers.[1] This is particularly true in the application of eco-industrial principles to brownfields.

Companies co-located in a park can share the burden of expenses for infrastructure and services, such as business services, waste management, purchasing, training and recruitment, recreation and childcare facilities, transportation and other common costs of doing business. Umbrella permitting and environmental compliance certification is another potential sources of savings.

Though not always possible, park-wide 'umbrella' permitting can possibly facilitate a park-wide, rather than a company-wide, certification process to cut down on paperwork and staff time (Cohen-Rosenthal and McGalliard 1996; Weitz and Martin 1995). Savings on production, disposal and regulatory costs can make companies more viable, with positive repercussions in terms of jobs and tax revenues for towns. EIDs subsequently help retain local currency and resources circulating within the community (Schlarb and Cohen-Rosenthal 2000).

The focus of EID has been on manufacturing, but it can benefit retail and other sectors as well. Eco-industrial approaches offer communities an opportunity to attract new businesses and jobs to previously undervalued areas. As businesses and their workforces locate (or remain) in a neighbourhood, markets for retail services are strengthened. Similarly, as businesses find new uses for their by-products, an EID strategy can help identify upstream suppliers and downstream customers by finding new opportunities to join in exchange relationships. Services in the financial, communications and administrative sectors may find new markets as local industries grow. Retail and small-scale enterprises can also be involved in eco-industrial exchange relationships. Springfield, MA, for example, has proposed engaging small main-street retail shops in a broader eco-industrial network through shared purchasing and services. Although the proposed networking relationships for these stores do not involve large by-product feedstocks, the ultimate goal is the same: the recovery and re-use of resources (Schlarb and Keppard 2001).

In practice, EID projects have derived strength from community engagement in EID planning and design to reach local economic objectives. Collaboration with community members can facilitate the collection of information, ranging from skill assessments of the local labour force to inventories of the regional resource base. Communities contribute valuable information about the local system, including its weaknesses, possibilities and resources, that can help EIDs operate more efficiently and productively in the local economy. The community also becomes an integral component of the eco-industrial system, supplying labour and forming a local customer base.

6.2.2 Job retention and growth

The industrial ecology literature has made little connection between EID and job retention and growth, with some exceptions (McGalliard et al. 1997; North and Giannini-Spohn 1999). EID supports job growth and retention in several ways. Its emphasis on building networks for resource exchange and recycling can help foster new businesses and new jobs. When companies reduce the cost of materials, waste disposal and fines for failing

---

1 For further discussion of the effect of EID on real estate value, see Chapter 14; see also RMI 1998.

to comply with environmental regulations, they can invest their savings in retaining employees and hiring and training new people. Savings resulting from greater efficiency and economies of scale particularly benefit small and medium-sized businesses. Many eco-industrial park projects have incorporated incentives for training and hiring minorities and women, salary improvement programmes and family-friendly policies. Emphasis on green design improves indoor workspace quality and therefore worker health and productivity. These factors result in higher employment for communities, better opportunities and working conditions for employees and a more skilled and productive workforce for employers.

6.2.3 *Community development*

Proponents of eco-industrial approaches point to a host of economic, environmental and social benefits for communities. The objective of these approaches is to add value to a municipality's economic base, strengthening its industrial, social and supporting institutions in a way that attracts new businesses and retains existing ones:

> The appeal of the concept is that developers and communities that create eco-industrial parks seek to build a foundation for industrial development that is more competitive, more efficient and cleaner than traditional industrial parks or regions. In addition, new business niches will be opened for recruitment or incubation of new companies that strengthen the local economy (Lowe 2001).

EID seeks to address environmental justice issues in the process by empowering communities to make their own decisions about what kind of industrial and commercial development will occur. Cases are well documented of economically distressed communities falling victim to the lure of quick income from landfills, hazardous waste dumps and rubbish transfer stations. In many instances, local governments accept these businesses into their municipalities because they see no other alternatives. Often, citizens living near these dump sites have no voice in the decision and are forced to live with the environmental and health consequences. EID emphasises the need to involve the range of stakeholders in identifying community assets, problems and alternatives and in planning and implementing economic development programmes.

Brownfield redevelopment poses a related environmental justice and community health issue that EID approaches may solve. Brownfields, or contaminated properties, are common blights in economically disadvantaged neighbourhoods. People residing near these sites potentially suffer the negative health consequences of toxins present in drinking water and soil (Meyer and Van Landingham 2000). They often also face acute challenges to attracting development projects and resources. The added costs and liability issues associated with brownfield clean-up and re-use make recruitment of industry especially difficult. The EID strategies for maximising resource efficiency, economic growth and community sustainability offer a set of mechanisms for redeveloping these sites without repeating the contamination of past industrial activities. Brownfields are often located close to existing industrial centres, and military bases in particular have accessibility to multi-modal transportation (Giannini-Spohn 1997b).

Sustainable redevelopment of brownfields is based on the common-sense notion that, in order to avoid the harmful health effects and high cost of remediation of contaminated land, new uses must avoid the mistakes of the past. By reducing the waste and pollution from industry, and by preserving green space, industrial land uses become better neighbours in mixed-use areas. As an advanced and efficient system, EID can attract to communities high-performance companies committed to international standards for environmental management (such as the ISO 14000 series) with the competitive advantages they derive from being in an EID network. Through co-ordinated planning of parking, shipping and warehousing and through administrative and support services EID projects are often more space-efficient, using less land and allowing urban areas to maximise the benefit from their industrial zones.

In fostering stronger partnerships between citizens, businesses, government agencies and non-profit organisations an eco-industrial park can enhance its host neighbourhood. By revitalising existing businesses, redeveloping brownfields and attracting new businesses eco-industrial projects can provide local residents with greater opportunities to work in their own neighbourhood and to walk to their workplace. Several current eco-industrial projects provide incentives to businesses for hiring local workers (i.e. the Green Institute in Minneapolis [see Chapter 17] and the Sustainable Technology Park in Cape Charles, VA [see Chapter 18]). Individuals and groups who are already networked in business and other partnerships can add the element of emergency response to the scope of their relationships. In Minneapolis, for example, as a result of eco-industrial networking, local businesses have combined resources to hire a security officer to prevent vandalism and theft (Krause 2001). Eco-industrial parks have also offered or shared services with the community, including day care, recreation and transportation. Although these community development benefits are not guaranteed outcomes of EID implementation, an eco-industrial framework can guide the process of strengthening communities.

Government can also find advantages in being involved in EID beyond tax revenues. Stronger neighbourhoods are easier to govern than are fragmented neighbourhoods. As Bracken Hendricks and Suzanne Giannini-Spohn discuss in Chapter 4, the federal government confronts the challenge of co-ordinating an array of complex and competing interests and priorities in the current context of increasingly devolved decision-making. EID offers an approach to developing public–private partnerships to optimise resources and improve economic and environmental performance. Communities that have strong interpersonal relationships between citizens, businesses and their supporting agencies already possess the channels of communication and interaction necessary for bolstered public education and decision-making. Eco-industrial approaches encourage individual firms to adopt measures to improve their performance voluntarily rather than in response to regulatory mandates. Businesses that voluntarily adopt more ecologically sound practices do not need to be regulated as closely, leaving government agency financial and staff resources available to address other pressing community needs.

6.2.4 Environmental stewardship

EID seeks to promote environmental stewardship at the firm, industrial park and community levels. The ultimate environmental goals of eco-industrial strategies are to reduce the use of virgin materials, decrease pollution, increase energy efficiency, reduce water use and decrease the volume of waste products requiring disposal in landfills. This approach encourages companies to adopt innovative processes and technologies that reduce waste of energy, water and materials. At the park level, park managers aim to minimise negative environmental impacts by basing siting, infrastructure and recruitment decisions according to ecological carrying capacity. Businesses linked in eco-industrial networks form material exchange relationships to decrease the amount of waste going to landfills and incinerators. EID encourages tenants and management to collaborate with the community to identify and support community-wide resource exchanges and recycling, re-use and remanufacturing opportunities. Some EID projects have looked to environmental restoration of open space, reforestation and riparian repair, both as a park amenity and for environmental improvement. A corollary benefit to firms is an enhanced environmental image.

As is the case with the economic aspects of EID, the environmental advancements of EID can be enhanced by the community's contribution. The community's knowledge of the local ecosystem, resources and environmental impacts and risks assist the EID project in creating innovative and pragmatic solutions to mitigate environmental stresses and promote environmental quality.

6.2.5 Building community capacity and pride

Social equity is achieved by realising the needs of local people, empowering the community and providing services equitably. Each of these components requires community involvement to reach their potential. Community members themselves are best equipped to assess and prioritise their needs and concerns. The more actively the community participates in forming the eco-industrial project, the more their concerns and ideas will be integrated into the project.

Community participation plays a large role in empowering citizens and creating a more equal forum for sharing opinions among diverse interests and stakeholders. Often, the most distressed of our communities bear the negative effects of industrial activities and growth. EID can empower neighbourhoods to reverse these trends. Through education and training, the public gains capacity and influence, not only in designing the EID project but also in future community decision-making. Building capacity enables communities lacking organisation and leadership to turn concerns into action. It helps citizens better process information and provide input into decisions by increasing their ability to leverage additional resources and capitalise on existing civic assets.

Lowe (1998) highlights how EID planning in particular can build additional skills and capacities to communities. Eco-industrial developers may subsidise waste disposal costs and offer more environmentally preferable methods of disposal. They may also share their information resources and expertise in areas such as green design, energy efficiency and resource conservation, which will help diminish the ecological footprint of the community overall. Furthermore, the eco-industrial park has tremendous potential to promote the community's awareness and pride in a sustainable community (North and Giannini-Spohn 1999).

6.2.6 Building social capital

In many respects, EID and the community participation process it supports help to increase the social capital of the community. Social capital, as defined by Szreter (2000:

56), is 'the additional productive benefits to the society or economy as a whole that result from the synergy of a set of mutually trusting social relationships'. In Chapter 11 of this book, for example, Ed Cohen-Rosenthal describes the social capital created among companies in Silicon Valley and other areas, suggesting the importance and benefits these networks of information and co-operation have had.

The social capital created within the community is similarly integral to EID. The community participation process in EID broadens traditional social networks to include other community residents, local businesses, industry, government agents and economic developers. All these sectors come together in the name of EID to improve their community, economically, environmentally and socially. As Lin (2001) suggests, social capital facilitates the flow of information among stakeholders, increases the influence of participants either directly or indirectly in decision-making, improves the social credentials of individuals, such as their accessibility to resources, and reinforces the value of participants in terms of their entitlement to resources and equitable status. The trust and capacity building fortified through the social networks and connections of community participation enable participants to have greater influence on the process, achieve broader consensus around the development project and gain more extensive benefits from the development.

6.2.7 Community support and buy-in

Participation achieves far more than simply providing stakeholders a forum to express their own interests and concerns. The process urges discussion among diverse groups to brainstorm new interests and concerns, improve understanding of the pertinent issues and create effective and appropriate solutions. The views and opinions of participants perpetually evolve, and often the process can achieve increased support and acceptance for a mutually devised EID plan. Furthermore, the resulting plan will be a better fit for the community than a plan not subject to the participatory model.

The recent movement toward a knowledge-based economy accentuates the significance the dynamic flow of information fostered through social networking offers EID. Maskell (2000), observing this 20-year trend, notes the shift from static price competition to rivalry based on knowledge creation that has occurred among competitors. He points not only to the growth and adoption of leading-edge technology within industry as evidence but also to the 'low-tech' learning and innovation activities that have led to ground-breaking developments in areas such as resource management, logistics, production organisation, marketing, sales, distribution and industrial relations. The community can contribute substantially to several areas of knowledge creation. From the community, the EID garners information about valuable local resources, local business assets, potential material exchanges, labour-force information and environmental risk. In return, the EID provides information to the community regarding the details of the development and its influence and impact on the community. This reciprocal relationship bolsters trust among stakeholders and further increases the exchange of information. Trust, and networks in general, among stakeholders can reduce transaction costs thereby increasing the community's economic advantage (Putnam 2000).

6.3 Effective strategies for engaging community

The EID projects described in this book use several methods for engaging community in meaningful and effective ways:

- Community visioning
- Planning charrettes
- Surveys and interviews
- Community advisory committees
- Community organising

Each of these will be described in turn in the following sections.

6.3.1 Community visioning

Visioning provides the community with an opportunity to be a part of the creative and innovative thinking that goes into shaping EID. The process brings together a diverse group of people who work to forge a vision of how they would like to see their community in the future. Visioning primarily focuses participants on the 'big picture', which is a positive way to develop relationships and networks among individuals and groups within the community.

The primary objectives for conducting a vision exercise include:

- To create a specific idea of what is hoped to be accomplished that is unique to the area concerned
- To build awareness and commitment among the broad range of stakeholders
- To set out a long-term framework for a project that will take at least several years to unfold
- To help provide specific items for action planning
- To help create criteria for assessing the success of the effort
- To motivate the general and business community

Visioning early on in the process can serve as a crucial source of inspiration and mobilisation for the community. It renders a base on which future collaborations and developments can be founded. The product can readily translate into specific principles, actions and changes to be incorporated into the EID plan. In addition, visioning can publicly establish the preferences, desires and goals of the community. Further decision-making can be greatly improved, since these standards can guide the direction of future decisions and bolster the legitimacy of actions and decisions made in accordance with the vision.

There are several ways to implement a community visioning process. Two methods in particular, search conferences and design charrettes, have proven effective in many EID settings.

6.3.1.1 Design charrettes

A design charrette is typically an intense two-day or three-day workshop in which architects and other design professionals, community leaders, public officials and citizens work together to envision alternatives for a specific project. For example, a design charrette might focus on a local building programme, or a neighbourhood or regional community project, often with an emphasis on long-term economic and environmental sustainability. They are frequently used to bring together a group of stakeholders to work through issues that are best represented visually, such as site plans and building designs. The results of charrettes include concise vision statements as well as illustrative reports filled with visually stimulating materials such as computer drawings, simulations, sketches and photos to represent and support the workshop deliberations. The towns of Cape Charles and Front Royal, VA, each carried out design charrettes to establish common visions and options for eco-industrial parks.

6.3.1.2 Search conferences

Search conferences help communities develop strategic planning capacities and meet community goals among diverse stakeholders. Like design charrettes, they bring people together to establish common goals and to search for a desirable future for common activities. They are, however, less focused on design details than are charrettes. Search conferences concentrate on prioritising concerns and solutions and have been helpful in eco-industrial settings in developing alternatives for how to revitalise industry.

6.3.2 Planning charrettes

Although the visioning process may make great strides in creating and visualising a community's understanding and hopes for its future, it is often weak in forming planning outcomes. Utilising planning charrettes in conjunction with the visioning process can help efficiently and successfully realise the ideas and expectations of the community in a participatory way. Planning charrettes allocate responsibilities and tasks to those who can best see them through. Although many of the participants of visioning sessions will be involved in planning as well, planning charrettes reach out beyond this scope to include individuals and groups that may not have been present during visioning. Planning charrettes therefore ensure that the participatory process continues throughout the design and implementation of the eco-industrial project, beyond the initial stages of development.

6.3.3 Surveys and interviews

EID planners can use surveys and interviews to gauge public sentiment, identify areas of concern and isolate issues of importance within the community. They can collect information needed in decision-making to determine how the EID will fit into the community. Surveys may be administered by mail, telephone or in person, in one-to-one interviews or through focus groups. In the case of interviews and focus groups, an individual or small groups of people are brought together in a confidential setting to discuss an issue with the assistance of a skilled facilitator.

Surveys used to appraise how the EID would fit into the community measure the assets of the community in baseline studies. Assets may include geographical assets, availability of raw materials, labour, education, existing industrial base, transportation, market access, governmental regulations and quality of life. In addition to these assets, and a facet that industry commonly already considers, EID also considers environmental factors, such as inputs and outputs of the primary industries in the region, the possibilities for waste aggregation, waste transformation and the environmental constraints and opportunities of the area. Rather than just listing these assets, the potential synergy and interconnections are sought.

6.3.4 Community advisory committees

Advisory committees are relatively small working groups composed of representatives from various interest groups, including businesses, trade unions, agency staff and local citizens. The advisory committee is established in order to inform decision-makers of the views of the community by producing a set of recommendations, decisions and/or actions for addressing a specific problem. The representatives are chosen in order to best reflect the interests of the larger public. They should be well integrated within the community and knowledgeable about the local ecosystem. These representative bodies, if assembled well, can greatly enhance the ability to reach acceptable decisions that reflect community needs and concerns.

6.3.5 Community organising

Community organising is primarily about mobilising the local community into taking action. People mobilise for various reasons. They may stand up for a decision or cause they strongly believe in and would like realised, or they may fight against decisions or activities they oppose. In engaging a broad cross-section of the community, community organising can encourage even greater public participation, tap commonly overlooked leadership and express the will of the people. Often it can be the impetus for the employment of other participatory tools, such as community visioning.

The Green Institute, for example, was conceptualised by a group of community residents in urban Minneapolis seeking positive alternatives to the siting of a rubbish transfer station. Opposed to the station, but concerned about declining industrial activity in the area, these citizens used the proposed siting as an opportunity to form a strategy on how to maintain and attract businesses and new jobs for local people without compromising the health and environmental wellbeing of the neighbourhood. In Londonderry, NH, developers wanted to develop a site but the community opposed the siting of a new power plant, leading to an innovative 'green' power plant design.

6.4 The community involvement imperative

Involving community in eco-industrial planning, as with any decision-making, entails both opportunity and risk. In a general sense, the challenge of integrated public participation into EID planning is to balance the needs and realities of the environment, business and community members, articulated aptly by US EPA administrator, Christie Whitman (US EPA 2002):

> Democracy isn't always pretty; it's not always neat and tidy and it's often contentious. But it is, without doubt, the best form of government yet devised because it gives every person a voice in the system and a chance to change the course of events. Democracy gives real power to individual citizens, but with that power comes responsibility. Democracy is not a spectator sport.

EID strategies, when effectively implemented, bring a host of economic, environmental and social benefits to local communities and local governments. The planning process, in turn, is enriched through the combined knowledge and skills of those participating in the design process. Citizens, government officials and developers alike not only have the power to shape EID projects, they have the responsibility.

7
THE DEVELOPER'S ROLE IN ECO-INDUSTRIAL DEVELOPMENT

Mark Smith
Pario Research, USA

Urban development is tremendously important to our society. The effect of the quality and effectiveness of our urban system on the quality and efficiency of our private lives, business enterprises and civic systems has been under-appreciated. A new paradigm is emerging that increases efficiency and increases value for all urban development stakeholders. As catalysts of development, real estate developers are in an important position regarding eco-industrial implementation. Developers' leadership in the new paradigm will realise the benefits of more efficient investment; but, if developers do not embrace this new leadership role, benefits and value will continue to be lost.

A temporal view is helpful. Eco-industrial development is in the early stages of its life-cycle, and particularly in the mainstream development industry. Thus, the developer's role is largely conceptual, and the best approach is to accept where eco-industrial initiatives are today and then to work toward goals and strive to optimise in-process activities. Later, as mainstream development integration occurs and we learn more about successes and obstacles, the developer's role can more formally emerge from those experiences. In this chapter I discuss the emerging developer's role—including educational needs, obstacles and the important role of process redesign—and, importantly, the necessary interfaces and support needed to assist developers based on where we are in the integration life-cycle.

There are many forms of eco-industrial development emerging, including the retrofitting of existing industrial zones, virtual eco-industrial parks that involve users and infrastructure in disparate locations, and new business parks. In the case of new business parks, real estate developers reform industrial parks to include eco-industrial design and programming, and they sometimes link the park's operations to other businesses in the region. This chapter is aimed primarily at the real estate developer building a new park.

7.1 The developer is the catalyst in the traditional development story

Of all the participants in the urban (real estate) development life-cycle, the developer organisation is usually the agency that largely determines the personality of a project and 'starts' the transformation from pre-developed real estate asset to developed or redeveloped buildings and communities. In this traditional function of 'catalyst' the developer is a nexus of the guiding municipal code, the economy and real estate market, the neighbourhood and the investment—resulting in an asset with a service life of perhaps 50–80 years. This is a remarkable scope for any business or social enterprise, and integrating sustainability into this disparate process complicates it even further by suggesting new connections and process participants, and changes to design. The process is already complicated.

Regarding industrial development, real estate organisations have been one of the last of the stakeholders to strongly embrace eco-industrial design. There are three primary reasons:

- The industry has little familiarity with the topic and what it entails.

- Much of the risk of sustainable design occurs in the development role, such as the reliability of technology, the market value of sustainable design and the absorption of potential time and cost increases. The industry avoids adding risk and processes to its already volatile circumstances.

- The primary benefit is perceived to be the inclusion of process participants outside the developer's role, and thus there is no expected gain under current topic views.

7.1.1 The motivation for an expanded developer role

The primary reasoning that supports eco-industrial development is that there is higher economic value for all participants in its paradigm. Higher value includes reduced pollution and its associated costs, reduced resource use and cost, and higher-quality, safer communities that are less costly for municipalities to operate. Real estate and its larger business context, as well as social systems, all lose value in the many 'disconnects' that are structured into their design and delivery. In essence, the eco-industrial developer connects valuable organisations and assets (usually connected in one or more of the development process steps) and produces more value because of shared knowledge and resources accreting into a whole that exceeds the efficiency and value of the segmented components—removing a limitation of the current traditional model. This benefit is little understood, poorly quantified and presented by eco-industrial advocates and is not effectively communicated.

7.1.2 The developer's role

With this, then, the developer's revised process for adoption of eco-industrial development can be viewed in two categories:

- Enhanced philosophy and objectives. This is required because:
  - Urban development is inefficient and methods are ingrained.
  - Understanding and facilitating whole systems increases efficiency.
  - Value is gained by increasing connections related to design, programming, resources and capital.
  - Process change is required.
- Revised and expanded relationships and revised processes. Relationships and processes are required that:
  - Involve all stakeholders, as there is added value and support to be gained from their participation
  - Create new alliances and partnerships with participants, both earlier in the development process and later in the operations of building and communities; this may include new relationships with municipalities or service companies for planning and services, and with building occupants to create and share savings during operations
  - Allocate costs properly with a new full-cost view
  - Share benefits and savings

This is relatively new thinking for real estate, though a revised approach to business is more common in most other industries, most of which, ironically, are the users of products of the real estate industry. The real estate industry presents some extraordinary obstacles to change that contribute to its relatively slow evolution. These fundamental reasons for resistance to change also inhibit the implementation of eco-industrial development.

Given this, the developer's role is, mostly simply stated, to accept that by striving to create more connection more value can be realised by all stakeholders, and their assets and investors are among the beneficiaries. With this acceptance as a foundation, the developer then must strive to re-engineer the development process to achieve eco-industrial development possibilities. Although the changes require effort, the result is higher value.

In other key stakeholder categories—such as government, municipalities, business and academia—participation with eco-industrial systems began many years ago, and topic understanding and support has been increasing. However, these advocacy efforts and organisations are not, for the most part, in a decision-making position to affect development. Real estate developers, those that are in the decision-making role, have been largely missing in this dialogue. Figure 7.1 indicates the main movements in urban development that are promoting better development patterns. The major movements are Smart Growth, New Urbanism and Green Building. Each has originated from a different sector and each has had unique as well as overlapping objectives. Importantly, these movements are moving toward the same goal and, as we learn more about the interconnections between design, resource systems, government and building and community occupancy, the dialogue between each of the spheres increases. Of note is the remarkable success with resource efficiency that is being achieved in businesses of all types as they try to find ways to reduce cost and increase profitability. These 'successes' in the business community may have a future influence on urban development if businesses translate their efforts into demand for more efficient urban systems, includ-

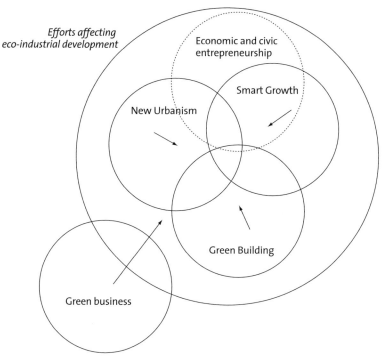

Note: Smart growth, Green Building and New Urbanism are paradigms relating to urban development that attempt to create better communities. Eco-industrial development includes many objectives and participants from each of these paradigms. Additionally, the business community is making notable progress at increasing efficiency and reducing cost. Many accomplishments of the business community are consistent with and/or driven by environmental protection objectives or regulation.

Figure 7.1 **Sustainable design and programming: allied movements in urban development**

Source: Pario Research, from Pario's Passiflora ProModel for Integration

ing even residential real estate, because companies are increasingly aware of the need for employees to have quality homes, neighbourhoods and schools. To reflect the emerging understanding by business, Figure 7.1 shows a 'green business' sphere entering the urban development movements, working toward more sustainable building.

When approaching eco-industrial development, developers should acknowledge that these are suggested changes in philosophy and implementation processes. In terms of eco-industrial advocacy (for the non-developer), sponsors must help with education, overcoming obstacles and providing good case studies that prove and show a path to success. Importantly, these advocacy tools should be placed in the syntax of the developer. All of these are important, and advocates can use these tools to achieve implementation success commensurate with the large forces and resources studying eco-industrial development and the significant resources committed to its implementation. The role of the developer is discussed in the following sections:

- Understanding the subject: the barrier before the obstacles
- The obstacles: missing connections and the influence of a limited span of control
- The approach: sharing risk and reward
- Process understanding: the 'as is' and the 'should be'
- Progress, natural systems and will

7.2 Understanding the subject: the barrier before the obstacles

> Good topic education is the first requirement in most potential applications of sustainable design.

Because eco-industrial development is poorly understood, education on the subject is the first step. Developers need to understand the relationship between resource availability, resource cost, preventing compared with managing ecology-related problems, and community—and the link between all of these and business profitability.

In its robust form, sustainable design touches on every functional department of an organisation, thus initiating a need to communicate to many people about something that will in many cases suggest that they change what they do, often adding risk to their processes. This is a challenge. Pario's experience with providing education on sustainability is that people require many instances of exposures to the message over a long period of time. Often, this can entail many discussions over several months.

Complicating this lack of understanding of the subject is that in the beginning many people correlate the environmental subject matter of this subject directly with their past association with environment-related issues. The majority of those past associations have added cost and delay to organisational processes, and they are viewed as a cost and not a benefit. Thus, the developer needs to understand the broader benefits of sustainable (resource-efficient) business and its manifestation in the eco-industrial paradigm.

Eco-industrial development can affect a development organisation in the functions relating to design, government relations, marketing, finance, construction and property management, among others. The time required for educating all of the people involved is an obstacle, and it is often difficult to get one targeted functional department to adopt methods that affect other departments. In this light, senior management endorsement is very helpful, providing a directive and leadership. This is the opposite situation from that of a functional department head who has solely adopted responsibility for the successful implementation of an initiative which requires tremendous change and endorsement from people not in his or her department.

Education is an important and under-appreciated issue even among eco-industrial advocates. There are usually several possible sustainable solutions for any given project or project component. 'Designer's bias' is common, and occurs when people or firms apply their specialisation as the solution when there may also be other, even more favourable, solutions. Usually, any system designed for sustainability is better than a non-sustainable alternative. However, designer's bias is an issue to be aware of and manage. It is much like the 'style' that many architects are known for and which they apply to almost any project, with little interest in more project-specific responses.

After this general education phase, the implementation phase brings similar educational challenges, only at a more detailed level of step-by-step organisational change. Further, during implementation, the development organisation adopting the changes needs to accomplish this same education–adoption–implementation sequence with many of the firms involved in the project. This can take considerable time.

Fortunately, once eco-industrial implementation efforts are under way, one sees considerable enthusiasm, commitment and business benefits at almost every step in the implementation process, and it is these benefits and accomplishments that propel eco-industrial efforts forward, even in the face of the new task structure. In the end, better assets are produced and a higher return on investment accrues to investors.

Developers need better case studies to rely on during the education and implementation process. Eco-industrial advocates need to better understand the development business and present information in a syntax that developers understand. Further, these cases need to present specific information needed by the developer to incorporate it into marketing plans, financial models and the material of other key functions. Currently, most case studies are merely project profiles and, though informative, are inadequate to truly aid decision-making. To some degree, an incomplete case-study presentation works against advocacy because an attempt to use this in a resource-committed adoption process yields a frustrating lack of adequate information. This may represent a disincentive for future adoption efforts, especially if a prospective adopter is already sceptical, perhaps because the subject is associated with the question of 'the environment', with which there is already negative experience.

The existence of suboptimal case-study presentations of is largely a result of the early stage of growth that eco-industrial development is in. Those that see benefit in the new paradigm and that are carrying out eco-industrial development are from a group of diverse professions—including academics in the life sciences and physical sciences, municipalities and government agencies, non-governmental organisations, architects and engineers and a growing number of businesses. Generally, these people and groups do not have in-depth understanding of the development business and are thus producing case studies that are, although well intentioned, not as effective as they could be.

Key information to include is the change to costs, revenues and time relative to case-study characteristics. Also important is a clear presentation of benefits to users, including a breakdown of benefits to various functional departments, such as marketing, finance, human resources, operations, etc.

7.3 The obstacles: missing connections and the influence of a limited span of control

> Lack of connection and lack of control are primary factors creating systemic inefficiency, and creating these resource and organisational connections is a

primary challenge of eco-industrial implementation. During implementation, the developer's lack of direct control over key process participants represents risk. The developer's role is to define value in the network of increased connection, and to derive benefit from creating those connections.

In the real estate development process, there are three major participant categories between which better connections can be created for mutual benefit. These include linkages between:

- Municipality and developer
- Developer and building occupant
- Building owner and lessee

The relationship between these participants is fundamental to urban development, yet the interaction is suboptimal on almost all levels. For example, the municipality produces codes and standards for development, but their guiding systems often do not produce the best results, in part because of a lack of an evaluation feedback loop. Overall, the connection and effectiveness of finance flows, resource flows and communication flows are suboptimal, and opportunity is lost to increase value and environmental performance.

When implementing eco-industrial design and programming, some key needs emerge that all participants should be aware of and embrace. These include the need for:

- Additional time
- Education on the subject
- Information on process change requirements
- Access to new materials and technology

and the need to:

- Transform existing partner organisations
- Identify and to work with new partner organisations

The boundaries of the real estate development 'system' are already broad, but broad in a way that threatens control and introduces risk. Major inputs currently come from outside the organisation, such as in building approval, finance approval and, to a greater degree than often realised, design decisions. Sustainable design suggests the need to widen these boundaries yet further, a move that is typically resisted both emotionally and systemically.

Changing mainstream real estate development to incorporate eco-industrial design and programming can be viewed in terms of two major categories: structural topics relating to change itself, and implementation procedure dealing specifically with eco-industrial requirements.

Structural changes require:

- High-level support for the process
- Time
- Resources

- Approval or authority for change, such as from a city council for code changes (outside the organisation)

Implementation procedure needs to:

- Define opportunity for eco-industrial design and programming
- Assess where economics or value are not optimal (by using whole-system thinking)
- Target designs, relationships and technologies where increased value can be captured
- Build an implementation team
- Manage the process
- Share the benefits

The developer is the one person (or organisation) that not only has a link to all participants of the urban development process but also has the largest direct stake (financial and in terms of reputation) in the outcome of the investment. Even when the business case for change toward sustainable design is made, the next task—implementation—introduces risk. It is difficult for the urban development industry to make changes. Most development organisations have much less control over individual processes than does, for example, a typical manufacturing organisation (see Fig. 7.2). The standard system of development is strongly fortified against change. New connections need to be established, and all can be generated and co-ordinated through the proponents of sustainability, adopted by organisational managers and thus can create a 'touchstone' purpose. With this, the reason for new working relationships becomes clear, and the creation of these new relationships can then be established as organisational goals. When the change is made in this way, many of the obstacles to integrating sustainability are minimised and an appropriate team is formed to increase process change efficiency.

Real estate development is an entrepreneurial business, even in its 'institutional' realm. In the development phase of real estate, many Wall Street investors require risk-calibrated rates of return that are equivalent to those of start-up businesses, indicating that for these investors each real estate project is uncertain, like a new business. Developers are assailed at all steps in the development process with factors out of their control. The industry copes with this relative uncertainty by being cautious. Change not mandated with clear benefits or for clear survival reasons is looked on as added risk and is often resisted. This is not to say the real estate industry does not innovate; it does, but usually within limited boundaries, especially compared with other industries. A primary reason is that, if a developer pushes for substantial change, it must involve and seek endorsement from many organisations outside its direct span of control; each of these organisations may put up resistance to the change from 'what works', and extra time will be required for the education and analysis process within those organisations. It is easier to make 'market returns' with existing product formats that are accepted by all process participants. These include equity financing, debt financing, municipalities for entitlement, the market, construction firms and property management. Thus, even though eco-industrial development presents opportunities for higher value, reduced cost and benefits for all process participants, the obstacles are important to understand—because

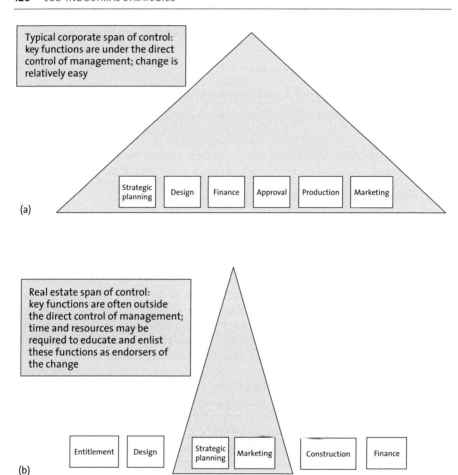

Figure 7.2 *Comparison of the span of control of corporations and real estate developers: (a) a typical corporate environment and (b) the environment of a typical real estate development organisation; the triangle represents the area of control*

understanding such obstacles is the first step towards solving them both for sustainability advocates and for developers looking to implement eco-industrial development.

7.4 The approach: sharing risk and reward

> The opportunity is realised in connecting things, and many of the connections are simple.

The role, and in essence the opportunity, of the developer in eco-industrial development is to expand on its current role of catalyst and conduct. In addition to the developer's current task—linking a municipal plan and the developer's desired project with current market forces—the eco-industrial developer needs to assess the factors affecting a project, translating them into benefits and specifications of sustainable industrial development. Major responsibilities are to:

- Capital investors
- Businesses
- Ecology
- The community

Risk is experienced in the form of task and activity changes, process investment and market acceptance. Reward is different for each project but, if properly defined and presented, can include favourable entitlement treatment, increased market acceptance, increased value, better-performing assets for all stakeholders and enhanced skill sets and reputations for the implementing organisations, which should increase future competitiveness.

7.4.1 Draw a bigger circle

Importantly, the developer can also step beyond the role of 'space provider' into a new role in the urban development process—that of synthesiser of participants, resources, information and capital—and thus better facilitate and participate in the management of buildings or communities. The developer should catalyse the new cast of characters, roles and responsibilities for implementation. Developers do this in their current role; however, to assess eco-industrial development opportunities and to implement them effectively the developer needs to expand interaction 'upstream' towards regional economic development, environmental planning and infrastructure planning—as well as to expand interaction 'downstream', with greater interaction with businesses in their operations and in real estate management. This is essentially 'drawing a bigger circle' around the existing relationships in the industry.

Even though many of these tasks overlap with the responsibilities of other participants, developers should initiate the new relationship formation for several reasons. First, the syntax of the development community is unique, particularly with regard to finance, marketing, sales and management. Much of the sustainable dialogue is now generated from non-developer participants—such as government, the design and engineering community and environmentalists. Although the message content is appropriate, the message is often not communicated in a syntax that allows the development community to hear and understand the information. And the communicators are often too busy to recode and repackage it. This same need for information packaging applies to proponents of eco-industrial development who are 'selling' it to the developer. By paying attention to the developer's information needs and giving help to overcome obstacles, proponents of eco-industrial development will enable integration of the principle at all levels.

Specifically, the developer can work in new ways with the public sector and that sector's upfront role in providing infrastructure and services (e.g. concerning roads, water [including management of storm-water and waste-water], power and solid waste) and can help the public sector with designing and implementing resource-efficient systems. Later, the developer can work with businesses that occupy space to provide better connectivity for resources, labour, waste, transportation, information and communications. Each of these has implications at different stages of the development life-cycle, and therefore various processes are impacted.

Ultimately, those who want change toward more sustainable design and programming need to do more to effect that change. During the 1990s, as awareness grew after the Brundtland Report (WCED 1987) and the Rio conference, considerable resources were applied to making society more sustainable. Overall, the success rate with regard to education and implementation has been poor, as only a small share of activity has been affected during the last decade. We need to learn from the failures in education and implementation. This is the responsibility of those who want change.

7.4.2 Enlisting process and cost support

Because of the forces trying to promote and support sustainability, there is in the short to medium term an opportunity to utilise resources to reduce implementation cost, including offsetting any marginal increase in development costs. Sharing risk and reward begins with adopting a philosophy that there is more value that can be realised in all stages of development by conducting business differently. And this can be a premise when reaching out to government, businesses and the community when designing and building an eco-industrial park. These desired process participants can be educated about and motivated by their gains as an incentive to assist the developer with the project. For example, an energy service company may reduce the capital costs of new power plants by assisting the eco-industrial developer with distributed energy integration—a common reason that such service companies now support sustainable initiatives. Similarly, a sanitation district may reduce the need for new plants by promoting and assisting with on-site waste-water treatment and re-use. The developer can assess each project for these opportunities.

Very little eco-industrial development has been adopted by developers relative to industry overall. Other industries have been aggressive in adopting sustainable development principles in response to changes in technology, consumer desires, communications and other trends.

Many businesses are already achieving resource-efficiency gains with use of such tools as technology improvements, process redesign, value chains and industry clusters. The idea of closer working relationships for mutual gain is becoming well established in business overall, but it is not practised to the same degree in the real estate sector. This 'gap' is an opportunity.

Development companies usually have a specialised understanding of certain types of real estate construction and building occupants, and associated with these specialised types of real estate go unique equity and debt financial structures for construction as well as for the subsequent ongoing ownership of the assets. Even though the developer catalyses this process, government, business and the community have an under-appreciated, longer-term and more financially involved role when all costs and benefits are integrated. Though some developers maintain ownership in the properties they develop, a large proportion do not. This lack of ownership interest in long-term operations is a key barrier to motivating developers to investigate better designs and to overcome the risk perceptions of implementing new systems.

In the early phases of the urban development life-cycle, municipalities create zoning and building guidelines as boundaries for key infrastructure, health and safety issues. And public agencies service real estate with resources and services for operations such as power, water and sanitation. We are learning how inefficient many infrastructure and resource-delivery systems are and about the increased costs of this suboptimal development. Key investment sinks include:

- Traffic costs, including the costs of pollution clean-up and healthcare and the consumer burdens of commuting time and travel cost
- Energy costs, including pollution clean-up and pollution-related healthcare costs and the impact of rising energy costs on businesses and consumers
- Water costs, including the cost of pollution related to tourism. Much water-related cost is probably not yet realised.
- Air pollution costs, including costs associated with the compromised health of citizens, lost productivity and the commensurate toll on longevity and family function
- Infrastructure costs. Here, inefficient and ineffective infrastructure costs more to build and maintain, and even creates allied problems such as ineffective storm-water systems causing a lack of local water absorption, depleted aquifers and downstream pollution.

To some degree, the very zoning and building codes mandated by municipalities preclude better development and sometimes promote inefficiency. Here, too, better design and incentive for change are needed, and many of the same needs apply to the municipal departments concerned.

7.5 Process understanding: the 'as is' and the 'should be'

Step-by-step process change represents the daily detail and the hard work of the integration effort. During implementation, obstacles to change occur daily, and, because eco-industrial development integration is new, the solutions are often creative new ideas. In essence, each real estate development task needs to have the requirements of sustainable design programmed into it, and process mapping is a useful means of exploring an organisation's situation in preparation for change. It is also a valuable teaching tool and has utility as a 'roadmap' for action. When people see the how their organisation operates today and have graphically demonstrated a direct process change comparison, many of the uncertainties are eliminated and a more educated implementation decision can be made. Figure 7.3 is an example of how process change is realised in real estate development. In this example, the 'detailed design' phase is shown in an 'as is' and 'should be' comparison, including the additional tasks and links associated with the integration of sustainable design and programming. The detailed design phase is a good example of how additional process participants are added with sometimes substantial interaction from each. Detailed design input often includes that from architects, planners and engineers and marketing information to inform building function. In the enhanced process model, new interaction may come from a variety of sources, such as tenants. Though mainstream real estate does design for target markets or specific tenants in build-to-suit relationships, these do not reflect the system integration opportunities that the eco-industrial paradigm suggests. For example, less costly, higher-quality and more reliable energy is being provided within business parks using distributed generation systems. These suggest that the developer, tenants, energy company and local utility collaborate—all new processes when compared with traditional development. This is only an example, and each project brings different influences and needs and different formulae of success and failure that must be determined by the development team and other stakeholders.

The 'design' phase of business process redesign represents a relatively small portion of the total change effort. So far, what sustainability advocates have done is to create guidelines indicating what developers should do, without commensurate support in the key implementation stages of the redesign effort. In other words, complications have been added without solving perceived obstacles. The advocates may not even realise the depth of the predictive thinking of developers if the advocate is not truly aware of how development organisations make decisions.

Ideally, the eco-industrial development initiative should include a person or organisation with experience in business process redesign. This will allow an understanding of the 'as is' and 'should be' as it applies to the specific organisations involved. In addition to this project-specific skill there are some good sources of information available that can inform all initiatives, including the codes, covenants and restrictions set out by the Work Environment Initiative at Cornell University (WEI 2000), the *Eco-industrial Park Handbook* by RPP International (Lowe 2001) and the LEED (Leadership in Energy and Environmental Design) design and building rating system of the US Green Building Council (USGBC 2002). These tools are not only educational but also adaptable to specific projects. To some degree, there is a misallocation of effort to provide a 'path' to more sustainable

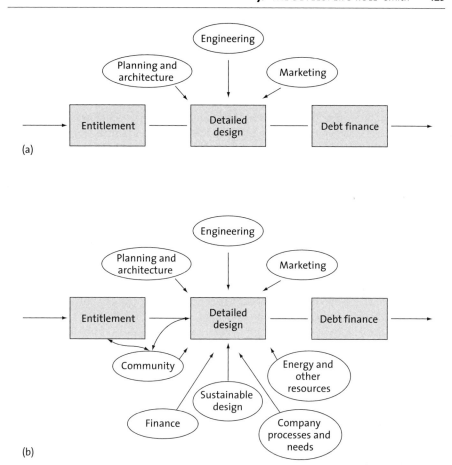

Figure 7.3 **An example of process change involving the detailed design phase of the development life-cycle: (a) the phase 'as is' and (b) the phase as 'should be'. The picture is similar for all phases of the development life-cycle. It can be seen that the level of interaction can more than double during the detail design phase, requiring additional time and resources. Outside participants, such as community, are often involved in several phases.**

development in that sustainability advocates often develop 'sustainable guidelines' to be followed by those the guideline sponsors wish to affect. Often, these guidelines are too prescriptive, though they are well intentioned. Two issues arise. First, many of the best solutions to sustainable design flow from project-specific design activity, and guidelines can preclude much of the exploration of issues appropriate for each site and user. Second, the time and capital investment for developing the guidelines would often be better used in creating local stakeholder teams that are ready to work with the developer or company that owns the project to be affected. What this 'reallocation' will do is, instead of giving the developer a stack of new procedures and little true implementation help, will give that developer a supportive team in the form of a toolkit and network, to guide them along a path that is best experienced as a project-specific exploration instead of a too-detailed estimation of the best result. Again, the intentions of guidelines are good, but in many cases the resources can be better employed.

Determination of the 'should be' version of an eco-industrial initiative involves four major phases: (1) assessment of project-specific and local opportunities; (2) outreach and alliance-building efforts to incorporate required skills and participation; (3) implementation; and (4) management of the sustainable design and programming for the life of the asset. A systematic approach to each of the resource-saving and labour-saving tools will yield unexpected positive results and considerable return on investment. A means of harnessing these tools to assist organisational transformation is what Pario Research calls 'resource engineering'. A resource engineer would take on a new position (or added responsibility, where appropriate) in an organisation, with responsibilities for integration of sustainable business methods. Specifically, responsibilities would include education, formation of internal linkages and external linkages (such as with suppliers, customers, government and other businesses) and consideration of energy, water, waste and other components of sustainability relevant to the organisational mission. A resource-engineering toolkit could be developed for each organisation. The position should pay for itself within a year and generate very favourable returns from savings obtained from medium and large projects. The position of resource engineer would not necessarily be permanent, being eliminated once sustainability is integrated into the organisation.

7.6 Progress, natural systems and will

Though developers are the catalyst for the development process, they have been the last of the major participants to meaningfully adopt the opportunity for eco-industrial development. In other words, the hub has not formed to connect the satellite elements. Indeed, government, business and community interests are promoting or making significant progress in understanding how resource-efficient business can improve quality and value. The integration effort is somewhat powerless unless embraced by the developer-catalyst. Eco-industrial development has been created largely in the academic and planning communities and has been promoted by those agencies along with government. Importantly, however, companies have made some of the greatest progress in the effort to save money and time, and found resource-efficient methods of operating and new technological solutions that preserve natural and financial capital. The group least involved is developers—those that create the space that all of this either takes place in or links with. As mentioned above, developers resist change in part because of the unique circumstances of their operating environment, but the operative factor resisting change is not circumstances but human will—the unique characteristic that sets humans apart from other living beings. It is common in the sustainability dialogue to make reference to natural systems as models for real estate development, urban and social systems and companies, suggesting that if they were to work on the same principles as natural systems they would operate more efficiently. It is important to realise that in the processes of education and the design and management of a sustainable community human will is a primary force behind 'obstacles' identified: learning about and implementing sustainability is often hampered by resistance to change.

The good news is that will is also the factor that motivates proponents of eco-industrial development. Transforming the will to resist into the will to integrate is key. Education, meaningful case studies, process understanding and new roles and responsibilities are the tools of transformation. The very good news is that the transformation from will to resist into will for change can be accomplished and has been achieved in an increasing number of successful new projects and communities.

Although the development community needs to become more receptive to learning about and implementing sustainable design and programming, much of the responsibility falls on sustainability advocates. Considering the resources committed to creating more sustainable communities during the past decade, and considering the knowledge of the benefits and successful examples, the actual change affected is disappointing. Simply put, sustainability advocates must become more effective. A big part of the challenge is to get sustainability advocates to present information in the syntax of the development community, so that real estate decision-makers can incorporate data into their organisations in the form of useful financial, marketing and management information, so that they can assess the impact of the change on other parts of the development process.

In other words, those who want to promote sustainability, eco-industrial development and more sustainable business operations need to increase efforts to promote an understanding of the topic, its benefits and the implementation process. With these as input to developer education, the developer's role in this new paradigm is to discover the reduced costs and higher value available through the integration of better design and resource linkages and to take a leadership role in conducting the new, expanded, team toward better development patterns.

8
THE ENERGY OF ECO-INDUSTRIAL DEVELOPMENT

Daniel K. Slone
McGuireWoods LLP, USA

Energy cascading—the use of multiple energy products (e.g. thermal products such as steam or hot water) derived from the generation of power, is a key part of many eco-industrial projects.[1] The opportunity for energy cascading occurs because of the inefficiency of most combustion processes. The burning of fossil fuels or biomass for energy can result in large amounts of heat loss. Energy production through non-combustion processes such as by fuel cells or digestion can also release heat. Many industries over the years have taken advantage of this relationship by setting up 'inside-the-fence' projects that generate energy for the facilities and utilise the heat in the industry's manufacturing process, either directly or as steam or hot water. The co-generation industry grew out of this same concept as larger power facilities were built to sell power to the power-distribution grid and to sell thermal products to nearby industries. Related types of developments, that may convert the by-products of power production into chilled water, air conditioning or compressed air, as well as steam, heat or hot water, have been called tri-generation or power parks.

Although energy cascading stands out as a pivotal opportunity for eco-industrial parks, it is by no means the only energy consideration. Some parks have sought to integrate alternative power sources such as photovoltaic cells or windmills. Other parks have focused on the energy efficiency of the buildings in the park. The embodied energy of goods and services is also a frequent focus of eco-industrial discussions. Numerous stories exist of businesses shipping materials from distant suppliers when inquiry shows that the original manufacturer or a comparable source is located nearby.

1 Kalundborg, Denmark, includes in its network of exchanges, in addition to the refinery, pharmaceutical company and plasterboard manufacturer, a coal-fired power plant and a district heating system for the adjoining town (Lowe et al. 1997: 131). US parks that have integrated power production facilities include Londonderry, NH, The Redhills EcoPlex in Choctaw County, MI, and the Cabazano Indian project located in southern California near the city of Indio.

Even more fully integrated approaches such as 'energy harvesting' or 'omnitilities' have been proposed. These approaches tie surrounding non-industrial uses into energy production or create master utility arrangements in which one utility deals with energy production and use as well as other utility functions.

8.1 Energy cascading

As a by-product of the inefficiency of combustion or chemical processes, energy-cascading opportunities are most robust when efficiencies of the underlying process are least. Thus, when electrical facilities move from burning coal to gas, or from gas turbines to combined cycle facilities, the available heat and, thus, thermal by-products decline. Consequently, as the energy production business becomes more sustainable by becoming more efficient at energy harvesting, the opportunities for energy cascading from this source decline and become marginal. Alternative energy sources such as wind, hydroelectric and photovoltaic, leave no opportunity for cascading (and, of course, geothermal energy concerns thermal output without power generation).

Nonetheless, energy cascading is an important part of eco-industrial design, for several reasons. Even new power plants, though more efficient than their predecessors, leave significant efficiency gaps. Older power plants, with even greater efficiency gaps, are being kept in operation much longer then originally planned. These plants also face significant costs for retrofitted pollution controls. The availability of significant wasted thermal energy and the need for new capital make these facilities attractive potential eco-industrial possibilities. New technologies, such as fuel cells, although extremely low in pollutants, have heat as a by-product of their operation and create new combined heat and power (CHP) opportunities. Finally, although most of the large-scale industries with heat as a major by-product of their process have already moved to co-generation or inside-the-fence arrangements that convert the excess heat to power, many smaller industrial operations have not done so and, thus, represent opportunities for eco-industrial integration.

Energy-cascading opportunities are 'scalable' and can be derived from the largest industries and power plants or from small-scale distributed generation facilities. Several companies have been developing CHP packages for single buildings. Others have optimised small power-production units to provide chilled water. These designs can be adapted to clusters of buildings or even whole new towns.

In adjusting scale, however, several other factors must be addressed. One is pollution control. Distributed generation has been labelled 'distributed pollution' by some organisations, particularly when it is accomplished with diesel-fired reciprocating engines or even gas-fired micro-turbines. Whereas large-scale power-production facilities may trigger rigorous pollution controls, distributed generation may fall below triggering levels. Thus, when analysed from a perspective of amount of pollution per megawatt of power, a pure distributed generation facility may be more polluting. The equation changes, however, if the facility increases efficiency by integration into surrounding buildings through energy cascading. Of course, if large-scale power-production facilities are similarly integrated, the equation changes yet again.

Another important factor related to scalability is economics. Energy cascading typically creates more opportunities for diverse economics. In one circumstance, the price of energy may drive economic success whereas in another the price of steam or some other thermal product may be more important. The equation may get even more complex if the fuel is generated from chipped tyres, biomass, black liquor or some other substance that a power producer might be paid to utilise in lieu of sending waste to landfill or deep-well injection. As scale declines, other relative costs such as permitting, dispatch, maintenance or operation typically do not decline at the same ratio, and the smaller-scale unit may have much tighter economic margins than have larger facilities. The cost of providing certain thermal products may not be justifiable. For example, because steam generation may require a full-time boiler attendant, a smaller facility may be able to provide hot water but may not provide steam economically.

Significantly less capital is required as scale declines and, thus, less is at risk than with large-scale facilities. Nonetheless, the risk of technological displacement may be higher for smaller facilities, particularly because these technologies are currently in a period of rapid development. Large-scale power-production facilities may be planned for a 20–30-year life. As a CHP micro-turbine is placed in an eco-industrial park, does its owner need to worry that it be displaced by the much cheaper energy from employees' parked fuel-cell-driven automobiles?

Other economic issues arise in connection with scale. Large facilities may be able to generate various environmental credits for carbon-use reduction, greenhouse-gas reduction or other reductions of air emission impacts. The value of these credits is expected to rise over the coming years. Individual smaller-scale projects may not cross threshold transaction cost barriers (though methods for aggregating these into a system for comprehensive assessment are possible).

8.2 Cannibalising your neighbours

In nature, opportunistic users respond to every energy source. A tree that has grown tall into the canopy of sunlight carries with it vines and parasites. A heat vent on the barren ocean floor is surrounded by sea cucumbers and other creatures drawn to the thermal energy. As every creature excretes or dies others are lined up to harvest not only elemental components but also the energy. So too is the design of the engineered, parallel world of eco-industrial development.

If a business has wood waste or agricultural waste, there is potential energy in the digestion or combustion of what would otherwise be waste. Black liquor, the by-product of paper pulping, contains small wood fibres that can be burned, producing energy and creating re-usable white liquor. Automobile 'fluff', the waste of car recycling—the non-metal portions of the vehicle—can be turned into energy. A company may need groundwater for cooling, but geothermal activity makes the water too hot to use immediately on extraction. Another company can use the heat first and return cool water. Similarly, a company may use steam in its manufacturing process or large amounts of water for cooling. Perhaps the scale is smaller than in other industries but still significant. In each case, the heat is often dissipated in a cooling tower or pond. Sometimes the energy is released up a stack as heat. In each case, other options to use the energy are available.

Not all opportunities make economic sense. In some cases, combustion or digestion creates energy, but at a price that greatly exceeds current competing energy prices. Temporary fixes such as collection of the tipping fee or state-sponsored fee for disposal of a waste product may create profitability, but such arrangements are often economically unstable. Interestingly, even a stable product such as wood waste can go through unstable cycles as prices become so low that suppliers go out of business. Consequently, users close to a source that is simply a by-product of another operation (such as a lumber mill) may be more stable, albeit at a smaller scale, than a large power-production facility that imports wood chips.

This co-location is where the energy aspects of eco-industrial development are at their best—when they are taking advantage of some business circumstance that already exists. Eco-industrial approaches are most unstable when used to try to take unprofitable undertakings to profitability. They work well in adding efficiency and thereby increasing profitability. Two vines in the rainforest cannot reach the sun, even when leaning on one another. Even if they did, they would not last long. They need a strong tree. This is not to suggest that every eco-industrial undertaking must have a large energy anchor. It works quite differently from that.

If a large power plant exists or is being built, it is an opportunity for eco-industrial development. It is likely that the power plant's business plan works whether or not there is secondary use of waste energy. But the facility may make more money if it can sell its waste heat. Or it may be more acceptable to the surrounding community if it can generate jobs at a co-located greenhouse or shrimp farm that utilises its waste heat. Or the facility may be more acceptable to the environmental community if it displaces the coal-fired boiler that has operated at the local chemical plant by supplying the chemical plant with steam.

But, by the same token, the scale of possible energy facilities could be, and often should be, quite different. Assume a lumberyard that is operating profitably. A power facility is built that consumes the waste wood and sawdust, reduces the base load electrical purchases of the lumber mill and generates heat, which is used in a kiln at the mill. The participants also bring in numerous small craftspeople who use the cheap waste wood in local crafts so that the first priority is to examine the best balance of avoiding waste or creating other products before energy harvesting. The kiln has the joint benefit of moving value-added payments to the mill and avoiding the embodied energy of transporting the wood for drying. Into this beginning cluster other related industries might join.

Whether scaled up or down, the opportunistic approach to energy harvesting is stable when based on a sound underlying ecology. As we change the underlying ecologies of energy production, developing technologies that are more efficient, these opportunities will fade away. As the ecology of human 'waste-streams' change and society reduces or eliminates various wastes, these opportunities will also disappear. It seems unlikely that there will be serious threats along these lines for the lifetime of the industrial parks being designed today.

Even as the opportunities disappear, new ones emerge for energy cascading. The roofs of buildings are bombarded with energy in the form of sunlight. Wind blows without regard to property lines. The bodies and machinery in office buildings build heat that must be removed. As the efficiencies of alternative energy sources become greater, the issues may become available exposure and movement of accidental energy.

Suppose you have an industrial building. Even with highly efficient photovoltaic cells the building does not generate enough energy to run your machines. Now suppose your neighbour has a tall office building with photovoltaic cells up the sides and on top. Your neighbour uses the heat generated in her building, her building is very efficient and, even before her employees' fuel-cell cars have been plugged into the parking deck to increase available energy, she has energy to sell you. The only way she will be able to capitalise this system of energy harvesting (because grid prices are so unstable) is with a long-term contract with your plant.

8.3 Omnitility[2]

One concept that has emerged from the energy and utility discussions in connection with eco-industrial facilities is the 'omnitility'. The omnitility is a utility that aggregates utility functions to overcome the system conflicts that lead to inefficiency. For example, in a particular jurisdiction water may be sold by the local utility. The charge for sewage treatment may be set according to the amount of water consumed. Thus, even if grey water is re-used and never sent for treatment, and even if the total consumption may be reduced, participants will pay for the capital costs of enabling the collection and use of grey water and for water treatment that never occurs. In the omnitility, one utility sells the potable water, the grey water and the sewage treatment. This company makes money by optimising the system. Redundant charges are eliminated and a means of off-balance-sheet financing for the grey-water component is created for system users.

On the energy side, several distinct functions may make more sense in combination. If the energy company also provides gas, this may create incentives for energy sources that use natural gas such as micro-turbines or fuel cells. If the company also sells energy management and energy efficiency (for a portion of the savings), then it is interested in and makes money from system optimisation rather than from system inefficiencies. In this setting, the sale of thermal products is consistent with the omnitility's profit-making goals.

Any number of utility aggregations are possible, constrained perhaps by that point in the process at which efficiencies are overwhelmed by complexities of combined operation. These combinations may even create strong enough financial incentives that they become magnets, drawing businesses into the eco-industrial park. The key to their effectiveness, however, is that they must counter the tendency of eco-industrial settings to require industries to negotiate multiple agreements to manage what were simple services in a conventional park. In a conventional park, the industry paid the electric bill

---

2 The omnitility concept was developed by this author after officials in the New Mexico State Land Office (Harry Relin, Assistant Commissioner for Commercial Resources, and Thomas Leatherwood, Manager for Planning and Development [Commercial Resources]) suggested in a charrette regarding development of Mesa del Sol (in Albuquerque, NM) that providing an easy outlet for waste management might create a 'magnet' for businesses. Thomas Leatherwood is now a private planning and environmental consultant at Thomas Leatherwood Associates, Commons on the Alameda #A3, Santa Fe, NM 87501, USA (tomsantafe@att.net).

and the water bill; it did not have to negotiate complex power-purchase or grey-water contracts. The omnitility must make its services economical and easily accessible.

8.4 Issues

Numerous issues go along with energy use and management in eco-industrial facilities. Years of co-generation development have stimulated solutions to these issues in connection with large-scale projects. Some of the available solutions scale down easily into smaller distributed generation settings, but in some cases new solutions must be developed.

Often, downtime—scheduled or unscheduled—is an issue in energy-cascading situations. A business needs power to operate. Though it may increase costs to be able to draw from the power grid, this is often a practical backup in case there is some disruption in the operation of the primary generation. If, however, a business is dependent on thermal energy for steam, hot water, heating or cooling, the consequences may be much greater if the source of these elements must shut down for repair or because it is uneconomical to run. If the power source is providing only a supplement to other systems when providing these thermal elements, the consequence of downtime might only be slightly higher costs. If, however, it is the primary source of necessary thermal elements, the consequence could be a significant loss of business. While a supplemental boiler may be an economical way to provide backup steam or hot water in large-scale systems, such redundancy may render small systems economically impractical. The other side of the same coin is the economic dependence of the power or thermal provider on the consumer of its output. If the business that consumes the power or steam shuts down for maintenance or because of adverse business conditions, the loss to the generator can be significant. Unless the electrical grid is available to absorb output (and the money has been spent on interconnection), there may be no alternative purchasers. Even if the grid is available, payment to the generator may be significantly less than planned. Several solutions to this type of risk have been developed, including take-or-pay agreements, partial load arrangements where the generator is never designed for full service, mobile generating facilities that can be moved if needed or peaking facilities that are designed to be cycled on and off as needed.

Another issue for some energy-cascading situations is the need to tailor equipment design to one type of energy product or another. Thus, the owner of a power-generation facility that sells power and steam may optimise the equipment for this purpose when it is installed. If the business climate later changes and more money can be made from steam than power, or from hot water rather than steam, then the generating equipment may not be able to adjust to the new business circumstances. The generating flexibility does not always match the business needs of an eco-industrial setting. One response to this issue is to utilise multiple smaller generating sources offering flexibility rather than a larger more limited source.

Similarly, infrastructure in an eco-industrial setting must be flexible. The costs of pre-installing all the pipes that might ultimately be useful in an eco-industrial setting is likely to be wasteful. Will steam pipes be extended to properties that will never utilise steam?

Will district heating or grey water be utilised? Although there are some designs of parks in which this may make sense, in many it will not. In industrial complexes such as chemical plants, open racks are maintained so that new pipes can be added at any time. This may not be practical in all eco-industrial settings and may be particularly difficult in those with multiple owners. In some settings the racks can be moved to re-openable trenches, but these can be expensive and difficult to maintain on sites with substantial truck movement. Another alternative is to use a conduit, which allows pipes and wiring to be added or removed as necessary. Various issues may still accompany this approach as maintenance concerns, mandatory separation of potable water, sewage and grey water or of power lines and other utilities are addressed.

Land-use issues may also arise in dealing with power generation. Local zoning codes often do not distinguish power generation by size, and, thus, a power generator, big or small, may stimulate special siting statutes or may require special or conditional use permits. In some cases, localities may be willing to characterise the generator as an accessory use, but when it is serving multiple properties or tenants and is independently selling its output this definition may become strained.

Perhaps the largest cluster of issues for energy cascading revolves around the energy regulatory structure of the jurisdiction. As a monopoly activity the sale of energy has been regulated for decades. Although the late 1990s saw the emergence of deregulation, the California energy crisis of 2000–2001 slowed this process and in some cases led to re-regulation. Consequently, in much of the USA, power can be sold only in retail transactions by a public utility. In some states this limitation applies to thermal energy as well. Many states, though not necessarily all, allow power produced on-premises by a non-utility to be consumed on-premises, but sometimes there are issues if ownership of the generating capacity is different from that of the consuming customer. In each instance, the consequences of becoming a public utility must be carefully examined. Different circumstances may present different opportunities for sale of energy as part of the lease or for a condominium approach to the generation facility. Each opportunity must be examined in light of the applicable state laws.

Even where retail sale of electricity is permitted for distributed generation, distribution itself may still be a monopoly. Consequently, something as straightforward as serving three separate parcels from a CHP micro-turbine may require the co-operation of the region's distribution company (the owner of all power lines) and some payment for such transmission.

Another area fraught with controversy has been interconnection itself. Safety concerns and system capacity limits require that all interconnections with the grid be done with the right equipment and procedures. Sometimes, these complex systems legitimately require careful study and expensive equipment additions. Some believe, however, that some utility companies require long, expensive studies and high fees to thwart distributed generation. Even the regulations for some undertakings such as 'net metering' (where the generated power at a site not only can be used at the site but also can be sent into the grid, causing the meter to spin backwards) may have consequences. Some statutes require the generation device be owned by the building owner, thereby foreclosing some decisions on capital risk or multi-tenant platforming.

8.5 Conclusions

Energy is a key subject of all eco-industrial discussion. It may arise only in the energy efficiency of buildings required pursuant to the restrictive covenants of the eco-industrial park. It may arise in the recruitment of tenants, in the design of the park or in park policies intended to reduce the energy embodied in products through their manufacture or transportation. Energy may be the key shared product, as power and thermal products are utilised with a goal of 100% efficiency.

Every boiler is an opportunity; every co-generation facility is a mere beginning and every power plant is a prospect. There is no reason why energy-cascading concepts should be constrained to industrial settings. The large, big-box retailers and the warehouses, offices, hotels and hospitals of a community are other considerations to be brought into the relationship. As new towns are built, the challenge for their designers will be not only to integrate industrial uses but also to create efficient relationships between all users. As existing communities are retrofitted, the challenge will be to find a mechanism of sharing the vision, risks, capital requirements and rewards of the efficiencies that can be achieved.

Part 3
KEY ISSUES IN
ECO-INDUSTRIAL DEVELOPMENT

9
LEGAL ASPECTS OF ECO-INDUSTRIAL DEVELOPMENT

Daniel K. Slone
McGuireWoods LLP, USA

Eco-industrial development, like all industrial development, involves legal issues (Frej *et al.* 2001; NCUED 1995). Typically, industrial uses stimulate more land use and environmental permits than do other human endeavours. Innovation in 'waste' management increases the number of legal issues. So too does the creation of legal relationships among industries or between industries and utilities. In conventional industrial parks, the codes, covenants and restrictions (CC&Rs) generate legal issues in their regulation of architecture and common-area maintenance. Some eco-industrial parks add the regulation of environmental and social sustainability to this mix. Although all business creates intellectual property issues, eco-industrial development creates new challenges because of its shared processes.

Although the combination of issues is new, many of the basic legal mechanisms utilised in eco-industrial development were worked out long ago in the electrical co-generation industry and in the co-product agreements of major smokestack industries. New challenges are presented, however, such as integrating evolving standards for environmental and social sustainability so that the legal documents can be as alive and flexible as the projects they represent.

9.1 Creation

Creation of eco-industrial parks or relationships involves several distinct actions and processes. First, the creator must determine the nature of the undertaking and the appropriate management entity. Then the creator must obtain appropriate land-use and environmental permits. Last, the creator must set up the rules of the undertaking either in the form of CC&Rs or contracts.

9.2 The management entity

To date, eco-industrial development has been stimulated through several different types of undertakings. In some instances the undertaking has been the development of a single park. In other circumstances, the goal has been a series of related parks.[1] Another version has been the creation of eco-industrial relationships between existing industries throughout a region, with actual parks for recruitment of new businesses as an after-the-fact option.[2] The management entity for an eco-industrial development may be different from the development entity. Thus, a park may be developed and the parcels sold by a private developer but be managed by a property-owner association.

Consequently, the next choice to be made, after choosing the type of undertaking, is to choose the type of relationship to be structured and managed. If the project is going to be a single park or multiple parks, will the park participants own their parcels, will they lease them or will the park have a combination of both types of occupant? If the project is going to be fostering regional relationships, will all of the relationships be direct or will there be any continuing role for an intermediary after the initial relationships are brokered? Various business concerns may affect all these choices. Leaseholds allow greater control of property than outright sales but, because the developer pays for the buildings and recovers its money over time, they may lock up capital needed for other projects. In a given market, one form of occupancy may be more marketable than another. Risk and potential liability may be higher when a developer takes a continued project role. For example, a continued role as the underlying property-owner with tenants may leave the developer liable for contamination of soil or groundwater.[3] A role as intermediary may carry liability for off-specification co-products or business interruption if a participant fails to fulfil its obligations.

After the nature of the undertaking has been defined and the types of relationship identified, then the management entity can be created. If the project is a private development and will be leased, not sold, then this may be as simple as creating the corporation to serve in the management role or as hiring a management company. Similarly, a public project may create an industrial development authority (IDA), which may manage a project directly or may utilise a management company.

---

1 Cape Charles Sustainable Technology Park is operated by an industrial development authority, the membership of which is made up of representatives from around the county, in anticipation of the eventual operation of multiple eco-industrial facilities with different levels of industrial intensity (www.sustainablepark.com). Similarly, Chattanooga has contemplated various eco-industrial parks, each with a somewhat different focus (www.csc2.org/index.html).

2 The Triangle J Council of Governments Industrial Ecosystem Development Project provides information to businesses and institutions in the Triangle region of North Carolina and helps to identify potential partnerships for the re-use of materials, water and energy. The project information can also be useful to other regions developing their industrial ecosystems. Currently, the project is working to promote by-product exchanges with a few existing and planned industrial parks (www.tjcog.dst.nc.us/indeco.htm).

3 See United States v. Monsanto Co. [1988]. In this case, the owner that leased land to a chemical manufacturing company that recycled chemical waste at the site was jointly and severally liable for the removal of hazardous waste under the Comprehensive Environmental Response, Compensation and Liability Act (CERCLA) of 1980.

The IDA or private developer may elect to create a tenants' association to advise on, or even make, key management decisions. The spirit of eco-industrial projects requires that there be some collaborative body to continue to explore efficiencies and to encourage capital investments that enable greater efficiencies. These groups may be created through the leases or through the CC&Rs.

Ongoing management is somewhat more difficult when the industrial parcels are to be sold. In these circumstances there is no lease to utilise for control of the occupants, and when all the property has been sold many private developers have little reason to continue with their involvement in the project. In most cases, a property-owners' association (POA) is then the appropriate manager. Conventional industrial parks have utilised POAs to provide for the continued enforcement of park rules and architectural requirements as well as to provide for maintenance of common areas such as private streets, landscaping or open spaces. In eco-industrial parks POAs can also oversee the environmental and social sustainability commitments. The public and regulatory agencies frequently express concern about this self-policing. Some of the mechanisms for continued public involvement are discussed below in the section on CC&Rs (Section 9.5), but one of the most important control devices is the economic self-interest of the parties. They are interested in increasing the efficiencies of the park and its participants because this increases their savings, and they are interested in maintaining the park's 'green' reputation because they have often paid a premium to be associated with that reputation and its benefits for the purposes of marketing.

9.3 Zoning

At one level, the zoning issues accompanying eco-industrial projects are no different from those of conventional industrial parks. The developer of any industrial park typically desires that the facility's zoning be in place and that no further consents be necessary before marketing begins. Because most localities already have industrially zoned property, the developer is often concerned that a new business will prefer the site that is already zoned over the site needing additional approvals. Developers express similar concerns regarding subdivision approval. They usually try to create a situation where no zoning or subdivision approval will be necessary. Many industrial parks are created so that only site-plan approval (which in many states is a ministerial, non-discretionary act) is required.

This desire for a speedy land-use approval process often encounters concerns by the host locality that the incoming industry will have some attribute needing specific regulation. Unless the locality has reserved some discretionary approval, it may not be able to legally regulate the attribute of concern. Eco-industrial facilities can sometimes address this concern through careful structuring of the CC&Rs. The regulations of the CC&Rs may give the locality comfort that any concerns will be addressed through the environmental review process and architectural approval. As an additional compromise, the locality may authorise a wide range of industrial uses to obtain a conditional use permit.

Eco-industrial facilities sometimes present unusual land-use approval issues. The facility may provide for private waste-water treatment in which the waste-water of several different participants is treated for re-use. Because this use is not strictly an accessory to any single user, separate land-use approval may have to be obtained. Similarly, utilisation of distributed energy facilities, if they are serving multiple participants, may require separate land-use approvals. Some jurisdictions have special siting provisions for these types of utility. Another unusual characteristic of eco-industrial facilities is that they often require extensive utility piping for elements not typically part of a conventional industrial park infrastructure such as for grey water, steam, heat and compressed air. Because these systems serve multiple parties, some jurisdictions may require that these utilities obtain separate approvals. Additionally, if the utilities are to be located in public streets, special approvals may be necessary.

Several eco-industrial projects have attempted to integrate industrial facilities into a 'New Urban' design.[4] New Urban designs are pedestrian-oriented new towns or infill developments characterised by mixed uses. The mixing process is modelled on the integration of uses that occurs in older towns rather than the separation attendant to most zoning since the Second World War. Most New Urban projects mix single-family, multi-family, civic, office and commercial uses but have not really dealt with the integration of industrial uses. Because eco-industrial facilities typically have tighter controls on their by-products and buildings they are the leading candidate for integration. As is the case for some of the larger-scale commercial and civic uses (e.g. big-box retail uses and hospitals) it may not be appropriate to scatter industrial uses through the neighbourhoods and town centres of these projects. There may, however, be an appropriate district within which they can locate in proximity to the town centre. The proximity creates opportunities for more shared resources among the different uses. Excess heat and power from commercial uses in the town centre, distributed energy or grey water from residential uses may be utilised in the industrial facilities. Heat or excess power from the industrial uses may be distributed in residential or commercial settings. The pedestrian orientation and proximity allows workers in the industrial facilities to use the restaurants and shops of the town centre without getting in their cars or having to access centrally provided transportation facilities.

New Urban designs often create unusual zoning and land-use requirements that must be taken into account in the design and permitting of related eco-industrial facilities (see Slone 2001). New Urban projects use build-to-lines principles, bringing buildings to the street edge—a practice common in older industrial neighbourhoods but abandoned in industrial parks. Parking is typically behind buildings in New Urban projects. Many projects utilise alleys and have narrow streets in order to encourage pedestrianism. Loading zones and truck access must be carefully thought through in the design and permitting of eco-industrial facilities in New Urban settings.

9.4 Permitting

Typically, eco-industrial facilities require fewer traditional environmental permits than do conventional industrial projects. Almost all true manufacturing processes require

4 On New Urbanism, see Duany *et al.* 2000; Katz 1994; see also www.cnu.org/about/index.cfm, www.newurbanism.org and Takahashi's discussion in Section 5.3.2 of this book.

some kind of environmental permitting. For the most part, goals such as zero discharge remain aspirational even for aggressively green manufacturers. Permitting may be necessary even for facilities without discharges (e.g. some states may require state-issued discharge water permits that regulate waste-water treatment even when there is no outfall for the discharge other than the public waste-water system).[5]

Eco-industrial design may help solve local air issues, but such resolution is fraught with complexity. Energy cascading or combined heat and power projects may be much more efficient than individual uses, preserving more of the local air increment. Nonetheless, if an area is 'non-attainment'[6] the required technology for a new source and the necessary offsets will still be daunting in some parts of the country. Retrofits of existing industrial facilities that replace ageing boilers or utilise existing waste heat often generate concerns that the project will stimulate new, higher levels of air regulation.[7] If these upgrades will eventually be necessary in any event, or if sufficient offsets are generated to create attractive value, then the eco-industrial approaches will typically have advantages over conventional approaches, even for existing industries.

Similarly, grey-water strategies may eliminate the need for a discharge permit. When grey water is furnished from a central facility to individual users, the legal issues are relatively few. Projects have run into difficulties in some cases, however, when they have tried to daisy-chain waste-water, taking process water from one facility and using it in another facility. In at least one instance (in the Port of Cape Charles Sustainable Technology Park) the regulating agency has required that water be treated to relatively high standards between each use. Other jurisdictions have taken a more comprehensive approach to eco-industrial facilities, focusing only on the final discharge.

Storm-water management also offers an opportunity for innovative permitting. As eco-industrial facilities use 'green' roofs or infiltration strategies to eliminate storm-water ponds and offsite disposal, their permitting often becomes easier. Key to permitting success is the ability to document calculations of capacity so that engineers can predict the consequences of different storm events (e.g. the 10-year and 100-year storm events; for an excellent manual on alternative approaches and the resulting calculations, see MWCG 1997). As storm-water permitting for individual sites becomes more prevalent and as jurisdictions begin to charge for storm-water management, economic incentives for these approaches should increase, provided that new ordinances and permitting approaches allow credit for on-site management. If they do not provide such credit, on-

---

5   For example, see State of Florida 1998; State of Mississippi 2002; State of South Carolina 2002; Commonwealth of Virginia 2002: 25-452-50.
6   The Prevention of Significant Deterioration (PSD) programme applies to major new sources with respect to Clean Air Act (CAA 1990) criteria pollutants in areas that have attained the National Ambient Air Quality Standards (EPA Office of Air Quality, Planning and Standards NAAQS; i.e. in attainment areas; see United States of America 2002: Section 52.21). The Non-attainment New Source Review (NNSR) programme applies to major new sources that result in potential emissions in excess of major source thresholds of criteria pollutants for which the area has not achieved the NAAQS (i.e. in non-attainment areas; see United States of America 2002: Section 52.24).
7   See United States of America 2002: Section 52.21 (indicating that major modifications of units are subject to best available control technology in PSD areas) and Section 52.24 (indicating that major modifications of units are subject to the lowest achievable emission rates in non-attainment areas).

site management will disappear as a practicable option because property-owners will refuse to pay twice for storm-water management.

Eco-industrial facilities sometimes pose other permitting issues. Some of these are common among industrial facilities. For example, there are federal and state laws that address the situation where a material is held for recycling and therefore does not have to be treated as hazardous waste, with all the regulations that such a designation invokes.[8] Other issues are more unusual, such as the permits necessary to annually burn prairie grasses near a facility, as part of a prairie restoration effort. Many of the eco-industrial facilities proposed to date have included components of preservation or restoration.

9.5 Setting up the rules

The rules for constructing and operating an eco-industrial park are usually set forth in CC&Rs recorded in the property records and applicable to any owner or tenant of the property (see WEI 2000). CC&Rs have been used for many years in conventional industrial parks to set forth rules for such things as architectural design, building height, outdoor storage limitations, maintenance of common areas and use limits. Many industries, particularly large-scale operations, typically do not locate in industrial parks and consequently they have no experience with CC&Rs. These are usually the businesses that need the most assistance in understanding and accepting any restrictions, not just those common to eco-industrial facilities. Eco-industrial facilities typically have distinctive regulations in addition to those common in conventional industrial parks.

CC&Rs for eco-industrial facilities may include standards for green buildings. An increasingly common practice is to refer to the LEED™ standards published by the US Green Building Council (USGBC).[9] The USGBC has a certification process to confirm that the building has met the required standards and to allow the posting of a confirming plaque. Some areas offer grant or tax incentives for LEED™ buildings.[10]

Some eco-industrial parks use screening standards for businesses, requiring that businesses meet environmental or social sustainability standards in order to locate in the park and that the businesses continue to meet these standards in order to remain there.[11]

8   See United States of America 2002: Section 272 (regarding the universal waste rule); Commonwealth of Virginia 2002: 20-60-1495 *et seq*.; State of North Carolina 2002.
9   The LEED™ (Leadership in Energy and Environmental Design) Green Building Rating System is a voluntary, consensus-based, market-driven building rating system based in existing proven technology. It evaluates environmental performance from a 'whole building' perspective over a building's life-cycle, providing an definitive standard for what constitutes a 'green building' (www.usgbc.org/LEED/LEED_main.asp).
10 So far, the State of New York has taken the lead in awarding tax credits for 'green building' (see www.dec.state.ny.us/website/ppu/grnbldg/index.html).
11 The Port of Cape Charles Sustainable Technology Park abides by certain covenants and restrictions (C&Rs). All occupants must meet certain minimum standards of environmental, social and financial sustainability. For example, applicants for occupancy must submit applications setting forth their commitment to sustainability and must satisfy basic eligibility requirements. The applicants must also maintain certain minimum standards of performance on the

In order to address business concerns, these standards must be flexible and allow businesses to change the mix of strategies from year to year in order to respond to market conditions. Eco-industrial CC&Rs may also require co-operation in identifying more efficient use of resources or in facilitating forums for discussing continued improvement of the park.

The CC&Rs can be a useful means of addressing stakeholders' concerns. The Cape Charles Sustainable Technology Park, for example, used its CC&Rs to address the architectural concerns of an adjoining world-class golf course and to address use concerns of the adjacent town of Cape Charles. In brownfield areas, where communal mistrust of industrial projects may be high, the CC&Rs may be key to communal acceptance.

Projects must deal with a wide range of issues in crafting the CC&Rs. Standards usually become the rules for operation and, thus, aspirational elements such as 'zero discharge' have to be clearly identified and separated from those elements that are immediately required. Not only do sustainability requirements have to walk the tightrope of satisfying third parties that the requirements are real and not 'greenwash', without creating business constraints that destroy economic sustainability, they must also address the consequences of cyclical economic downturns. Often this issue arises in the discussion of enforcement when a business is not able to meet standards. The documents should allow flexibility to work with a business that desires to comply but cannot because of temporary economic conditions but to evict or otherwise punish a business that simply decides it no longer desires to be environmentally responsible. As stated in one community meeting during the creation of the Cape Charles CC&Rs—'We don't want to kick out businesses and jobs just when the economy is bad and businesses and jobs are hard to come by.'

Similarly, developers must determine whether the standards and the inherent increased profitability will drive all sustainability efforts or whether other incentives such as rent reductions are useful in encouraging a continued increase in sustainable approaches. One of the most important issues for CC&Rs is whether the standards will be mandatory requirements or merely guidelines. An eco-industrial park might be essentially a normal industrial park in which innovation and efficiency are encouraged and facilitated but not mandated. In the alternative, standards might be flexible but mandatory. Often, the circumstances in which the park is created affect this decision. If the eco-industrial form is adopted by a private or public developer as an option for improving the marketing of what could otherwise be a conventional industrial park, the standards are often guidelines. If the park would not be approved except for the eco-industrial design, the standards are usually mandatory.

9.6 Relationships

Risks and rewards define the relationships created in eco-industrial arrangements, just as these elements define almost all business relationships. A business may seek the

> 'sustainability criteria' and pass a 'sustainability audit'. Occupants must also be able to meet the financial obligations, and applicants that have not taken adequate steps to maximise economic sustainability are not accepted.

reward of lower energy prices by taking advantage of an opportunity to buy energy from a distributed facility. The business takes the risk that the distributed energy source will be unavailable. As rolling blackouts in California showed,[12] any business may already face the risk that loss of electricity will interfere with its operations. If the power grid remains available as a backup, then the business will save money and, with two energy sources available, it will have reduced its risks. Similarly, if a business replaces its traditional supplier of benzene with a local business that has benzene as a by-product of its manufacture, its reward is reduced prices and its risks are supply interruption or a poor-quality product. Again, it previously faced these risks, though there may have been a record of good performance, making the occurrence of problems unlikely. Now, however, the business has lower costs and if anything interferes with its new supply of benzene it has the ability to return to traditional supplies. Thus, its risks are lower.

9.7 Risks

The principal service of law and insurance to business is the mitigation of risk. Like all businesses, those involved in eco-industrial projects take on new risks when they form new relationships. Tainted grey water may destroy a manufacturing facility's production run. A company accepting a co-product may mishandle it, resulting in regulatory liability for not only the accepting company but also the original manufacturer. A dedicated third-party source of steam may need unscheduled maintenance or go bankrupt, leaving a manufacturer with a costly loss of production.

Legal documents, particularly leases and contracts for energy products and co-products, allocate these risks. They typically assign the responsibility for loss or damages, describe the steps agreed on to minimise risk and mitigate loss and describe any financial protection required such as bonds or insurance, which must be maintained to protect against default. These contracts also describe how disputes are to be resolved.

A wide variety of relationships may be created in connection with eco-industrial development. For example, a power company burning coal may contract to sell its ash for use in concrete products. The power company may be concerned that economic or regulatory conditions may cause it to need to change its fuel type. It knows that it will have plenty of lead notice, but because of this possibility it does not want a contract for more than five years. The concrete manufacturer is seeking financing for its new facility and the lender may be concerned that the life of the loan exceeds the term of the contract. The power company might solve this problem by committing other ash from its other plants, but transportation costs will change the economics of the deal. The concrete manufacturer may be able to manufacture its product without ash and take the risk of a more expensive process. The contract can make it clear which course is the basis for the agreement. The contract should also deal with the consequence of less ash being produced because of reduced operations or some improvement in boiler technology. It

---

12   See the time-line of the California energy crisis at www.sfgate.com/cgi-bin/article.cgi?file=/news/archive/2001/04/06/state1705EDT0232.DTL; see also www.pbs.org/frontline/shows/blackout.

would also typically specify that the ash will be within a certain range of parameters for such constituents as heavy metals. Because there have been periodic threats to characterise coal ash waste as 'hazardous waste'[13] the contract should address the consequence of such a designation.

For most co-product agreements, amounts and quality, supply interruption and co-ordination of scheduled outages, consequences of regulatory changes and damages for off-specification products are the most important issues. Related issues, such as responsibility for products that are potentially hazardous waste may be handled by designating the point in time when ownership transfers. Dispute resolution provisions, ranging from mediation by park management boards to formal arbitration, should be included as consistent with the philosophies of eco-industrial developments and as a means of reducing legal costs.

In structuring their legal agreements the parties must have a thorough understanding of a wide variety of background utility issues as well. For example, an eco-industrial park may be charged for waste-water treatment based on its intake of municipal water, even though it discharges no waste-water, essentially consuming it all through re-use. Although such re-use does reduce the water intake and therefore the waste-water charges, the park is still paying for the infrastructure for pre-treatment and re-use as well as paying for treatment services it does not use.

In many parts of the country interconnection costs—the cost of setting up to send power into the power grid—are high and the amount paid for unsolicited power is low, so, often, notions of net power sales to the grid are aspirational. Even the more achievable goal of project net metering—the consumption of generated power within a multiple tenant or participant project—may have severe restraints. Retail sale of electricity is still not available in much of the country. Some net metering laws, although allowing self-generated electricity to spin the meter backwards as energy is conveyed to the grid, limit this ability to a property-owner who also owns the generating assets. In some jurisdictions, although generation has been deregulated, transmission has not, and even the service of three properties from a combined heat and power micro-turbine would require involvement of and payment to the authorised transmission utility.

Similarly, the provision of voice, video and data services to park participants may have to take into account local cable franchises and FCC (Federal Communication Commission) regulations on telecom providers entering into exclusive agreements.

9.8 Insurance

Insurance can be used to reduce some, but not all, of the risks associated with eco-industrial development. Environmental insurance can be obtained to address liability

---

13   US Congress adopted the Bevill Amendment in 1980, which required the US Environmental Protection Agency (EPA) to study the effects of coal ash and other by-product materials on human health. On 24 April 2000, US EPA announced that such waste will be regulated as non-hazardous waste under the 1976 Resource Conservation and Recovery Act (RCRA). However, several members of Congress and environmental groups were pressing to have these wastes characterised as hazardous waste (see www.corn.org/web/cc_0500.htm).

exposure from locating on ground potentially contaminated by hazardous substances.[14] Business insurance may cover contractual indemnifications. Business interruption insurance may be used to address gaps in the parties' risk agreements. The cost and terms of these policies must be carefully examined. Insurers frequently require reports or studies and impose conditions before providing such coverage.

9.9 Innovation

Because eco-industrial discussions often involve innovative participants and discussions, it is not unusual for issues of intellectual property to arise. As new processes are created that sometimes save large amounts of money, the parties should develop a strategic plan for intellectual property. Such a plan would involve monitoring for identification of protectable property, a determination of whether the parties desire to protect the concepts and then implementation of the protection plan. In some cases it may be useful for someone on the team to become familiar with the process of obtaining and maintaining such items as copyrights, trademarks and patents (on patents, see Chisum 1978; on trademarks, see Gilson 1974; on copyright, see Boorstyn 1998). It may also be useful to learn the time-frames and rules of disclosure for various types of protection.

The process innovation of eco-industrial development also frequently requires various impediments—regulatory, financing or simply practice—to be overcome. The team should include participants experienced at implementing innovative approaches and anticipating such issues.

9.10 Conclusions

Eco-industrial development involves familiar legal issues in unusual settings as well as new legal and regulatory issues. All the risks and legal concerns are manageable with proper documentation and thoughtful legal support.

---

14 Pollution legal liability policies can be obtained to cover unknown releases of petroleum, hazardous substances and other contaminants as well as future releases from operations. Only government-required actions or third-party claims are covered. Voluntary remediation efforts where the insured party cleans up a release where not mandated by an agency or third party are not covered by such policies. Known releases will be expressly excluded; however, cost cap policy coverage providing protection for remediation cost overruns can be obtained. These policies generally will provide coverage after the insured party has spent 120% of the estimated cost for the remediation. Owing to numerous claims in the industry with regard to cost cap policies where the remediation costs were estimated at less than US$1 million, some insurers decline to provide coverage or are providing coverage only once the clean-up exceeds 200% of the estimated remediation costs.

10
ENVIRONMENTAL AND RESOURCE ISSUES
An overview

Raymond P. Côté
School for Resource and Environmental Studies, Canada

Although it is tempting to simply list the ever-growing litany of environmental issues attributed to industrial production and consumption, the emphasis from a developmental perspective should be on those materials or processes that are likely to cause synergistic or cumulative impacts. These are the issues that have ecological rather than the more narrowly described toxicological repercussions. For example, we know that there are approximately 100,000 commercial chemicals in production and use today (OECD 2000). Many of these are used in small quantities and in limited applications where they may be toxic to the workforce in the industrial facility in which they are used but have little or no anticipated ecological impact.

This is not to suggest that these chemicals do not present any risk. In some cases, they may be transported and stored in larger quantities and serious accidents can occur at these stages of the life-cycle. Parts of the life-cycle occur in the more than 20,000 industrial parks or estates around the world. The proximity and concentration of industries and workers in these parks also concentrates the environmental and health hazards. Since the disaster in Bhopal, India (Côté 1991), chemical companies in particular have devoted considerable effort and resources to reduce the risk of chemical accidents. However, accidents will continue to occur. In other circumstances, the chemicals have been used in many products or as propellants, both of which will result in their dissipation into at least one environmental medium. Our experience with chemicals such as polychlorinated biphenyls (PCBs) and chlorofluorocarbons (CFCs), which were presumed to be quite innocuous in terms of ecotoxicological effect, should cause all users to take extra care with any chemical products released into the air, water or soil.

PCBs have been used in many products since their initial manufacture in 1929 (Paehlke 1991). These uses have included carbonless copy paper, paints and as dielectric fluids in transformers. Because they are persistent and bioaccumulative, these chemicals have found their way into organisms around the planet. CFCs have been used in fewer products, but these uses have resulted in the dissipation of these compounds into the atmosphere. They have found their way into the upper atmosphere, wherein they take part in chain reactions with ozone, destroying the ozone molecules and thus affecting the ability of the ozone layer to block UV rays. The dissipation of anthropogenic chemicals should be curtailed as much as possible to guard against unanticipated side-effects, many of which we will not quantify for many years.

The points being made here are that (1) everything is interconnected and (2) humans do not always understand the consequences of their actions, and some of these actions have potentially serious consequences. Increasingly, societies believe that taking a precautionary approach is the responsible way of managing chemicals, waste and products. In this chapter I will address the environmental and closely linked resource issues involved with the various stages in the life-cycle of a development project: in this case an industrial park or estate. In the chapter, the terms 'industrial park' and 'industrial estate' will be used interchangeably.

10.1 Environmental and resource issues in the life-cycle of an industrial park

10.1.1 Clearing the development site

Chemicals are not the only threat to sustainability. Industrial development normally begins with setting aside or purchasing a plot of land, leading to physical and biological changes. The first step may be the purchase of land that may be forest, agricultural land or a 'brownfield' site having been previously used for industrial development in the recent or distant past. In the case of forested land, the site may have been an old-growth forest with ecologically valuable species of plants and animals. The loss of prime agricultural land is of concern because of the reduction in food production capacity. Although forested and agricultural land continue to be converted for urban and industrial development, attention in recent years has turned to previously contaminated, sometimes abandoned, industrial properties in urban areas. These 'brownfield' sites, as they are known, present opportunities, but they also present financial, legal, engineering, social and ecological problems associated with their remediation.

The nature of the contamination will determine the degree of remediation necessary to satisfy regulatory requirements. The sites may have been old munitions factories, gasification facilities, paint manufacturers where lead and arsenic may have been utilised or hat manufacturers where mercury was extensively used. The excavation and disposal of contaminated soil will be of concern to several groups. Risk management is the name of the game and involves satisfying a range of stakeholders, including the lenders, the regulators and the local community.

Fresh and coastal waters and ecosystems are under increasing threat from a qualitative as well as a quantitative perspective around the planet. If a development site requires clearing and levelling, erosion during construction is a distinct possibility, with resulting siltation of nearby and downstream water bodies.

As mentioned above, the site itself may have been forested or previously converted to agricultural use. In the former case, there may be impacts on biodiversity through loss of habitat of threatened or endangered species, or by interfering with historical corridors for wildlife. In the case where the land has previously been converted to agricultural use, the original transformation to such use may have occurred decades if not centuries ago. At this stage in the life-cycle of the development, several cumulative impacts are becoming evident. For example: there is a potential contribution of the conversion of forest to the further loss of biodiversity; conversion of forest and productive soil also contributes to a reduction in carbon dioxide sink capacity, with implications for climate change; erosion associated with clearing and levelling of the site could exacerbate water-quality problems downstream.

As described above, part of the site preparation may have involved levelling the land. One type of habitat that has often disappeared during this activity is wetland (Fig. 10.1). These are typically considered swamps and wasteland. Their ecological and economic value is usually underestimated. Wetlands serve as natural flood-control structures, as filters and as water recharge areas. All of these functions have economic value for the developer if ways can be found to integrate wetlands into the development plan. They also have ecological value as habitats for many species of plants and animals as they bridge terrestrial and aquatic habitats.

Figure 10.1 **An engineered wetland in Burnside Industrial Park**
Photo: Eco-efficiency Centre, Burnside

10.1.2 Installing the infrastructure

Today's development is supported by various types of infrastructure including transportation, energy, water supply and waste disposal. Once the area is cleared, transmission and distribution lines are installed to deliver electricity to the customers of the industrial park. In most instances, the energy source may be a long distance from the development. In many areas, there may also be gas transmission and distribution lines carrying natural gas as an energy source for heating purposes. Furthermore, there are communication cables that must be installed. Trenches are dug for the installation of water supply, sanitary sewers and storm-water sewers.

One aspect of land clearing and infrastructure construction that requires particular attention is the nature of the bedrock in the area. Some geological formations are heavily mineralised with ores such as arsenopyrite, for example. When these formations are disturbed, the interaction of a naturally occurring bacteria, *Thiobacillus ferrooxidans*, oxygen and water release acid known as acid mine drainage. The acid solubilises metals, resulting in a 'brew' that is highly toxic to aquatic life downstream (Chiaro and Joklik 1998). In addition to the environmental costs, the financial costs of remediation can be very high.

The materials used in the pipes may have been extracted from quarries and processed into cement at kilns located tens if not hundreds of miles from the site. Roads are constructed into and through the area by using materials transported from quarries and asphalt plants, also located elsewhere. There may be opportunities to connect the development to a rail line through one or more spurs. The use of rail, especially for the transportation of goods into and out of the development, will reduce wear and tear on the roads and reduce greenhouse gas emissions on a per ton transported basis.

Water and energy are two aspects of industrial park development in which cumulative impacts are presented. Water supply is an increasingly important issue in many countries and communities. The supply may be constrained in terms of quality or quantity. There are techniques for using water more efficiently and effectively. Industrial parks, with their extensive parking areas, roadways and flat roofs, funnel large volumes of water. Some of this water could be diverted into tanks, cisterns or surface ponds and used for toilet flushing, washing and landscaping. Energy production, distribution and use are relatively inefficient, especially where fossil fuels are involved. Energy infrastructure that supports co-generation and district heating systems should be encouraged. Other infrastructure such as thermal energy storage systems with use of groundwater or rock formations can also be considered to reduce energy consumption.

It should be clear by now that, although consideration is given to the environmental impacts associated with the life-cycle from land clearing through to operation, there are also upstream and downstream impacts associated with individual components at each stage (Fig. 10.2). In the context of an eco-industrial development, the nature and scope of these impacts may influence decisions about the design and selection of materials and equipment.

Impacts can be local and regional as well as cumulative. There are an estimated 20,000 industrial parks around the world. Some of these are very large, involving 25,000 acres, whereas others are small, involving 1–2 acres. Some parks involve large numbers of employees, in the tens of thousands, whereas others involve only a hundred. Fig. 10.3 is an oblique view of a portion of Burnside Industrial Park, which is located on Halifax Harbour, Nova Scotia. The park has approximately 1,300 businesses and 17,000 employees, with many vehicles moving through the park daily. The scale of the park will influence the scale of the infrastructure and the magnitude of the impacts (UNEP 1997).

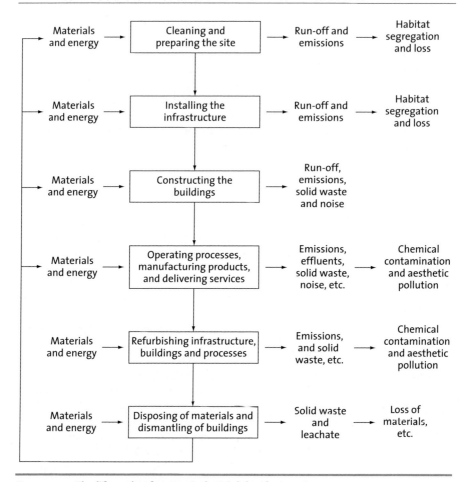

Figure 10.2 *The life-cycle of an eco-industrial development*

10.1.3 Construction of buildings

The nature of the industrial park will also directly influence the type of industry that will be attracted and consequently will affect the nature and size of the buildings. Some industrial estates are designated as heavy industrial sites, with the intention of attracting oil refineries, petrochemical plants and related industries. Others may attract mixed heavy industrial industries, including steel mills, aluminium smelters, ammonia facilities, methanol plants and so on. The nature of these facilities will differ extensively from an estate designated as light and medium manufacturing, for example, attracting companies in the electronics, computer and telecommunications sectors.

In the case of the refining and processing of petrochemicals, buildings will require large amounts of space because of the nature of the raw materials and the processes

Figure 10.3 *An oblique view of part of Burnside Industrial Park, Dartmouth, Halifax Harbour, Nova Scotia*

Photo: Sundancer Photo, Halifax, Canada

involved. Oil refineries and petrochemical complexes typically have large tank farms with pipelines and railway spurs leading into and out of the complex. These generally require extensive land to be set aside around the complex to protect adjacent facilities and communities from noise and accidents. Transportation to and from these facilities tends to occur 24 hours a day, seven days a week, with all the related impacts. As can be seen in Figure 10.4, refineries and petrochemical plants are well-lit facilities, for safety reasons. In some cases, the lighting is so bright as to influence local residents. The resources on which the materials are derived for the construction of the many complex aspects of these plants may be sourced from anywhere on the planet. The iron ore could be mined in Australia, whereas the steel could be manufactured in Taiwan and carried on ships built in Japan.

The fabrication of the plants might also involve the construction of pollution-control facilities for air and water pollution as well as facilities for the storage of solid and hazardous waste. Such facilities may also be large, depending on the nature and level of treatment required. Biological oxidation systems can occupy several acres of land.

In the case of electronics and telecommunications companies, the buildings tend to be of a standard nature, resembling office and laboratory buildings (Fig. 10.5). Their exterior facade is often constructed of stone or concrete, whereas the interior may include stone, concrete, gyproc (plasterboard) and wood. Increasingly, buildings are being constructed of materials that can be readily assembled and disassembled, such as steel and aluminium siding.

The largest estates have several hundred, if not a thousand or more, buildings, as can be seen in Figure 10.3. Burnside Industrial Park in Dartmouth, Nova Scotia, Canada has more than 1,300 businesses, although approximately half of these are in industrial malls of 2–10 businesses.

The magnitude of the raw materials that must be extracted, mined and harvested, processed, manufactured and then transported to the site of the development can be measured in millions of tons. Attempts are often made in environmental impact assess-

154   ECO-INDUSTRIAL STRATEGIES

Figure 10.4 *The lights of Imperoyal Oil Refinery, Dartmouth, Nova Scotia, at night*
Photo: Eco-efficiency Centre, Burnside

Figure 10.5 *An example of a modern telecommunications company building in Burnside Industrial Park*
Photo: Eco-efficiency Centre, Burnside

ments to identify and even quantify the environmental and social impacts associated with these large developments. These impacts are frequently underestimated and simplified to those directly resulting from the construction of the infrastructure itself.

10.1.4 Operation of the park and companies

The companies that locate in an industrial park vary considerably from place to place (see Box 10.1). It is therefore difficult to discuss the resource and environmental implications of the operation of the industries except in a general way. All industries require electricity and water. In the case of water, the uses may be limited to cleaning and sanitary purposes or they may be an integral part of the industrial process. If any goods are processed, services performed or administrative operations undertaken, the company will need inputs of various types. All of these result in some wastage, be it liquid, solid or gaseous.

Business sectors found in Burnside Industrial Park are, in alphabetical order:

- Accommodation
- Advertising
- Air conditioning
- Automobile parts and supplies
- Automotive repair
- Bakeries
- Building materials
- Carpeting and flooring
- Chemical manufacturing and sales
- Cleaners
- Clothing
- Computer manufacturing, sales and service
- Construction
- Door and window manufacturers
- Electrical contractors, sales and services

- Electroplating
- Food processing
- Furniture manufacturing and sales
- Heating and ventilation
- Industrial equipment
- Marine and fishing products
- Printers
- Packaging manufacturers and sales
- Office equipment sales and service
- Paints and supplies
- Photofinishers
- Sign manufacturers
- Telecommunications sales and services
- Transportation
- Waste disposal and recycling

Box 10.1 **Business sectors represented in Burnside Industrial Park, Dartmouth, Nova Scotia**

For example, electronic and telecommunications factories have become very attractive to industrial park developers and managers because they are viewed as being 'high-tech', with highly trained and salaried employees. In addition, the types of processes that take place in these plants generally require very high standards of cleanliness. Although the buildings will be pleasing in appearance and clean from a production point of view, the manufacture of their products often involves metals and organic chemicals that result in substantial toxic and hazardous waste. This waste not only represents problems for the companies producing them but also presents environmental and health risks for the industrial park if the waste is not stored and transported carefully. The waste may receive some kind of pre-treatment on site but would generally be transported to a hazardous waste treatment facility located away from the park. Some parks, especially in South-East Asia and in Europe are being built with their own common waste-treatment facilities, but these are in the minority.

10.1.5 Refurbishing and remanufacturing

Facility owners and managers are continually upgrading facilities in order to attract new tenants or customers. In some cases, the exterior and interior facade of buildings is modified whereas in other cases all materials in the interior are torn out and new equipment and furnishings are introduced. Until recently, this process of refurbishing resulted in a considerable percentage of the waste discarded to landfill. In some cities and towns, construction and demolition debris sites, used building material stores and renovator resource centres have been established. Some of the material extracted from the waste-stream is now re-used one or more times. Entire buildings can be refurbished and re-used (Fig. 10.6).

Although large corporations such as Xerox have had 'asset recovery management' programmes in place for many years to recover and remanufacture copiers (Azar 2001), entrepreneurs in small businesses have also entered the marketplace to remanufacture products, which heretofore would have been sent to landfill, where they would contribute to the leachate problem. As indicated in Box 10.2, Burnside Industrial Park has seen the

Figure 10.6 **An old brewery transformed into a new office and manufacturing facility**

Photo: Eco-efficiency Centre, Burnside

Companies are performing functions in the following areas:

- Recovery of
  - Fine paper
  - Cardboard
  - Bottles
  - Metals
  - Batteries
  - Chemicals (through chemical separation)
- Remanufacturing of
  - Toner cartridges
  - Furniture
  - Printer and typewriter ribbons
  - Tyres (retreading)
  - Computers
  - Automotive parts
- Re-use of
  - Used building materials
  - Tools
- Rental of
  - Construction equipment
  - Tools
  - Trucks
  - Pallets
  - Communications equipment
  - Photocopiers
  - Uniforms
- Repair of
  - Computers
  - Electronics
  - Lorries and cars
  - Furniture
  - Buildings
  - Communications equipment
- Environmental services, providing
  - Soil decontamination
  - Environmental site assessments
  - Environmental audits
  - Energy audits and efficient equipment
- Environmental products, such as
  - Solvent recycling
  - Cleaning products
  - Specialised composting and recycling containers
  - Recycled cardboard
- Recycling of
  - Waste oil
  - Solvents

Box 10.2 *Companies performing scavenger and decomposer functions*

evolution and the colonisation of the park by new businesses that re-use, rent, refurbish, remanufacture and recycle materials. Such businesses can reduce the impacts associated with extraction and transportation of resources from beyond the park boundaries as well as the impacts involved with the disposal of materials within the park. In some parks, a material exchange could be launched by estate management or a not-for-profit organisation to accelerate the re-use, repair, remanufacture and recycling of materials.

10.2 Cumulative resource and environmental impacts

In this chapter I have attempted to draw the inevitable conclusion that every stage in the life-cycle of the development of an industrial park results in resource and environmental impacts associated with inputs and outputs. Some of the impacts are associated with site and infrastructure preparation through such interventions as fragmentation, removal of biotic soil cover, soil compression, fertilisation, desiccation, contamination and habitat destruction (Kollmer 2000). These are multiplied thousands of times as new developments occur in and around cities and towns. There are also impacts associated with the many products used and manufactured in the factories and businesses. In addition, other services that support businesses in industrial parks also use natural resources. The physical, chemical and biological disturbances result in impacts that are cumulative on soil, air, water and other biota, not only at the local level but also, in some instances, regionally and globally. Locally, the various point and non-point sources of pollutants contribute to the deterioration of air and water quality. For example, the health impacts of deteriorating air quality in our cities cannot usually be attributed to a single source of pollutants. Rather, it is the many sources of emissions, both stationary and mobile, as well as drift of pesticides and the long-range transport of pollutants that exacerbate the problem.

Regional impacts such as acidification and excessive water extraction and global impacts such as climate change and the loss of biodiversity are of increasing concern to governments, environmental organisations and international agencies. It is the causes of these impacts that must receive more consideration from public and private planners and developers. One of the perspectives that must be adopted by government and industry is that there is no such thing as waste, only misplaced and misused resources. Reaching sustainability demands that we use the Earth's resources more wisely; we must 'do more with less' (WCED 1987). This will necessitate devoting more effort to understanding the metabolism of materials and energy within industrial parks, communities and regions (Ayres and Simonis 1994).

10.3 Managing resources not waste

Management of resources within the context of an eco-industrial development takes on a very different perspective than is the case in a standard industrial park. In most parks, individual companies are responsible for the procurement of raw materials and the disposal of waste from their operations. In the case of waste, they must satisfy environmental regulators from the relevant jurisdictions that the waste is being handled appropriately, often at a high cost. Although some companies have recognised that waste-prevention opportunities exist in their industrial plants, the majority of waste is still treated with use of end-of-pipe technologies and processes. These wasted materials are often trucked away to an off-site treatment and disposal facility. Each business or property manager in the park negotiates with waste-management companies that offer their services in the area.

In the case of an eco-industrial development, the materials coming into the estate as well as those in the estate such as building materials and equipment are looked on as resources to be husbanded for continuous use. As such, these resources should be managed in an environmentally and economically responsible manner. The Worldwatch Institute (Brown et al. 2000) has reported for many years that renewable resources such as forests, fisheries, soil and water are being depleted at rates in excess of the levels necessary for sustainability. Among other things, the eco-industrial development planner and manager should assist tenant companies in identifying by-products first within and second beyond the boundaries of the industrial park that could be used by other companies. In this way, resources such as water, energy, gases, packaging and building materials could be used more effectively, reducing the demand on renewable and non-renewable resources and the overall 'ecological footprint' (Wackernagel and Rees 1996) of the park. Furthermore, this approach may create new business opportunities, as illustrated by Box 10.2, while reducing resource and environmental impacts.

A number of industrial ecologists and organisations (see Côté et al. 1994; WEI 2000; Erkman 1998; Erkman and Francis 2001; Lowe et al. 1997; Orée and UNEP 1998; UNEP 1997) have described strategies and tactics for improving the use of these resources in eco-industrial developments. This chapter concludes with a set of guidelines (Box 10.3), proposed by my co-workers and myself at the School for Resource and Environmental Studies, Dalhousie University, based on applied research on Burnside Industrial Park, in Nova Scotia, Canada (Côté et al. 1994). Every eco-industrial development will have its own geographical and ecological characteristics and guidelines that must be selectively chosen in each case. There is no 'one size fits all'. These guidelines do provide the reader with a sense of the types of practice that would have to be adopted to establish an eco-industrial development.

The guidelines found in Box 10.3 should be viewed as indicative rather than comprehensive. In addition, a cursory review of these guidelines will indicate that each subsequent category is influenced by the category that precedes it. Site selection influences the nature of the decisions about the development of the site. The choices made about the site and changes in the landscape during development can affect the design of the infrastructure. There are various infrastructure and support systems that can influence planners and architects of the industrial facilities in the estate. Decisions about the layout of the facilities will affect the efficiency of the operations. Finally, selection made in the design of facilities and materials used will determine the costs and impacts associated with refurbishing facilities and revising and recycling materials and by-products.

This chapter has clearly underlined the fact that planning and design are critical to the reduction of the cumulative impacts associated with industrial estates and the success of an eco-industrial development. Recognising the reality and potential of cumulative

- **Site selection**
    - Take account of ecosystem functions and attributes in the area under consideration. In particular, water flow and impoundments should be mapped and protected.
    - Encourage flexibility and foresight in site planning to consider how use of the site may change over time.
- **Site development**
    - Maintain wetlands to provide habitat areas, to filter surface run-off and, where feasible, to treat waste-water.
    - Replant endemic vegetation throughout the park for aesthetic value, wind protection and shading; where possible, replace large unused portions of lawns with local ground-cover varieties.
    - Ensure all structures have access to southern exposures for passive solar gain.
    - Design streets so that building orientation maximises solar access, on the one hand, and winter wind resistance, on the other.
    - Maintain land forms and other landscape features that support ecological functions and energy efficiency.
    - Protect some wild spaces as parkland and corridors for wildlife.
    - Reduce the amount of land disrupted for development, in terms of buildings per hectare, infrastructure, parking area and so on.
    - Designate sites for vegetable gardens in the park, to be used by employees of businesses in the park and, particularly, by park restaurants.
- **Planning and operating infrastructure, including support systems**
    - Where natural wetlands are not available, build engineered aquatic ecosystems (living machines), which use sunlight, bacteria, plants and other aquatic life to break down toxic materials, concentrate metals and treat organic material in sewage. These can be developed for larger industrial facilities or clusters of smaller businesses.
    - Bio-treat grey water from restaurants and food-processing facilities in organic filtration beds, which are solar aquatic water purification systems.
    - Use ground-source heat pumps or heat-recovery ventilators for space heating and air exchange.
    - Collect rainwater for firefighting, irrigating plants, flushing toilets and, where possible, process water.
    - Adopt economic instruments that will encourage clean production and penalise waste generation; disposal fees and tax relief must be of a magnitude to make a financial difference to the business.
    - Set up information systems and incentives to attract businesses and groups that can make use of wasted materials.
    - Support co-operative efforts in procurement and waste management.
    - Develop local, small tri-generation facilities in industrial parks to produce electricity, steam and hot or cold water for heating or cooling.

Box 10.3 *Guidelines for the establishment of an eco-industrial park development*
(continued over)

Source: Côté et al. 1994

- **Construction of industrial facilities**
  - Where possible, co-locate buildings and businesses to make effective use of waste heat, water and other resources.
  - Consider the nature and composition of building materials to reduce air emissions into the work environment.
  - Insulate buildings for local weather conditions but utilising technologies and practices that do not create 'sick-building' conditions.
  - Design buildings to reduce heat loss. For example, design large transfer bays with hoods or ensure they are away from winter winds.
  - Use waste heat from major facilities for district heating purposes within the park.
  - Encourage small businesses to use solar panels and photovoltaic systems to heat water for small businesses.
  - Install low-flow shower heads and taps and low-flush toilets to reduce water consumption and waste.
  - Standardise building materials as much as possible to reduce waste during construction, and encourage re-use of materials.
  - Build with manufactured products and techniques such as the force-fit, no-nail technique to facilitate the moving of interior walls and the re-use of building materials.
  - Encourage re-use of building materials within the park by setting standards that require durable, re-usable materials.
- **Operating industries**
  - Use electric machines and vehicles as much as possible. (assuming renewable sources of electrical energy are being used).
  - Encourage recovery, re-use and recycling of chemical and metal waste.
  - Avoid use of hazardous substances where possible, and otherwise reduce volumes stored and used at any one time.
  - Reduce the use of highly toxic and persistent chemicals.
  - Encourage the use of non-toxic, non-hazardous cleaners and supplies within the park.
- **Facility refurbishment and materials recycling**
  - Encourage the use of materials that can be readily recycled first within the park, second within the urban area and third within the jurisdiction.
  - Encourage the establishment of maintenance, repair and reconditioning businesses such as companies that refurbish furniture and laser printer cartridges.
  - Require separation of waste to encourage repair, re-use and recycling.
  - Encourage composting and other uses of organic waste. For example, mulching lawnmowers could be required.
  - Install recycling centres into buildings to facilitate collection and transfer of materials.
  - Install water-management systems that recycle water, extract harmful materials and provide cleaner releases to water sources than at intake.

Box 10.3 (continued)

environmental impacts is a necessary beginning. However, addressing these issues from a life-cycle and an increasingly ecological or systemic perspective will require the adoption of a range of policy and regulatory instruments and management tools. Park managers, whether public or private, can begin by adopting environmental or sustainable development policies. Economic instruments (such as water rates, sewer use fees, recycling fees and landfill tipping fees) and legal instruments (such as landfill disposal restrictions) will also influence the re-use, remanufacturing and recycling of materials. Environmental management and information management systems, monitoring and reporting will assist estate managers and tenants in their quest for a sustainable approach to industrial park development and operation.

11
MANAGEMENT OF ECO-INDUSTRIAL PARKS, NETWORKS AND COMPANIES*

Edward Cohen-Rosenthal
Work and Environment Initiative, Cornell University, USA

11.1 The need for management systems in eco-industrial parks

All industrial parks require a management structure. They may come from the developer, an industrial estate authority, an economic development organisation, a property management firm or the co-located industries themselves. In many ways, the management requirements of an eco-industrial park (EIP) are even more critical to success than they are in traditional industrial real estate. Although there are much more active industrial parks where the management provides other functions, a park where recruitment is based on placing a sign on the perimeter or link solely to attract commercial real estate brokers that are looking to locate prospects often requires administering minimum codes, taking care of the roads and collecting the fees. Industrial park managers have enhanced these roles by providing common sewers and rubbish collection, organising social events and providing canteens, among other strategies. What an eco-industrial approach requires is stretching the boundaries of common activity by actively seeking connections and beneficial opportunities among participating companies, pro-

* This chapter may seem a bit choppy as it is the compilation of a good deal of work by different people over the years associated with the Work and Environment Initiative. Tad McGalliard and Bruce Fabens and I collaborated in thinking through management structures when working with Chattanooga in the design of a proposed eco-park; Astrid Petersen, most recently a graduate student at Cornell and accomplished on these issues in her native South Africa, worked with us to think through the relationship to the ISO 14000 series; and the last section on ecological workplaces was initially developed as a presentation for a seminar for the Johnson Graduate School of Management. Judy Musnikow helped with the research for this chapter as well. The chapter draws from this work with editing and changes to incorporate new ideas, and I take responsibility for their inclusion and editing as part of this chapter.

actively working with the community and looking to necessary and value-adding functions at the central level.

The notion of how to manage in an environmentally responsible way is the subject of an increasing number of books and manuals (DeSimone and Popoff 1997; Dunphy and Griffiths 1998; Piasecki and Asmus 1990; Scanlan 1998; Stead and Stead 1996). Overwhelmingly, these are about how to 'green up' individual companies or the particular challenges of a particular sector. These can be helpful guides for individual companies or for particular processes. Even those that are derived from broad environmental frameworks such as The Natural Step tend to focus on each individual company (Anderson 1998; Nattrass and Altomare 1999). In eco-industrial activity, the system level is 'bumped up' at least one level, where higher value is sought through networked or diverse connections.

11.2 Traditional management roles in industrial parks

Although management of eco-industrial developments may present new and unique challenges to the realm of industrial park management, in many ways eco-industrial managers must assume many of the same roles held by traditional park managers. Both eco-industrial and traditional parks must strive to maintain a manageable and practical size. Managers must take account of, and adjust accordingly to, the fact that the park may not be able to accommodate the needs and requirements of every industry that wishes to locate in the park. Although eco-industrial development requires deeper consideration of the interconnections and integrative possibilities of various industries, traditional industrial parks must also heed the relationships among tenants and the size of their estate in order to ensure a collaborative environment conducive to productivity and growth.

The establishment and operation of eco-industrial and traditional industrial developments are both reliant on the skills and input of various fields of expertise. The creation of industrial systems requires technical know-how, ranging from the mastery of engineering to a deep understanding of marketing. It is infeasible for many industrial parks to employ such a diverse and specialised staff on a permanent basis. Park managers must decide how specialised a staff the park can support and for which tasks and expertise they should defer to outside experts. Eco-industrial developments may necessitate assistance from even more diverse sources than traditional sites since they integrate a larger variety of disciplines within their bounds.

Several other universal issues exist for park managers, including the necessity to uphold sufficient and accurate data. Keeping comprehensive records of such topics as the industrial site, park tenants and industry needs as well as of other areas can improve the management of existing occupants and help attract additional tenants. Another issue relates to the aesthetic appeal and reputation of the park. Managers must maintain the physical look and 'good name' of the park in order to retain current tenants and draw in future tenants as well.

The creation and enforcement of standards in traditional industrial estates, although not necessarily done in the same manner as standards promulgated in eco-industrial developments, is also an issue park managers must contend with. Standards in many ways shape the principles of the park and can dictate many other aspects of park management. Financial management is another concern of the park managers. This requires agreeing on common services, fair fees for these services and collecting proper amounts and accounting for their use. Enforcement and monitoring, possibly more so for eco-industrial developments than for traditional estates, are often 'touchy' responsibilities that managers must work out how to handle. Managers must also balance the above-mentioned responsibilities and other expenses with the overall objective of maximising the long-term cash flow to developers (SIR 1984).

11.3 Business clusters and networks

Business clustering is a popular concept in economic development. By getting certain sectors to concentrate in an area, support services, suppliers and trained workers are more likely to be available. Clusters are composed of related companies that have certain commonalities such as customer–supplier relationships, markets, research and development (R&D) needs or product lines. Figure 11.1 illustrates a cluster of industries in Southern California. states, municipalities and businesses use cluster analyses in order to target economic development policy for greater success in business retention, expansion and attraction.

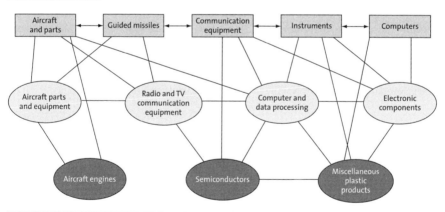

Figure 11.1 *Aerospace and advanced manufacturing cluster in Southern California*
Source: Anderson 1994

A prominent example of an advanced business network has evolved in Silicon Valley. In the early 1990s, the Joint Venture Silicon Valley Network was organised in order to 'bring people together from business, government, education and the community to identify and to act on regional issues affecting economic vitality and quality of life'.[1] The network works on a variety of issues, including, among others:

1 JVSV (Joint Venture Silicon Valley), www.jointventure.org, 1998.

- Business retention, expansion and attraction
- Business entrepreneurship
- Public–private partnerships for cost reduction and regulatory streamlining

11.4 Business networking

One means of applying cluster analysis understands the active and potential relationships in a business cluster. For example: do the companies in a cluster realise that they are part of a larger system and work together for mutual gain, or is the group of companies latent or underachieving in its interactions?

Competition is the defining buzzword of the free market; however, it is also true that businesses more frequently find it necessary to co-operate. The most obvious means of co-operative activity is the trade association. Others include strategic partnerships, short-term alliances and contractual relationships. Most often, these types of inter-firm collaborations are for harnessing a competitive advantage. Also, increasingly, with the globalisation of the world economy, maintaining competitive advantage requires flexibility, responsiveness and adaptability to rapidly changing conditions.

Business networking or flexible manufacturing is a strategy whereby small and medium-sized enterprises (SMEs) come together for joint activities (Chisholm 1998). Examples of such collaboration may include R&D, product manufacturing and assembly, training, purchasing, and marketing. Since most firms are SMEs, networking provides a means of increasing economies of scale, competing with large integrated conglomerates, lowering costs of doing business and creating continuously 'learning' organisations. Larger companies have now 'got into the act' and are using 'partnerships' as a way to shed non-core functions to be able to reduce overheads and increase flexibility.

11.5 Using networks to facilitate eco-industrial development

Eco-industrial development is a means of creating a network of companies in a number of areas. In essence, the goal is to facilitate awareness of the collaborative possibilities that can emerge from closer co-operation. The following sections outline the primary areas in which eco-industrial collaboration can occur. Many clusters are proposed within single sectors. These have emerged in various parts of the world, such as the textile cluster in the Emilia-Romagna region in Italy. Through self-organisation, thousands of companies—small and larger—have a highly adaptive and successful interrelationship. Although this kind of co-operation is possible, the eco-industrial tends to deal more with diverse systems. This diversity provides latitude for complementary production schedules, material (resource) needs and product innovation. Reliance on a single sector means that a slump in that sector brings the whole system crashing down, competition is more of an issue and peaks and valleys in needs are accentuated.

The management system associated with the eco-industrial business network must mirror the values of system intelligence and value-adding integration. The goal is to create a management system that preserves the dynamism of market processes with the synergy of strategic alliances.

Several principles are important to emphasise as overarching goals:

- Participation in the eco-industrial network aims towards achieving lower overall costs of doing business, improved revenue yields, better environmental results and an increase in responsible community interactions.

- Beyond minimum requirements set by the eco-industrial park or network (EIP/N), constituents or local ordinances, participants are independent companies that create or do not create alliances in areas that make sense to them. Meeting particular needs for quality, quantity, timeliness and cost are among the parameters that will govern relations among all, or subsets of, participating companies.

- The goal is to stretch each participant's practice of mutually beneficial interactions.

- Flexibility in system response is the key to adapt to changes in the environment of individual enterprises and the EIP/N as a whole.

- Consensus-based action is the preferred mode of operation. A clear dispute-resolution process should be developed. The goal of the EIP/N is meeting goals, not meetings. Facilitation is required to ensure that group time is used effectively.

- Specific performance indicators should be publicly known and measured. These help to inform and create accountability to a commitment to continuous business and environmental improvement.

11.5.1 Management through continuous improvement in important business drivers

The companies involved in an eco-industrial network have to view themselves as businesses first. This is not an abandonment of environmental commitment but a way to ensure that they can continue to produce environmental gains in the future. My presumption is that the eco-industrial network has to impact at least one of the top three to five drivers for business success of the company (Cohen-Rosenthal 1998b). If this is not the case, at some point the participation in the eco-industrial network will be marginalised and possibly abandoned. But each company has a diverse set of such drivers such as market access, equipment uptime, quality, waste minimisation, time to customer, cost and so on. Although a set of drivers is outlined here, certainly there may be other categories or configurations developed.

Two strong warnings are provided. One, a management approach has to be systematic and cover the significant drivers—otherwise major opportunities are lost in 'silo' thinking. Second, there is a tendency to 'cherry-pick' project-by-project or, worse yet, to pick something easy to prove it can work. Gearing up for and carrying out each project abandons the power of focusing on improving the overall driver—not just 'pussy ant' steps in that direction. It takes more time, and reviewing other opportunities becomes more difficult. It does not institutionalise systematic thinking. This does not mean that groups may not prioritise and undertake discrete activities or projects, only that they connect with a broader sense of purpose. Our goal is to make significant change, and that rarely if ever comes from such 'wishy-washy' tactics. If you are going to invest your time and money, make it worth doing.

In the following sections I outline a number of potential areas in which eco-industrial connections can be made within a network of companies. In each major area, the steps listed form a progression from an early stage through later, more developed, stages. For instance, material connections between companies, at the simplest level, could include buying supplies in bulk to meet several companies' needs. With continued focus on material issues over time, waste to raw material exchanges might develop between compatible companies, possibly followed by the development of altogether new material markets. The factors or drivers of eco-industrial development form an illustrative array of possibilities for connections between companies, leading to the growth of business, jobs and the community.

The premise behind eco-industrial development is that business and environmental performance can improve simultaneously and that the use of eco-industrial strategies will improve business performance beyond other strategies. The progression of steps in each of the areas represents the implementation of a continuous environmental and business improvement strategy. To some participants, the initial attraction may be cost savings, or organisational improvements that lead to higher productivity. As these are implemented, further strategies will become apparent that can build on the initial successes. Not all companies will move forward in all areas, nor do they need to start at the beginning. Differences in products, services, production techniques or locations will affect the choices of participating companies and organisations and will lead to new imaginative possibilities beyond those listed below.

Industrial development agencies and communities can view the factors of eco-industrial development as focused topics around which to promote their locations. Certain factors can be promoted as themes from which to recruit businesses, or they can be used to contribute to the targeted marketing of eco-industrial sites. For instance, some groups may promote a material exchange network, others a trained workforce or a common product focus, such as has been the case with semiconductors in Silicon Valley. These become themes, not only for attracting like-minded companies but also for creating synergies between companies. Thus one of the services that would be attractive to companies would be the strength of the vision and the success of the implementation of these factors among participating companies.

The factors may be viewed in two ways: (1) for evaluating company growth and development goals and (2) as a recruitment tool for communities using eco-industrial development principles to inform their strategies. The factors represent fundamental aspects affecting many businesses. Individual firms are more greatly affected by some factors than they are by others. Any given firm, however, will tend to address most or all of these factors at different times. Within each factor are a number of steps that can be described as progressing towards greater eco-industrial symbiosis. Companies interested in improving their business and environmental performance will view the steps as strategic directions toward these ends.

A discussion of each factor follows (see also Fig. 3.1 on page 59). Examples are presented to show the benefits for participating companies. Recruitment strategies can also be developed to demonstrate the advantages of the eco-industrial development approach in growing local businesses, attracting new firms and creating new jobs.

11.5.1.1 Materials

Material flow is at the core of industrial ecology as a way to reduce cost, toxicity and net volume within business operations. Different organisations invest different amounts of their budgets in the purchase and processing of materials. Further similar needs will generally be clustered within various sub-groups. In all cases, money amounts can be reduced and quality can be increased, thus adding value to each organisation involved. Cost, inventory, quality or quantity flexibility and improvements can be made on a variety of levels, from common buying, to setting supplier specifications (including environmental requirements), to material exchange connections of products and by-products. A material once considered waste or a by-product with little value could become a valued input in a new process. In all cases, from a business perspective, the goal is find the most effective and safest use of materials (see Table 11.1).

Who?	Topics	Outcomes
■ Purchasing managers ■ Inventory managers ■ Operations managers ■ Process engineers ■ Waste and environmental managers ■ Municipal solid waste manager ■ Materials brokers	■ Recycling ■ Common buying and alternative sourcing ■ Environmentally preferred products ■ Customer–supplier relations ■ Take-back and disassembly ■ By-product connections and re-use of materials ■ Setting common material standards ■ Creating new material markets	■ Lower material costs ■ Higher-quality materials purchased ■ Less-toxic materials ■ Availability of more reliable materials ■ A minimised redundant inventory ■ Less waste

Table 11.1 *Material flow. Mission: to maximise resource productivity and ensure quality, safety, timeliness, environmental stewardship and cost-effectiveness.*

At the simplest level, co-operating companies could order products jointly to get volume pricing and to reduce the shipping costs incurred with separate deliveries. For example, many hospitals already participate in buying co-operatives. From there it is a short step, conceptually, to evaluating customer–supplier relations at any number of levels, including, but not limited to, shipping procedures, matching production and purchasing cycles, and product design for customer needs. Each of these proven strategies can reduce the waste of materials and possibly how the material is used and in what quantity in the final products.

Incorporating the use of environmentally preferred products is another area of market development. Increasingly, companies are shifting their purchasing requirements to include the sourcing of products that have a demonstrably lower environmental impact than comparable items. As a purchaser, a company would want to evaluate existing patterns of buying to become part of this emerging wave. As a producer, a company would want to position its products to qualify for this growing sector of purchasing. This would include documenting not only the content of the products but also promoting other activities that the producer is involved in that benefit the environment. This strategy not only contributes directly to cumulative environmental improvement but also directly supports the companies active in those areas.

Another example characteristic of an eco-industrial strategy is to review production, based on a regional material flow and the creative cycling of materials between companies. This might include reviewing possible by-product connections between companies as well as production methods with an eye to alternative material use. Other options might include reaching out to a broader range of regional companies. Communications through these business networks maximise the possibility that connections can be made between companies that could constructively use one firm's by-products for another's benefit. This strategy is not new. Opportune business networks or industry groups have always traded information concerning the development and use of new materials.

When taken to the fullest extent eco-industrial networks have the potential to create new material markets. Numerous recycled material markets such as for paper, plastic and metals have developed and stabilised over the years. Other recycling and reprocessing industries have developed over time in response to changes in production arising from the need to reduce a particular by-product for liability or cost reasons. Companies that keep abreast of these changes will position themselves for new markets. Communities and networks of companies that keep abreast of these changes will create new jobs, within sustainable companies and communities.

11.5.1.2 Energy

Energy represents a cost of production for all individual firms as well as a global environmental consideration. Access to adequate energy supply is requisite for accomplishing the goals of any organisation. At the firm level, attention to energy-efficient design and operation can reduce costs and ensure greater reliability of supply. Networking of energy supply can increase overall energy efficiency, adjust for peak load variations and drive down costs by reaching across various companies. Economies of scale may provide other means to experiment with renewable sources of energy. When energy is being cascaded it requires integration of measurement and delivery systems. As an environmental system, the net effect is to lower overall emissions that result from energy production, transfer and use (Table 11.2).

At the level of the firm, reduction in energy costs can come from use of lights, motors and heating, ventilation and air conditioning (HVAC) that are more energy-efficient and from reduced energy use in production processes. Green buildings and green retrofits are the next step. These include energy-saving designs and technologies as well as

Who?	Topics	Outcomes
▪ Facility managers ▪ Energy managers ▪ Local utility representatives ▪ Other experts, as needed	▪ Green buildings (e.g. use of energy-efficient lighting and HVAC) ▪ Energy auditing ▪ Energy cascading between firms ▪ Co-generation facilities ▪ Spin-off energy firms	▪ Lower energy use ▪ Lower energy costs ▪ Lower emissions

HVAC = heating, ventilation and air conditioning

Table 11.2 **Energy. Mission: to identify means for overall energy supply, efficiency, conservation and cost-effectiveness.**

planning for employee efficiency and comfort in the workspace. Green designs represent greater overall savings than conventional designs, but they may have a longer payback time because of their greater initial capital cost. An example of the green building strategy would be a flexible manufacturing building constructed to green standards. The developer is able to present a more attractive building on the market while at the same time energy-saving benefits go to the tenants (or the developer) in terms of lower energy bills. Many companies are worried about assuring quality and consistent energy flow in case of glitches in the system or the grid. These are expensive backups. Sharing such a precaution can lead to lower costs and greater reliability.

Energy cascading represents an opportunity for an energy user to sell excess or waste energy in the same way that companies have looked for ways to market by-products from their production. Connections consisting of, for example, steam lines, excess cooling water or chilled water can be made between companies. Low-pressure steam can be taken off after high-pressure use, or hot water can be drawn from a cooling tower. Another aspect of energy cascading is co-generation, which allows various forms of energy to be made available from one producer. This can involve use of turning gears in one process to power a turbine elsewhere and use of flare gases for adjutant uses as well as other creative solutions.

The end goal for energy use based on sustainable development strategies would be to promote the use of renewable energy sources. Many people predict that in the long run the price of non-renewable energy is set to rise over time. Technologies to produce solar, geothermal and wind power are possible candidates to be matched with co-generation to produce energy. This goal for energy production may provide fertile ground for the development of spin-off companies in new and more efficient energy technologies.

11.5.1.3 Transportation

The ability to move people and goods in a timely and efficient manner is one of the keys to business success. Transportation in all its aspects provides various means to achieve these goals. For individual firms, flexible transportation options allow an ability to ensure that customer requirements of timeliness and cost are met. A comprehensive look at transport seeks synergies between different transportation modes. Synergies are also possible that enhance the ability of SMEs to deliver to their customers through common shipping activities, vehicle maintenance, fleet management and facilities. In businesses with a just-in-time approach to inventory, transportation is critical, whether as the customer or the supplier.

Who?	Topics	Outcomes
▪ Company shipping-and-receiving representatives ▪ Representatives of the air, rail, truck or water-vessel shipping company ▪ Public-sector transport officials	▪ Shared commuting ▪ Access to local mass transit ▪ Shared shipping of materials and products ▪ Common vehicle maintenance ▪ Packaging reduction and disposal ▪ Alternative fuels ▪ Intra-park transportation ▪ Integrated logistics ▪ Warehousing and depots	▪ Lower shipping costs ▪ Lower impact on surrounding community ▪ Reduced emissions ▪ Cost-effective transportation and flexibility

Table 11.3 *Transportation. Mission: to ensure that materials and people are able to move within and outside the eco-industrial park or network with highest reliability, lowest environmental loading and best cost-effectiveness.*

Transportation also has a strong impact on community relations by the volume and timing of truck traffic in residential neighbourhoods and can mean the difference between good and bad relations. Access to quality employees and employee timeliness is a function of connection to transportation, whether mass transport, car pools or private vehicular traffic. By seeking the lowest impact one can also better meet local air-quality commuting requirements.

Some companies have a consistent shipping volume and can maintain a fleet of delivery vehicles. Other companies ship in cycles and choose to rely on contract shippers for their transportation. Energy use and environmental impact can be addressed in each of these cases. Addressing transportation issues in the context of eco-industrial development can result in strategies for improved performance. I find intriguing the notion of dedicated logistics facilities run by those whose reputation is based on excellence in this area. By partnering with such a group, companies in an EIP can reduce their inventory and shipping costs while assuring higher levels of performance.

Employee transportation issues concern commuting and use of company vehicles for business. Car pools, company vans, walking or bicycling are strategies for reducing energy used in employee commuting. Commuting incentives can be instituted by the network. Alternatively, groups of companies closely located or that draw employees from common neighbourhoods can participate together in cost-saving commuting services.

Regarding the daily use of vehicles for business or employee emergencies, clean cars can be available for short-term rental to meet those demands. Reducing the frequency of trips and re-routing travel for greater efficiency can also be evaluated. The use of clean-fuelled vehicles and maximising the effectiveness of vehicle mileage are also important steps in reducing environmental impact and the cost of transportation.

The evaluation of the methods used to ship products from place to place and the overall efficiency of distribution represent the next level in the area of transportation. This includes reviewing how products are shipped and how long they remain in storage. Can the loads of two unrelated companies be combined so that one joint load travels for less than two separate loads? The area of integrated logistics has grown to include warehousing and depots for improved materials distribution through multi-modal hubs. Improved evaluation of production requirements and shipping needs can make better use of the transportation required to move products between producers as well as on to consumers.

Transportation opportunities span the range between preferred commuter parking and van pooling, links to local mass transit, and the development of a comprehensive regionally integrated logistics system. Integrated logistics requires that transportation of people and materials be seen as an integral part of the product life-cycle. As with the other factors of eco-industrial development, the notion of integrated logistics management requires evaluating the use of resources on a cross-company basis, possibly even regionally. This perspective comes from evaluating material flows and methods of least environmental impact. Some preliminary examples of this include improved predictions of shipping and receiving needs, full loading (and return freight) of vehicles and just-in-time production techniques that minimise wasted delivery trips.

11.5.1.4 Marketing

No business makes money without revenue. An exclusive focus on costs or operations ignores the importance of a profitable outcome. Markets are the primary driver of company success. As such, association as part of an eco-industrial network should enhance market access, presence, penetration and sales. Independent of an environmental focus, common marketing efforts allow smaller and medium-sized firms to get recognition for their products and services from a broader audience than they could achieve alone. Further combining different capabilities allows firms to bid on a broader range of projects and on larger projects than they might be able to do separately. The environmental focus provides a special advantage when it comes to developing and exploiting 'green' markets at the retail or supplier levels in the private or public sectors. Different firms may select different approaches towards how much to display their 'green aura', but marketing research confirms that when consumers have a choice between two products with similar costs and features the more environmentally sound product wins. The need to be in a similar cost and quality ballpark with other products underlies the other drivers discussed in this section.

Marketing strategies can focus on both green labelling and expansion of markets in the initial stages of eco-industrial development (see Table 11.4). Individual companies have used green labelling as a marketing strategy. It is also possible for a network of companies to use green labelling based on its eco-industrial connections. Green labelling would be used to set the products from member firms apart on the basis of product

Who?	Topics	Outcomes
■ Sales and marketing managers ■ Chamber of commerce ■ Economic development agency	■ Green labelling ■ Accessing green markets ■ Joint promotion (e.g. advertising, trade shows) ■ Joint ventures ■ Attracting new, value-added, EIP/N partners	■ Increased sales ■ Increased market access ■ Increased market visibility ■ Broader business opportunities

EIP/N = eco-industrial park or network

Table 11.4 **Marketing. Mission: to expand the market recognition and revenue base of constituent companies and the eco-industrial park or network (EIP/N).**

content, corporate environmental performance or an overall strategy that stresses sustainable development principles. By promoting the network of companies, the labelling symbolises a broader strategy of interactions between a number of companies and a commitment to overall environmental performance, not merely the recycled content within an individual company's own products.

This broader strategy can lead to joint promotions among firms and an increased credibility that the activities are more than just claims. Exposure can be gained through websites and trade show promotions as well as through print advertisements and direct-mail advertising. A group of companies can market themselves and their products under the umbrella of eco-industrial strategies that reduce waste and thus bring higher value to the products. They can also stress the reduced impact of their products over their entire life-cycle. These goals will focus participating companies on the value of implementing these strategies as well as attracting customers.

These marketing strategies are likely to lead to further joint ventures between participating firms as well as attracting new firms to enhance existing product lines. This occurs through the development of business networking between the firms. Marketing departments may make the first contact but the overall strategy of business networking can lead to creating broader communication between participating companies. As marketing strengths are shared, in a growing atmosphere of co-operative problem-solving, joint solutions will be created that enhance the participating businesses. This could include joint ventures between participants or the recruitment of new value-added partners that could bring further capacity to the group.

Narrowly speaking, the results of this growth of partnerships are similar to those of other business growth strategies and joint ventures. However, the mechanisms used to bring the companies together suggest a major difference. With a persistent focus on environmental improvement and efficiency, business networks are created with conscious intent. By creating these forums the opportunities for co-operation, creating efficiencies, joint ventures and marketing can be maximised. Contacts among businesses for market access now become front-burner items for all firms involved. Previously unrecognised connections are discovered between companies and partners are added to the network. The contacts are soon recognised as winning situations for all members of the eco-industrial network by providing broader market access and increased revenue growth through the partnerships created.

11.5.1.5 Human resources

Another potential area of overlapping management function is that of human resources. In many instances, companies may go through three or four people to find a steady employee. The expense of this process could be minimised, especially in seasonal businesses, by having a common hiring pool where workers could be known to the eco-industrial association as quality workers and maintain steady employment even when shifting from one company to another.

The cost of employee benefits and services could be spread over a number of businesses in a network (see Table 11.5). Multi-company health plans could be negotiated. Employee wellness programmes could be initiated across multiple companies to reduce the start-up costs and provide more options by including a broader population of workers. Employee services such as daycare, eldercare or commuting services could be initiated more easily with broader participation because the overhead costs could be borne by a broader employee base if multiple companies were included.

Who?	Topics	Outcomes
▪ Human resource managers ▪ Union representatives from unionised companies ▪ Local training providers	▪ Human resource recruiting ▪ Joint benefits (e.g. multi-company health plan) ▪ Employee wellness programmes ▪ Shared services (e.g. daycare, commuting services) ▪ Common business skills (e.g. payroll, maintenance, security) ▪ Training programmes ▪ Flexible employee assignment ▪ EIP/N orientation programme	▪ Larger pool of employees ▪ Higher skill base ▪ Fewer seasonal adjustments ▪ Lower insurance costs ▪ Increased benefit availability and cost management

EIP/N = eco-industrial park or network

Table 11.5 *Human resources. Mission: to increase the range and quality of opportunities for recruitment, career development, occupational flexibility and benefit availability.*

At the next level, a common services pool could be created. Companies that make different products may use similar skills to run the company. Very often, these skills, though not needed full-time, are essential. In some cases, this provides an opportunity within the community for a start-up subcontractor. However, if the volume is not big enough or is not continuously needed, there exists a possibility for shared service position. There are many possible sets of services that could be maintained and scheduled by several companies to minimise overheads for any one company and to provide flexibility for all.

Another possible example includes cross-company training options, the costs of which could be shared by the group of companies or with government training assistance. This would serve to enhance the flexibility and skill level of the overall workforce. It would be aimed at increasing worker productivity and performance for participating employees. Examples of training programmes range from basic maths and workplace skills to higher-level production or engineering skills, to hazardous waste disposal training or education in environmental excellence. As the benefits from co-ordinating these types of programmes are demonstrated the next possible step is the realisation by management that there are a number of other functions that could be shared co-operatively. Payroll systems have been subcontracted out for years based on this type of reasoning.

Some or all of the companies may be unionised. I urge that the unions in the area also meet together and participate actively in the various cross-company activities by creating quality workplaces characterised by active labour–management co-operation. Valuing natural resources also means valuing human resource and employment in a company affiliated to an EIP should bring decent wages, benefits and working conditions.

11.5.1.6 Information and communication

Information and communications systems are areas that can benefit from business networking. Individual companies can improve their internal communications and training. At the same time, they can work with other partners to create more productive external connections, such as access to useful databases, World Wide Web resources and fast and accurate electronic communications between customers and suppliers (Table 11.6). Information related to enhanced environmental performance could be shared over a network or on a website, which would allow participation by more firms. One example of this could be the posting of a waste materials exchange accessible by all participants or linked to other regional resources. Another example could be making available the results of a company's energy auditing to other participating companies. Through the use of modern electronic means both of these examples would reach a broader audience and have the potential to effect greater change.

Ultimately, the creation of environmental information systems to track compliance or to work with government agencies creating beyond-compliance standards would be a logical next step after creating management information systems. Rather than reacting to individual regional and national standards for air, water or ground contamination, beyond-compliance goals would be set for a firm or a group of linked firms. From a management point of view, this could lead to higher performance, lower compliance costs and reduced future liability exposure by minimising possibly polluting inputs. Joint regulatory permitting would allow the distribution of expenses across numerous participants. It could lead to higher environmental performance by all parties and would have the added strength of being based on maximising efficiency of production by reducing waste.

In today's world of business-to-business (B2B) computing the issues of compatible computing platforms and the ability to integrate remote sensors or triggers for needed materials or energy come to the fore. Although certain information will need to be protected by individual companies, firms will have to work through how information will be shared. If an EIP were to include a variety of components in a value-chain strategy,

Who?	Topics	Outcomes
▪ MIS managers ▪ Network administrators ▪ External systems consultants ▪ Communication managers	▪ Internal communications systems (e.g. bulletins, videos, e-mail) ▪ External information exchange (e.g. database and Internet-access vendor communications) ▪ Common communications facilities (e.g. video conference centre) ▪ Facilities and production-monitoring systems (e.g. flow monitoring and purity) ▪ Compatible measurement-and-evaluation systems among EIP/N companies ▪ MIS system to track key EIP/N performance indicators and optimise software applications	▪ Improved company information for employees ▪ Improved opportunities for information exchange and discussion between employees and firms ▪ Easier access to external research and information sources ▪ Improved customer and supplier communications ▪ Development of overall EIP/N facilities and energy-monitoring systems

EIP/N = eco-industrial park or network   MIS = management information system

Table 11.6 **Information and communication systems. Mission: to ensure the broadest and most efficient use of, access to and transfer of information.**

information with pull inventory systems would need to be established. Although this integration is easy to write about, in reality it requires a great deal of face-to-face interaction between system designers and users to make it work well.

11.5.1.7 Environment, health and safety

Although most of the items mentioned above also affect environmental concerns, there is a discernible measure under the environment, health and safety (EH&S) banner. In many companies, safety and achievement of environmental goals are part of the formula for payouts for bonuses and pay for performance schemes. There are a number of low-hanging fruit in this area, especially when it comes to training on health, safety, hazardous waste handling and environmental stewardship. For those companies interested in wellness activities as a way to improve and monitor employee health, such activities can be combined into a common facility or be jointly staffed (Table 11.7). The same is true with employee assistance programmes that help with drug, alcohol and family stress issues; a common plan may provide better services with more anonymity for the individuals using those services.

Who?	Topics	Outcomes
▪ Designated environmental officers ▪ Health and/or safety managers ▪ Union health and safety representatives ▪ Representation from environmental and occupational safety agencies ▪ Community representatives	▪ Accident prevention ▪ Emergency response ▪ Waste minimisation ▪ Compliance management ▪ Air, water and land multimedia planning for pollution prevention and hazardous-waste management ▪ Design for environment ▪ Shared environmental information systems ▪ Environmental management systems (e.g. the ISO 14000 series) ▪ Joint regulatory permitting	▪ Increased flexibility and lower costs of environmental management ▪ Beyond-compliance environmental performance ▪ Reduced worker-compensation and environmental liability rates ▪ Integrated accounting of community impact

Table 11.7 *Environment, health and safety. Mission: to promote continuous improvement in all facets of environmental performance and in community and employee health and safety.*

This can be extended to emergency preparedness to protect against accidents and natural disasters that could affect the EIP/N companies and community. Common co-ordination with police, fire and emergency medical services can be done better and in less time than it would it take if each company were to carry this out on its own. There also may be strategically placed equipment that may be needed in an emergency. By sharing safety or environmental walk-throughs, firms may receive the benefit of fresh sets of eyes to identify potential hazards and solutions to avoid violations before they occur.

I have long been a fan of electronic reporting as a way to cut down paperwork. Furthermore, such reports can be aggregated and compared across the EIP. Just as firms hire accountants to deal with their reporting needs, an EIP can hire an environmental resource that can make sure that forms are filled out correctly and that can look for areas of opportunity for improvement at the firm or EIP level. Continuous improvement on aggregate emissions in all environmental media can be a valuable goal and an assurance to the local community. Those who have done well in source reduction can mentor those having problems with particular substances or processes.

Further integration is possible if the participating companies want to use common permitting for the whole park as a means for providing flexibility. There are significant implications for current regulatory practice and for liability for major violations that would need to be addressed in such a plan. All companies within the EIP can apply as a group for certification within the ISO 14000 series. Those companies that are already certified can provide assistance to other participating companies and help them develop effective and recognised environmental management systems. Such a system could be attractive to supplier companies whose customers require environmental certification. The EIP can issue an annual report for the community about progress and problems in the past year and host open houses for neighbours to explore ways to avoid environmental hazards at home.

11.5.1.8 Production processes

Each of the drivers discussed in this section represents an attempt to look at whole processes within an organisation to help re-engineer connections to maximise performance and resource utilisation. Although the production process incorporates many of the other drivers discussed here, it has integrity in itself. The goal is different within and between firms. Within firms it is to ensure an optimal connection between design, technology, skill requirements and quality management that leads to reduction of unnecessary steps, to operational flexibility and to outcomes that are of a measurably higher quality. It seeks alignment within the manufacturing or service delivery process. In order to maximise interaction between firms, process integration or alignment would also be required. At the system level, utilisation of resources, technology and people can be optimised still further in mutually beneficial ways that reduce cost, increase flexibility and enhance capability.

The development of co-ordinated and measurable quality systems within a company has been a successful first step to improving productivity. Quality of the product and, in many cases, the environmental standards of the company have also shown improvement through these programmes. Although these programmes are usually initiated by individual companies, there is often overlap in the methods used to initiate the programmes and train the participants. This type of activity would benefit from the connections between networked firms.

Another very basic type of business networking is the use of common subcontractors. The local machine shop, machine calibrator or the welder are examples. If the service is specialised or the volume is not large enough, a subcontractor network develops. If two or more companies could use the technology, facilities or skilled contractor on a part-time basis, economies would result. These and other opportunities to share under-utilised assets represent a first level of business networking around common aspects of the production process. The savings from shared equipment could be financial or could provide the ability to purchase better equipment.

Shared skills are another area in which economies could be realised as a result of networking. From the office to factory-floor workers, development of training and skills leads to more capable and productive workers. With developed networking between companies the application of that skill base could be spread over more than one company and yield both more productivity from the skills and higher job satisfaction for the workers. In an example such as heating, ventilation and air conditioning (HVAC), several companies needing only part-time services could share the skilled employee. In an area such as environmental monitoring, skills may also be readily transferable between types of production and could be implemented in companies where a full-time dedicated position would not be affordable.

As networking develops, more information can be transferred between companies and the potential for interaction, technology sharing and integration of product lines increases. The implementation of modern computer networking can also enhance this

Who?	Topics	Outcomes
■ Operations managers ■ Production engineers ■ Facilities managers ■ Shop-floor workers	■ Pollution prevention ■ Scrap reduction and recovery ■ Production design ■ Common subcontractors ■ Common equipment ■ Shared skills ■ Technology sharing and integration	■ Reduced production costs ■ Increased production flexibility ■ Reduced waste in processes ■ Improved working conditions ■ Increased quality of products ■ Increased customer satisfaction

Table 11.8 *Production processes. Mission: to ensure the most flexible and effective use of capital stock and production processes within and between participating firms, leading to excellent customer satisfaction.*

process. This is not to say that the cabinet company, for example, will always have something in common with the electronics assembly company. However, the trend in international business is to have smaller companies team up on a project to compete where once only larger more fully integrated firms maintained the capacity.

In production processes, the environmentally better outcome usually is to eliminate unnecessary processes, and the same is true from a business perspective. Much of the energy and material waste remains embedded in the system and the product. The pollution or 'waste by-products' are often the tip of the iceberg. Environmentally beneficial strategies also transform extra cost for control technologies and process steps into very significant bottom-line savings. By taking a systems view of the production or service process, one can identify the potential for such savings or one can hold redundant capabilities in reserve for emergencies.

11.5.1.9 Quality of life and community connections

Even the most connected system has boundaries that must be managed to ensure its functionality and survival. On a practical level, community issues and concerns must be part of the mix. But the reason for action in this area is more than seeking ways to avoid community interference or censure. For the eco-industrial project itself, community engagement provides a means to gain insight into areas for improvement and to integrate its activities, opportunities and facilities with those of its neighbours in a win–win way. As such, availability of quality services in the community enhances access for the network and its employees as well as for local residents. Safe streets and good fire departments benefit everyone, as does support of local restaurants and shops. Good relations with the community can lead to recruitment possibilities when the EIP/N is viewed as a desirable place to work, and raising the skills and educational level of the area is a community and business service. Common recreational opportunities are mutually enhancing. Seeking ways to contribute to community improvement is also part of a larger vision of responsible corporate citizenship (see Table 11.9).

Who?	Topics	Outcomes
▪ EIP/N representatives ▪ Community advisory board representatives ▪ Union representatives ▪ Mayor's representative ▪ Other local resources as needed	▪ Integrating work and recreation (green ways, recreation programmes) ▪ Co-operative education opportunities and school–work transition ▪ EIP/N-initiated volunteer and community programmes ▪ Environmental awareness activities in the community and employee homes ▪ Involvement in regional planning ▪ Co-operative relations with public-sector services	▪ Increased services for local residents, employees and their families ▪ Career opportunities for local residents ▪ Improved integration of EIP/N activities with local community services

EIP/N = eco-industrial park or network

Table 11.9 *Quality of life and community connections. Mission: to ensure excellent interactions between the EIP/N and the surrounding community in ways that enhance the community, strengthen companies and provide benefit to employees.*

Finally, and not because it is any less important in the framework of eco-industrial development than the other factors, there is the link between business and the community that promotes quality-of-life issues. Most communities are strongly committed to the idea that companies contribute to the local quality of life. The first steps of this are, for example, to create 'green ways' and hiking trails or to support community sports activities by integrating work and play within the community.

At the next level, businesses can work with the educational community, at all levels, to develop co-operative educational opportunities. Co-operative education benefits both the company and the community by building the skills base and providing school-to-work transitions for participating students. At this level, firms interact with local educational institutions, creating a standard for participation in sustainable business in the community.

As business continues to strengthen and grow in the community, with the implementation of eco-industrial strategies, job skills are improved. This can attract better, higher-paying jobs because of the higher skill level of the community. This can generate positive changes in the community, such as improved community services and a greater variety of retail businesses. By this stage, the growth, based on the recently created business and community partnerships, will have also brought about discussions on how the direction and rate of growth will impact the community.

11.6 Eco-industrial park management structures

Though management approaches may vary from location to location, there will have to be formal structures that help the EIP manage itself. Lowe et al. (1998) argue convincingly that to the maximum extent possible these should be self-organising learning organisations to mirror natural systems and to reduce the fear of participants that their autonomy will be constrained too much. It may be a contracted firm that specialises in the variety of responsibility it takes to manage an eco-industrial park or estate. It may be the task of the initial developer or part of a local authority. Whoever is in this position, all the participating companies and the community have to be intimately involved.

Ernest Lowe et al. (1998) observe:

> An EIP is two distinct but overlapping business entities. It is a real estate development property that must be managed to provide a competitive return to its owners. At the same time an eco-park is a 'community of companies' that must manage itself to gain common benefits for its individual members. The latter is a looser association in business terms, but the owners of the member companies will be no less concerned with their return on investment. You will need to respond to the needs of both entities in designing a management system for your EIP. Fortunately, their basic goals are very complementary.

Common services will require a common funding source to do this adequately. If all companies in the park must or want to be involved in particular activities, then a service fee can be assessed. Again, there is little difference from traditional parks except in scope. If only a few companies will be using a service, then they must identify how to fund it. Penny-pinching on services by a management company to save money can backfire on the allure of the EIP. Prudent management and accountability to park ownership and tenants is important while remaining true to viewing the EIP as a locus of unfolding possibilities.

11.6.1 Leadership team

As an operating business connection, the primary responsibility for operating and deepening the EIP/N is vested in the leadership team, composed of senior executives from each of the companies involved. This requires a real commitment of time and energy on the part of these individuals that cannot, in most cases, be delegated. The purpose is to ensure that the EIP remains connected to the core business concerns of the companies involved and that decision-makers are at the table to avoid lengthy delays in obtaining agreement. The value of the interconnection will produce results in direct proportion to the degree that business leaders invest themselves in the connections. The leadership team would have representation from the community advisory group in an *ex officio* capacity, and it may add other *ex officio* members to the group based on the value that they would bring.

The team will require facilitation and some team-building training, at least in the beginning, to ensure the best possible interaction among the players and to test the boundaries of possible collaboration. The purpose of the group is to make the key strategic and operational decisions. The leadership team could hire staff or consultants to perform any of the tasks they deem necessary and can set a fee schedule to help pay for

Who?	Topics	Outcomes
■ Chief executive officers, chief financial officers, chief operating officers and officers of EIP/N companies ■ *Ex officio* community advisory group members	■ Strategic planning ■ EIP/N policy development ■ Delegation and review of proposals from and to the integration teams ■ EIP/N management oversight (e.g. staff and consultant hiring) ■ Budget development ■ Annual performance review ■ Formation of ad hoc groups as needed to advance EIP/N goals ■ Business retention, expansion and recruitment	■ Increased return on assets ■ Demonstrated environmental improvements ■ Improved MIS for participants ■ Excellent community relations

EIP/N = eco-industrial park or network   MIS = management information system

Table 11.10 **The leadership team in the EIP/N. Mission: to increase the return on assets.**

these common services. As such, they will make budgetary decisions for the organisation as a whole. This dynamic will ensure quality service and value that meets the expectations of those who are participating.

11.6.2 Community advisory board

A community advisory board (CAB) that provides community input and oversight should be organised to represent the interests of the local community. It does not make the business decisions for the companies involved but does have a voice at the table. It would meet on a quarterly basis, or more frequently, as deemed necessary. It may also recommend participants for the eco-integration teams, discussed in Section 11.6.3.

Community involvement and enthusiasm for the project is a must for success. The primary emphasis of the CAB should be on ways to contribute to the success of the EIP/N as an asset to the community. As such, ideas that it produces that would improve the performance of the EIP/N should receive serious attention. The CAB serves as a bridge to the local community in order to facilitate communication with, and involvement in, the local community. The EIP/N should serve to enhance the local community by increasing access to services and by generating job opportunities. The CAB also has an important role to play in ensuring environmental excellence. It needs to be involved in emergency management issues, in traffic planning and in carrying out overall reviews of and in providing feedback on environmental plans and performance. It should have access to all information necessary to carry out that role and have an ability to tour the facilities.

Who?	Topics	Outcomes
■ Representatives of local community organisations, educational institutions, environmental groups, political officials and others, as designated by a senior local elected official	■ Assessment of community impact ■ Review of EIP/N environmental performance ■ Feedback and advice on any area of EIP/N operation ■ Communication with the local community ■ Link to local education institutions	■ Improved community conditions ■ More harmonious business–community relations ■ Support for community initiatives

EIP/N = eco-industrial park or network

Table 11.11 **Community advisory board. Mission: to link the surrounding community with the EIP/N and encourage two-way communication on issues of concern and possibilities for mutual assistance.**

11.6.3 Eco-industrial integration teams

The primary mechanism used for networking is a series of integration teams. These are keyed to the core functions presented in Section 11.5.1. The logic is simple. If these are the factors that drive business success, then mechanisms need to be put in place that help drive those factors. Precisely what they do is a result of the local situation and the mutual needs, interests and imaginations of the local parties. Membership is also flexible, based on who can contribute to the best decision-making. As such, it provides a flexible mechanism for encouraging integrated responses among the key stakeholders in the EIP/N. It is anticipated that some issues will cross the boundaries of two or more groups and that they would form overlapping teams to work on these issues. It is the leadership team (see Section 11.6.1) that charters the various groups and sets their membership and mission. Each year, these groups should participate in the overall strategic planning process of the EIP/N. In Figure 11.2, the business network, consisting of the drivers described in Section 11.5.1 and the management structure described in this section (Section 11.6), is shown.

11.7 Managing actively and imaginatively

Finding common ground is not easy; not all solutions will not work out as imagined; and circumstances will change the aspect of what seemed wise. However, approach the issues discussed above we must, so that industrial networks mean more than random connections, so that past practice does not become the dictator of the future, so that the value of

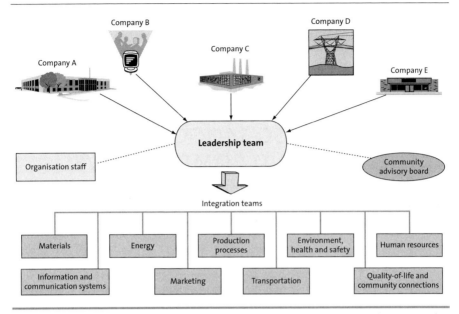

Figure 11.2 **The structure of the business network of an eco-industrial park or network**

interactions bears full fruit for all those involved. An eco-industrial system requires more than simply managing better; it means shifting to managing ecologically, where whole-system thinking is the norm. There can be a shift towards ecological thinking that helps to inform the organisational values reflected in organisational practice (Cohen-Rosenthal 1998b) and shifts away from mechanistic mental and organisational models. In each of these areas there is a challenge to stretch our notions of the possible by asking how can we do business better, nurture those who work in and live around EIP/Ns and enhance the environment.

12
LINKS TO THE ISO 14000 SERIES AT THE PARK AND COMPANY LEVEL

Astrid Petersen
Cornell University, USA

It has long been recognised that there needs to be a balance between industrial development and protection of the environment. In many countries, including the USA, South Africa and Asia, there has been an increasing tendency to group industries together as part of the land-use planning process. Various terms have been used to describe this phenomenon, including industrial development zones, industrial parks, industrial estates, commercial and business parks (Côté and Balkau 1999) and, in some cases, eco-industrial parks (WEI 2000). In this chapter, the term industrial parks will be used.

Industrial parks have emissions typically associated with industries. Such emissions result in air and water pollution and include the generation and disposal of waste, spills and accidents. In addition, industrial parks tend to magnify environmental impacts or potential risks:

- Pollution and other risks are concentrated in a small area, with potentially serious cumulative environmental impacts.
- There is an increased potential for impacting coastal ecosystems, as well as for impacts on other land uses, such as tourism (Côté and Balkau 1999).
- There may be an increase in accidents, emergencies and health impacts.

How can industrial parks be managed in order to ensure sustainability in the environment and communities in which they operate? Excellent environmental management of these environmental issues is essential for the sustainability of the park. One of the approaches to managing environmental issues in organisations is the development and implementation of an environmental management system (EMS). Many of the management frameworks have been developed for individual organisations rather than for industrial parks as a whole. The ISO 14001 EMS, developed by the International Organisation for Standardisation (ISO 1996), is an internationally recognised system and has been used worldwide in individual organisations for addressing environmental concerns in those organisations in a systematic, formalised manner. The purpose of this chapter is to explore how this ISO 14001 management system can be adopted and applied to industrial parks.

12.1 The ISO 14001 environmental management system

An EMS is defined as the 'organisational structure, responsibilities, processes, procedures and resources for developing and implementing the organisation's environmental policy' (ISO 1996). The main components of the system are environmental policy, environmental planning, implementation and operation, monitoring and corrective action and management review (ISO 1996), which translates simply to plan, do, check and correct, and improve (see Fig. 12.1).

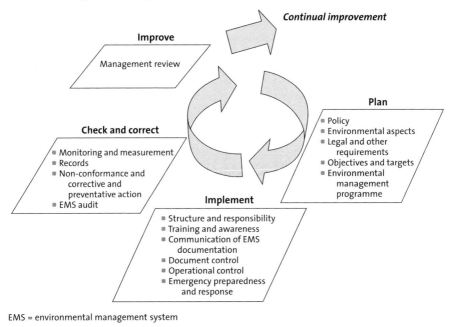

EMS = environmental management system

Figure 12.1 **Model of the ISO 14001 environmental management system**
Source: ISO 1996

Although initially regarded as time-consuming and costly during its establishment, many benefits have been associated with an EMS, including the following (Côté and Balkau 1999):

- Increased energy efficiencies
- Less wastage and use of hazardous substances (with a resultant reduced impact on the environment
- Consideration of prevention and remediation of environmental aspects
- Improved environmental awareness in employees
- Commitment to comply with regulatory requirements and continual environmental improvement
- Facilitation of permit approval
- Reduced accidents and liability
- Increased ability to satisfy investors, and improved trade relationships
- Improved marketing image
- Improved industry, government and community relations

An EMS enables an organisation to identify the environmental risks comprehensively, act proactively and focus resources and effort on managing the significant aspects. In essence, there are two primary reasons to pursue EMSs such as ISO 14001. The first reason is to systematise and improve the environmental management of a facility or operation to improve results; the second is to use attainment of such a status for marketing purposes to distinguish the operation, gain access to certain markets and meet environmental supplier requirements.

12.2 Components of ISO 14001[1]

12.2.1 General

12.2.1.1 Scope of an environmental management system

One of the issues that needs to be considered at the outset when developing an EMS is the scope of activities, products and services to which it will be applied. The EMS was initially established for organisations that have direct control over all their operations. The application is also flexible in that an EMS can be developed for all activities, products or services in a company or be applied to only a sector of its operations.

In an industrial park there are various activities, products and services produced by a number of different parties: namely, the park managers, the tenants in the park and, possibly, independent contractors. The managers of the park have direct control over their own activities, which are typically managerial, administrative, planning and design-oriented or relate to the supply of services and infrastructure. They are, however, not

---

1 For ease of reference the components of ISO 14001 will be discussed in the order in which they are described in the international standard *Environmental Management Systems: Specification with Guidance for Use* (ISO 1996).

directly responsible for the operational activities of the tenants in their park (Wheatley 1992).

The degree of control that park managers have over operations is an important factor when considering the scope of the EMS to be implemented. Côté and Balkau (1999) describe a number of variations in how this EMS model can be adapted to an industrial park (see Box 12.1).

An EMS can be applied to different stages of operation of an organisation. Typically, when an EMS is developed for an individual company, this development occurs when the company is already in existence and operational. In an industrial park, however, an EMS can be developed during the planning and design phase of the industrial park, during the operational phase (regarded here as the entering and operation of tenants in the park) and during the construction phase. These phases can also occur simultaneously in an industrial park as different tenants enter the park through time. The scope of the EMS developed for an industrial park needs to take all these phases of the park into account.

12.2.1.2 Environmental review

A key step when preparing an EMS is to prepare an environmental review. This requires a thorough understanding of the existing environmental issues, environmental or management practices and procedures (if it is for an existing operation), surrounding ecosystems, land uses and communities. A large number of environmental aspects need to be considered in a review. The environmental impact statement (if one was prepared when the park was established) or baseline environmental studies conducted in the area are useful sources of information.

The environmental review should enable management to identify environment and safety issues associated with the park. Consideration should also be given to concerns or issues raised by interested and affected parties (Côté and Balkau 1999). Positive environmental impacts that can be enhanced through the operation of the park should also be identified, such as water conservation and use of open space.

12.2.2 Environmental policy

An environmental policy sets the environmental framework within which the EMS operates. It gives the overall sense of direction of the industrial park and demonstrates management commitment to the organisation, tenants and other stakeholders in the environment. The policy is a public statement of management's environmental vision for the park and ensures that actions are co-ordinated and work within this framework. Some of the considerations in developing a policy include: compliance with laws, regulations and other environmental requirements, policies and strategies; clarification of the park's vision or values with respect to the environment; and incorporation of principles such as continual improvement, prevention of pollution and communication with stakeholders (Côté and Balkau 1999). The policy also needs to incorporate the significant environmental aspects identified during the EMS process.

The senior management of the industrial park is responsible for signing the policy and ensuring that it is adopted throughout the organisation. The policy signifies the scope of the EMS. For example, policies may include a statement of consideration of impacts throughout the life-cycle of products and work with suppliers in order to reduce impacts.

The enterprise option

The park manager simply encourages the adoption of environmental management systems (EMSs) by companies in the park. The environmental problems are then resolved on a company-by-company basis without any overall co-ordination. While having the advantage of simplicity, this option does not facilitate building of synergies, nor does it address cumulative environmental impacts pertaining to the park as a whole, nor is it efficient from a total cost perspective. The manager can facilitate this option by assisting individual companies in their efforts to implement their systems by, for example, providing seminars and offering training and technical assistance. Parks could also offer fee differentials and other incentives to encourage companies to develop EMSs.

The infrastructure option

The park may implement an environmental management system for its own activities and services such as roadways, landscaping, water supply, sewage treatment, solid waste collection and recreational facilities. Such an approach would set the example for companies in the park and could be a useful step in a process leading to the formulation of an EMS for the park as a whole. On its own, however, such an option is of limited use, as the park activities do not usually constitute the major impact. The exception would be where the park supplies major services and manages drainage, common effluent treatment, waste disposal and perhaps recycling plants.

The comprehensive option

In this option the EMS is developed for the park as a whole, viewed as a total interacting system. This option is only possible where the park has direct influence over the pollution activities of its tenants through, for example, permits or other contractual agreements. Such an EMS would probably have to be developed with tenants as primary stakeholders. In the case of new parks, the existence of a formal EMS would exert a selection pressure on future tenants in favour of those who have a similar systematic approach to environmental control.

The environmental charter option

The charter can be adopted in conjunction with any of the three above or developed as a stand-alone arrangement. In this option, the park, tenants and outside service suppliers prepare a contractual charter that specifies environmental responsibilities of each partner. While not formally an EMS, it can encompass many of the elements of an EMS, and the process could be a co-operative one also with the outside community. The particular model chosen will depend on the local circumstances and in particular the environmental objectives that the park wants to achieve. A critical aspect concerns the reasons for developing an EMS in the first place. It was mentioned earlier (Côté and Balkau 1999) that such a system allows a more cost-effective response to environmental problem-solving. But, if public relations, marketing of environmental quality, more favourable financing and good relations with government agencies are also a high priority, then one of the comprehensive systems involving a high degree of community consultation should be chosen.

Box 12.1 **Approaches to the environmental management system model**

Source: Côté and Balkau 1999

The policy is normally displayed throughout the industrial park and is also regarded as a public document that should be readily available to the public.

12.2.3 Planning

12.2.3.1 Environmental aspects

One of the critical steps in developing an EMS for an organisation or an industrial park is the determination of the significant aspects. An environmental aspect is defined as that aspect of the activities, products or services that interacts with the environment, i.e. causes or can cause an impact (ISO 1996). This is a key component of an EMS as it determines the management priorities. All other components of the system (i.e. objectives, targets, environmental programme and procedures) are focused on managing these significant aspects.

Typically, many, or even most, of the significant aspects and impacts on the environment may be related to the tenant's activities in the park. Information and knowledge of tenants' activities, products and services as well as the associated impacts during normal and emergency conditions are essential. There needs to be good co-operation between the park manager and the tenants, as the information needs to be obtained, where possible, from the tenants.

The procedure for rating the significance of the environmental aspects needs to be determined and documented. To facilitate the rating of environmental aspects, the description of activities and impacts in the park may be stated in a broader manner (e.g. by grouping similar activities in the industrial park) without losing meaningful details. The level of detail may differ from that in an EMS for an individual organisation, where the aspects can be described in a more detailed, process-specific manner. The criteria used for the rating of aspects may differ slightly between organisations; however, generally, they may include legislation, environmental impacts, stakeholders and, in some cases, business risk–benefit (financial considerations). Actual or potential environmental impacts should be considered in terms of severity, probability of occurrence and duration of impact. Determining management priorities is important so that resources and effort can be focused on addressing the most pressing issues first.

An important environmental issue in an industrial park is management of cumulative impacts. Such impacts include the effect on overall air quality, total volume of stormwater run-off and the impact of waste generated from the park. In this respect, it is essential that there be an emphasis on pollution prevention and continuous improvement. The park management should promote this by imposing pollution restrictions or use of economic incentives, particularly when new tenants enter the park. Environmental guidelines can be established as a framework for the industrial park to be cognisant of these cumulative impacts.

12.2.3.2 Legal and other requirements

One of the requirements of an EMS is for the organisation to comply with the relevant national and local environmental laws, policies and regulations. Implementation of such regulations can be quite varied in an industrial park. In some parks, the park owner may have delegated authority for the enforcement of environmental regulations and standards, whereas in other situations the national environmental agency may deal directly with the tenants on some or all environmental aspects (Côté and Balkau 1999).

Often, park managers develop their own set of legal instruments, such as codes, covenants, conditions and restrictions specifically for a park and have the power to regulate certain activities of their tenants (WEI 2000). It is important for a complete inventory of environmental regulations to be available as well as a mechanism for regularly updating the inventory. Park managers should keep a record of permits and authorisations that pertain to businesses within the industrial park (Côté and Balkau 1999). This inventory should be communicated to managers and employees of facilities and service organisations as well as to the park management.

The industrial park manager can be a source of information to plants and provide information on policy, permits and other park requirements, or direct tenants to the relevant authorities. However, all parties in the industrial park are responsible for ensuring that they have the required legal and policy requirements related to their own activities and that a mechanism is in place to ensure the compliance register is kept updated.

12.2.3.3 Objectives and targets

Environmental objectives need to be developed once the significant aspects have been identified and prioritised. Environmental policy objectives are goals for the EMS and apply to the industrial park as a whole. Objectives can be management-oriented (concerned with the functioning of the EMS) or operation-oriented (concerned with the environmental performance of particular aspects). There should be a set of measurable environmental performance indicators to measure the progress in meeting the objectives and targets.

Targets are the desired environmental performance levels to be fulfilled within a specified time-frame. Targets can be quantitative (measurable) or verifiable (auditable). It is appropriate to set cumulative targets for the industrial park as a whole. The tenants can then develop their own set of (process-related or service-related) objectives and targets for their organisation that recognise those set by the industrial park. Examples of objectives, performance indicators and targets for an industrial park are given in Table 12.1.

Knowledge of the operations and impacts of the activities and services should be known to the industrial park manager when developing these targets. Many of the environmental benefits of meeting these targets will be cost savings. Targets are normally set for a period of time, usually per annum or per month, and should be reviewed at least annually during the EMS management review.

12.2.3.4 Environmental management programme

The environmental management programme is the action plan for achieving the objectives and targets. For example, if the objective is to improve relations with neighbours, the action plan could include the publishing of newsletters or an environmental report and the communication of the park's environmental performance to stakeholders at meetings and other public forums. The programme should clearly identify the responsibilities of and the time-lines and resources used by the industrial park staff as well as

Objective	Environmental performance indicator	Target
Minimise the risk of accidents involving hazardous materials	Number of accidents with hazardous materials	0 per annum
Improve relations with neighbouring communities	Number of public complaints	Less than 10 complaints per annum
Reduce solid waste sent to landfill	Percentage of waste sent to landfill	30% reduction in waste sent to landfill compared with the base year (e.g. 2002)

Table 12.1 **Examples of objectives, indicators and targets for an industrial park**

the issues requiring action by tenants. Agreement should be obtained from tenants on the actions to be carried out as part of the programme.

12.2.4 Implementation and operation

12.2.4.1 Structure and responsibility

The structure of and responsibility for the EMS should be clarified not only within the industrial park management and staff but also between the industrial park and its tenants. Overall responsibility for meeting environmental standards and objectives for the industrial park lies with the senior manager or the board of directors of the industrial park. The day-to-day implementation of the programme is assigned to one person (or more) in the park. In the case of a comprehensive EMS, the park management and the tenant companies would share the day-to-day responsibility, each having well-defined responsibilities under the environmental management programme (Côté and Balkau 1999).

The environmental responsibility should be clear to all parties with regard to action plans and procedures. This responsibility is often defined in individual employee job descriptions or performance charters. EMS representatives responsible for EMS implementation may, where possible, be designated in each tenant organisation. The tenant representatives can report to their own organisations and also communicate environmental issues and serve as links to the eco-industrial park. The implementation and effectiveness of the EMS and environmental performance should be a regular review item at management, operational and environmental meetings. Several companies with similar issues may share the funding and time of a designated person to meet these tasks. The opportunity of an eco-industrial park is to share the learning process of what it takes to have an effective EMS.

12.2.4.2 Training, awareness and competence

The ISO 14001 standard differentiates between general awareness training and competency training. Competency training refers to the teaching of specialist skills required to implement the environmental programme and is provided to those personnel whose jobs impact on the environment. However, in most cases, all personnel in an organisation, at all levels, whether managerial, technical, environmental, operational or clerical staff, should be given environmental awareness training appropriate to their tasks. Personnel should be aware of the impact that their jobs have on the environment and be accountable for reducing that impact. This responsibility is often reinforced by being stated within employees' job descriptions.

The park management can directly influence the recruitment, training and skills development of only its own employees. However, tenants can be helped through information dissemination and by developing joint or common training programmes. The park management can, for example, implement a 'train the trainers' programme by providing EMS training to key tenant personnel; this can then be 'rolled out' by these key personnel to other tenant staff. For example, training courses can be offered on water and energy conservation, waste-exchange programmes or emergency preparedness and response, which may be of interest to many, if not all, tenants.

Training of contractors should also be included where relevant for reducing environmental impacts. Training to contractors can include methods of reducing impacts during construction activities and the promotion of certain principles such as water-wise gardening and the landscaping of open space areas with indigenous vegetation.

12.2.4.3 Communication

Given the various stakeholders involved in a comprehensive EMS, there is a need to develop an effective communication system. There needs to be a mechanism for internal communication (between the industrial park and tenants) and external communication (for communication with other stakeholders, such as interest groups, public and legal authorities and the media). In an eco-industrial park the range of those involved may be broader than that in traditional industrial parcels.

The type of information to be communicated and reported internally includes legal and policy requirements, environmental performance, public complaints, EMS monitoring or auditing results, procedures, incidents of non-conformance with procedures or legal and other park requirements, and other incidents and emergencies. This can be done through various methods, including environmental or management meetings, newsletters or bulletins, e-mail, and as specified by the park's EMS procedures.

Communication with external stakeholders could involve formal reports to government agencies, the development of a procedure for handling public complaints, the publication of newsletters and environmental reports, public meetings or press releases for outside stakeholders. Information communicated could include environmental compliance with standards, environmental policy, objectives and targets, environmental performance evaluation, opportunities for improvement and independent verification of the functioning of the EMS. A mechanism must also be implemented to allow for communications to flow from stakeholders to the park such as through a public complaints line, use of comments sheets, meetings or stakeholder forums. It must be noted, however, that, although the attainment of an ISO 14001 EMS can market the organisation, an EMS is generally used as an internal management tool for improving environmental issues within the business or park.

12.2.4.4 Environmental management system documentation

Key components of the system need to be documented to ensure everyone uses the same documents:

- As a reference
- As an educational tool for new employers or stakeholders
- As a historical record
- For audit purposes, to allow managers to check progress and communicate results with tenants and authorities

The documentation must include the environmental policy, objectives and targets; action plans, programmes and procedures; monitoring reports, incidents of non-conformance and corrective actions taken; and personnel responsibilities. In essence, the documentation must provide evidence that the EMS is being implemented.

It is also important to ensure that documents are periodically reviewed and that updates are available to key personnel (ISO 1996). Documentation should not be cumbersome but should be appropriate to the size and complexity of the organisation. The relevant EMS documentation should be available at the industrial park and be held by key tenant personnel in a manner that will ensure easy access to staff.

12.2.4.5 Document control

It is important to ensure that documents are updated as changes occur and that correct versions of documents are used. The ISO 14001 standard indicates measures to be incorporated for controlling documents. These include the identification of documents, indication of the revision status and the removal of obsolete documents from circulation.

12.2.4.6 Operational control

Operations that cause or that potentially can cause a release to the environment such as emissions, waste, spills, noise and effluents need to be controlled. Operational control is a key to reducing impact on the environment and is especially important during emergencies. Operational control could involve implementing appropriate technology to prevent or reduce emissions and implementation of procedures for managing environmental aspects. This is particularly relevant in an industrial park when emissions can occur from different sources. Implementation of control measures and monitoring of the effectiveness of these measures is an important step in managing environmental impacts and ensuring accountability within the park.

12.2.4.7 Emergency preparedness and response

Emergency preparedness and response is essential in an industrial park, where industries are located in proximity and there is an increase in potential for incidents and knock-on effects. However, emergency situations can arise in any industrial park. Accidents are serious environmental, health and safety concerns for industrial park managers and tenants and especially to neighbouring communities because of the concentration of industries and possible proximity to major roads and harbours (Côté and Balkau 1999).

The emergency plan, roles and responsibilities, service organisations, contact information, communication plans, training and exercise drills should be clearly designated for the industrial park.

Response to an emergency also depends on where the emergency originates. For example, if an emergency occurs on park-owned land, the industrial park manager is responsible. If it originates on a tenant's property, the tenant will be responsible; however, if the impacts extend beyond a tenant's property and affect other areas the industrial park management would assist.

The industrial park manager generally has a co-ordination role when an emergency arises in the park by ensuring that the necessary emergency services have been contacted and that the required assistance is at hand. Co-operation and communication between tenants, emergency services, the public and authorities is essential. It may also be useful for a central emergency centre to be established to facilitate co-ordination in the event of an emergency in the industrial park. In certain cases, emergency equipment can be shared between tenants and the eco- industrial park.

12.2.5 Checking and corrective action

12.2.5.1 Monitoring and measurement

Practitioners have often mentioned that, before one can manage environmental issues, one needs to measure them. Environmental performance needs to be monitored and compared with the objectives and targets set for the park. Environmental monitoring and measurement ensures that the performance of the aspects is evaluated and assists in identifying areas that have been performing well or those requiring improvement.

The industrial park typically will have a park-wide monitoring programme to measure, for example, storm-water run-off from the park, ambient air pollution and effluent discharge from combined sewers from the park or at an effluent outfall or to monitor construction activities or public complaints. The type of monitoring will depend on the local circumstances of the park. Tenants, in turn, will be responsible for monitoring the emissions from their individual properties.

12.2.5.2 Non-conformance and corrective and preventative action

As tenant non-conformance with regard to the environment may impact the overall performance of the industrial park, tenants should report incidents of non-conformance to the industrial park. The types of non-conformance to be reported, time-periods, responsibilities and procedures for reporting need to be agreed between the park manager and tenants. Non-conformance and corrective action can be identified through site inspections, monitoring, auditing and other mechanisms. The findings and recommendations should be documented and communicated to the relevant personnel in order for corrective and preventative actions to be implemented. Ideally, once implemented these actions need to be audited to establish whether they have been effective.

12.2.5.3 Records

The master copies of the EMS documentation and records should be kept with the industrial park EMS representative. Key tenant personnel (e.g. a designated EMS representative) should have the necessary EMS documentation relevant to their operations. Records are valuable for auditing purposes and as a historical record of activities, incidents, impacts and programmes to monitor changes in the system through time. They can also help newly locating companies move through the process quicker and better.

12.2.5.4 Environmental management system audits

An EMS should be audited periodically to establish whether it is effective and whether environmental performance has been improved. Audits can be conducted internally by park management or externally by independent consultants. In the case of industrial parks, an audit team could also include representatives of other parks from the region or country (Côté and Balkau 1999). An audit programme should include an audit of the industrial park's own activities as well as of tenants' performance relating to significant aspects of the EMS.

The purpose of an audit is to identify areas for improvement in the EMS and should be regarded as a beneficial process that will aid management rather that being used as a 'fault-finding' process. To obtain certification as an ISO 14001 company it is necessary to pay for an outside, ISO-approved, auditor and to register with ISO. In an eco-industrial park, these processes and costs can be managed in a co-operative fashion. Attainment of ISO 14001 status is a cause for celebration, and those inside the company and others in the park should join in the recognition of this accomplishment.

12.2.6 Management review

One of the main principles of an EMS is to continuously improve the implementation and environmental performance of the system. The aim of a management review is to assess the implementation of the EMS and the actual environmental performance of the industrial park and to identify areas for modification and improvement. This is the final stage in the EMS process and it enables the organisation to act proactively in light of changing circumstances.

The review should cover: the environmental objectives, targets and performance; environmental aspects; audit results; incidents; training and awareness; the environmental policy; and the continuing suitability of the EMS in relation to current and future legislation, changing technology, procedures, market changes and communication from stakeholders. The elements of the EMS should be reviewed to ensure it is adequate and effective and to allow for continual improvement.

A management review requires the collation of information from tenants on their environmental performance. This requires communication between tenants and the industrial park on environmental performance and the other EMS-related issues. The information can be obtained through completion of performance questionnaires, meetings, records, monitoring reports and so on. The review is conducted in a climate of trust and transparency between the organisations. The management review is an important step, as it establishes whether the objectives and targets have been met. This component is typically conducted at least annually by senior management and 'closes the loop' of the EMS.

12.3 The link to eco-industrial development

To an extent, an EMS is focused on identifying environmental management priorities and minimising pollution risks. However, there may also be environmental opportunities in this form of development when synergies between companies are explored. Eco-industrial development highlights the benefits of grouping or clustering industries together, where possible synergies created can give advantages over the traditional forms of industrial development. The benefits are to the environment, businesses and to the communities. Embedded in eco-industrial development are the following principles, which can enhance the effectiveness and reduce costs by looking for synergies and eliminating unnecessary redundancy (WEI 2000):

- Environmental planning and the precautionary principle, giving consideration to environmental issues during the planning stages of an industrial park (see Chapter 10)
- Ecological design, giving consideration to the type of materials used in the construction of buildings, energy and water efficiencies and use of indigenous vegetation
- Total quality environmental management, applying an EMS as described in this chapter
- Cleaner production, focusing on processes and technology that prevent pollution
- Resource recovery and industrial ecology, including the exchange of waste and energy and the provision of common services (UNEP 1997)

Eco-industrial development is regarded as an alternate model for sustainable development (WEI 2000).

12.4 Concluding remarks

In this chapter I have explored the adaptation of an EMS such as the ISO 14001 EMS to industrial parks. All the components of the EMS are relevant and can be applied to an industrial park, with modification where appropriate. The main objectives ensure that the environmental risks are managed and that the potential benefits are enhanced. Although an EMS does not stipulate environmental targets and performance requirements, if the EMS is implemented successfully there should be a visible improvement to the environmental performance of the park (ISO 1996). This has led to a strong criticism of the ISO 14001 system, because it is a process-oriented rather than an outcome-oriented activity. However, the industrial park and the tenant companies can supplement their process control with specific targets for improvement that are measurable and can report on how well they have done. A comprehensive EMS for an industrial park can provide a systematic and preventative approach to the management of environmental issues (Hopfenbeck 1993). It is a valuable tool for addressing cumulative impacts, enhancing opportunities and formalising management action when there are many parties involved. It is necessary to have a high level of co-operation and communication between tenant companies and the industrial park management, where all parties take environmental action seriously. An EMS differs from a project that has a finite end. The value of an EMS is that it is a continuous improvement process, integrated within the business, and contributes to the successful operation of the park.

In eco-industrial development, beneficial opportunities may be enhanced by the clustering of industries and the implementation of eco-industrial principles. The principles include, among others, the exchange of waste and energy between industries, cleaner production, provision of common services in the park, and integration with surrounding communities.

An EMS tends to focus on managing environmental risks and opportunities, whereas the eco-industrial development principles focuses on harnessing beneficial synergies within an industrial park and the larger community. Both approaches, however, complement each other and bring park managers closer to managing industrial parks in a sustainable manner.

13
MAKING IT HAPPEN
Financing eco-industrial development*

Dennis Alvord
US Department of Commerce

An organisation, whether a local government department interested in redeveloping a core industrial area, a non-profit organisation seeking to revitalise the community or a developer looking for a sizeable return on investment, will have a greater chance of success if it develops a well-thought-out strategy for securing financing for the project. This chapter is intended to help community, government and private-sector leaders to think through three sets of issues. First, how can the project be structured in ways that best sell the project to investors—both financiers and prospective tenants? Second, how can a developer strategically market project features to potential investors? Third, as with any real estate transaction, where are available public-sector and private-sector funds? The chapter concludes by offering recommendations for how national, regional and local government along with the private sector can develop policies and practices that create greater incentives for this innovative approach to industrial development.

13.1 A dedicated leader: an essential ingredient

Eco-industrial projects to date have varied considerably in terms of who assumes the leadership role in initiating and implementing the project: The Londonderry, NH, Ecological Industrial Park was initiated by a private-sector developer; the Endicott, NY, Aurora Project by a private company, International Business Machines (IBM); the Choctaw County, MS, Red Hills EcoPlex by a state government agency; the Burlington, VT,

---

* The opinions expressed herein are those of the author and not necessarily any entity with which he is associated. Joan Glickman of the US Department of Energy's Office of Energy Efficiency and Renewable Energy offered invaluable advice on the manuscript.

Intervale Food Center (formerly Riverside Eco-Park) by a local government agency; the Cape Charles, VA, Sustainable Technology Park by a quasi-governmental agency, an industrial development authority; and the Minneapolis, MN, Phillips Eco-Enterprise Center by a non-governmental organisation. But regardless of the organisational framework, an effective and committed leader who communicates well, perseveres and develops strong partnerships is crucial.

Strong leadership ensures that someone is invested in the outcome of the project and will tend to all the details of project implementation, including planning, community participation, financing, design, construction, marketing, leasing, and business attraction and retention. This local 'change agent' can set strategic project goals and make organisational decisions to ensure efficient and effective project implementation.

Whoever assumes this leadership role needs to be strategic about packaging the eco-industrial development (EID) project to attract investors, local, regional and national governments as well as business partners and tenants. The following discussion maps out a number of factors to consider when putting together an EID project to help promoters ultimately to sell the concept to public and private financiers.

13.2 Creating a viable project: eco-industrial development as real estate transactions

Eco-industrial financial deals are essentially real estate development transactions conducted within a community and economic development framework. To attract financing, one needs to consider many of the same factors that guide 'typical' real estate deals.

13.2.1 Site selection

Historically, a variety of factors have been known to influence business location decisions, including access to customers and labour, wage rates, utility costs and tax levels. More recently, increased attention has been given to the role of quality-of-life factors, such as recreational and cultural opportunities and the quality of the built and natural environment in influencing these decisions. As with all development decisions, the real estate mantra, 'location, location, location', holds true with eco-industrial projects and is the first factor to consider when putting together an EID project. Some location criteria are fixed (i.e. cannot be changed), such as the distance to customers; others are flexible and can be negotiated, such as taxes and wage rates. It should be kept in mind, however, that negotiations over flexible factors can delay projects.

In remarks at a recent meeting of Cornell University's EID Roundtable, Justin Bielagus, developer of the Londonderry, NH, eco-industrial park (EIP), noted that he and his partner followed conventional market trends when selecting a site and putting together their EID project. First, they looked for an opportunity to enter an emerging market early on. Next, they focused on population centres and attempted to anticipate market growth in these areas. Finally, they worked to develop in the 'path of progress' (WEI 2001). In the case of the Londonderry EIP, the stars were aligned. AES Corporation, an international power company and the park's anchor tenant, identified an opportunity for power generation in the rapidly expanding north-eastern US market.

13.2.2 Infrastructure

Infrastructure is a vital component of any successful local development project, regional and national economic development and global commerce. One needs only to look at the economic impacts of the past century's canal, railway, road and port facilities and this century's broadband telecommunications and regional airport systems to understand the enormous impact of infrastructure on a nation's economic prosperity. Among the most important characteristics of any industrial development project, particularly manufacturing-based efforts, is access to regional transportation networks.

Environmental infrastructure—such as water, sewer and solid waste management facilities—has become another important component of development decisions. Sites with existing infrastructure, or readily available access to such infrastructure, may offer a competitive advantage over sites that lack these amenities.

The eco-industrial model also calls for, when appropriate, 'next-generation' infrastructure to support activities such as materials, heat and energy exchanges. The model also facilitates shared conventional infrastructure to reduce system redundancy and overcapacity and thereby increase both the efficiency and cost-effectiveness of the infrastructure. By considering how best to use all types of infrastructure, EID offers new opportunities to tie together project infrastructure in a more coherent and compatible manner.

13.2.3 Market demand

EID, by its nature, deals with industrial development. Therefore, in order to attract financing and industrial partners, a project must provide good business opportunities for potential industrial tenants and enhance their financial bottom line. The greater the value of the EID model to these intended users the greater the potential for successful development.

Private developers that have been involved in EID, such as the developers of the Londonderry, NH, EIP, stress the importance of market fundamentals in advancing an EID project, particularly in securing financing for such projects. Generally, developers look to double their money on a present-value basis, charging developer fees, brokerage fees, financing fees and infrastructure development fees; so, a US$1 million investment requires US$3 million in sales over three years to obtain a present value return of US$2 million. In the case of the Londonderry eco-industrial park the developers were able to show banks a credible third-party appraisal indicating that the project would return 3:1 present-value money—demonstrating the project's strong market fundamentals (WEI 2001).

13.2.4 Site assembly

EID projects often require large tracts of land under single ownership in order to attract and support large-scale industrial projects. In cases where assembly of multiple adjacent parcels is required—or where smaller parcels that are interdispersed among larger parcels in a patchwork-quilt manner typical of early 19th-century spatial development patterns must be acquired—time considerations may become a factor in project viability. A number of factors (e.g. land speculators, homeowners unwilling to take buy-outs and mothballed sites) can delay or prevent developers from assembling a large parcel. In the most contentious cases, use of local eminent domain authorities may be required. This raises an entirely new set of political, community involvement and public relations issues that may further delay the project.

13.2.5 Financial constraints

Industrial construction is expensive in densely populated urban areas where property values are high. The economic, environmental and social benefits of the project should be weighed against costs of land, materials and labour.

13.2.6 Site preparation

Redevelopment projects may involve demolition, deconstruction or adaptive re-use of existing buildings or structures. Preparing a site for re-use, particularly when removal of underlying structures or foundations is involved, can impose significant costs on developers. Although conventional demolition, removal and disposal are the norm, many communities have started to recognise the benefits of alternative strategies. 'Deconstruction' refers to the selective dismantling of materials for re-use or recycling (NAHB 2000). 'Adaptive re-use' calls for the integration of existing structures and/or materials into a new development project. This might involve wholesale redevelopment of buildings for new uses, or salvage and re-use of structural components (e.g. beams, joists, etc.) in new facilities.

Although deconstruction and adaptive re-use strategies may have higher 'upfront' costs (e.g. deconstruction is a more labour-intensive process than is demolition), these costs can frequently be recouped through the sale and re-use of on-site materials as development proceeds. From a community development perspective, additional benefits include jobs for unskilled and unemployed workers, business development opportunities and the diverting of waste from landfill. The Green Institute in Minneapolis, MN, for example, has created a value-added deconstruction business coupled with a neighbourhood retail outlet to sell recovered materials. The Resource Recovery Center trains unskilled labour in the building and construction trades and thereby helps bridge the gap between marketable skills and employment opportunities. Furthermore, when developing the Phillips Eco-Enterprise Center, the Green Institute saved thousands of dollars by re-using donated construction materials and systems.[1]

---

1 Personal communication with C. Brinkema, E⁴ Partners Inc., Minneapolis, MN, USA, 21 August 2001.

13.2.7 Environmental characteristics

EID has frequently been undertaken in the context of urban infill and brownfield[2] redevelopment as a means to recapture public investment in urban areas over the years. EID projects and brownfield redevelopment projects often share many of the same objectives. In particular, infill development is often targeted to stimulate employment-generating growth in already developed areas where jobs are accessible to low-income populations, thereby promoting efficient land-use patterns and reducing pollution from urban sprawl.

Although brownfields offer many benefits in terms of community appeal, existing infrastructure and accessible labour pool, among other attractive features, developers need to consider how environmental concerns might pose risks both in terms of time-delays for site assessments and characterisations and in terms of assuming liability for pre-existing contamination without regional and/or national assurances. If significant contamination is discovered, land-use-based remedies may impact potential future uses of the site. In some cases, clean-up costs can be prohibitive and therefore a 'deal-breaker'.

In general, private lenders are extremely conservative. They are concerned about anything that may impact the ability of a borrower to repay a loan as well as potential lender liability. Sometimes these concerns can be addressed through a clear quantification of risk during the assessment and remedial design phases of a contaminated-site redevelopment project. At other times, environmental insurance products that cover expenses associated with future site liability and/or remediation cost overruns may increase lender comfort.

At the same time, in the USA brownfields can be sold as an attractive investment to lenders, as banks can receive credit through the 1977 Community Reinvestment Act (CRA) for supporting brownfield redevelopment projects. There may be a need, though, to target lenders that are proactively looking for opportunities to include brownfield transactions within their loan portfolios.

In addition to hazardous waste issues, developers may confront a variety of other environmental issues, including ecologically sensitive areas and habitat for endangered species; water treatment and discharge; wetland fill and restoration; and flood plain issues. All these considerations can delay projects and therefore affect costs and financing.

13.2.8 Historic and cultural significance

Although a historic or culturally significant site may provide a developer with a unique and attractive redevelopment opportunity for certain project types (e.g. housing or civic structures), such a site can just as easily be perceived as a barrier in the context of industrial development. Outmoded or incompatible structures may be protected under historic preservation statutes or have a strong community or cultural significance that limits re-use.

2 According to the US Environmental Protection Agency, 'brownfields' are 'abandoned, idled or under-used industrial and commercial facilities where expansion or redevelopment is complicated by real or perceived environmental contamination' (US EPA 2001a; www.epa.gov/brownfields).

EID projects have dealt with this issue in two ways: first, whenever possible, they have explored avenues to re-use existing historic or culturally significant structures as part of the development project; second, when this has not been possible, they have worked to maintain part of the site's historic identity even if structures are demolished. The Green Institute, for example, has named its project the Phillips Eco-Enterprise Center, after the Phillips neighbourhood in Minneapolis, MN, where the project is located. This has helped sustain strong community ties, established during the controversy over the future use of the site from which the project emerged, long after that controversy has subsided.

13.2.9 Zoning and compatibility with existing land uses

Existing zoning designations are the staring point for developers. A fundamental component of any real estate development is that the project must be compatible with existing and adjacent tenants. Historically, in the USA, 'incompatible' land uses have been separated in an attempt to protect public health and to mitigate nuisances (e.g. noise and traffic). Consequently, attempts to encourage mixed-use developments have sometimes resulted in community opposition and even legal action. EID projects often challenge this model by co-locating non-polluting industrial users with housing, retail and civic-purpose structures—diverse uses that historically have been considered incompatible.

As development patterns evolve and more people express an interest in Smart Growth and working and living in the same community, mixed-use development is becoming an acceptable and even desirable means for reconnecting the social fabric of communities. However, these trends evolve slowly, and zoning variances can still be difficult to secure and almost always take considerable amounts of time to enact.

13.2.10 Community and political considerations

EID projects often attract significant community interest. As with other types of development, conflicting visions may emerge. Opposition, including legal or political challenges, from an individual or group can delay development considerably.

Public concerns vary, based on the nature of the proposal. For example, some neighbourhoods may welcome a smaller-scale mixed-use operation but will pose vocal opposition to a large-scale single-tenant facility. Some community members may resist eco-industrial models based on the simple assumption that they propose something different from what has been proposed in the past. In economically distressed communities, it is not uncommon for community members to favour the same economic development models that have failed them in the past, initiating a 'cycle of decay' that can be difficult to overcome. Others may favour more forward-looking models that promote civic prestige but do little to assist the economy. The key to success, of course, is to balance these competing interests and develop a project that satisfies different stakeholders while remaining viable.

13.3 Ready, aim . . . pitch! Marketing the project to funding sources

A key to attracting financing is to understand the appeal of the project to various audiences and then to market the project strategically to funding sources that are most likely to be receptive. At this point, it is important to think about how the project is different from the average real estate deal. Remember that EID is a community-based economic development strategy that responds to local social, environmental, economic and quality-of-life goals. In developing the marketing pitch, one should consider how the project helps advance these goals. The following discussion outlines some of the reasons why eco-industrial projects are often attractive—not only to communities but also to public and private financiers interested in advancing broader social, economic and environmental goals.

13.3.1 Smart Growth

Today, communities have begun to recognise that unplanned growth sometimes has negative consequences, including long commuting times and distances, air pollution, loss of open space and community park land, and development that is isolated from work and community. EID can promote liveable communities through attention to ecologically sound design, the preservation of open space and community space, a healthy work environment (e.g. with enhanced 'daylighting' in buildings) and new approaches to transportation (e.g. the promotion of cycling and walking trails, transit-oriented development and use of alternatively fuelled vehicles). In the USA, as more states begin to implement Smart Growth programmes, state programmes can increasingly be used to leverage EID activities. For example, projects in targeted locations may be eligible for breaks on builder fees.

13.3.2 New market development

A number of public and non-profit institutions, as well as some private financiers, are interested in funding projects that advance new technologies and new markets. Commercialisation and deployment of advanced environmental technology is one way in which EID supports new market development. In his landmark study, *Sustainable America: New Public Policy for the 21st Century*, Benjamin Goldman (1995) explores the hypothesis that emerging clean technologies will create the next industrial revolution, the way that cotton gins, trains, automobiles and computers transformed the economy in the past. He also notes that 'economists widely agree that improvements in technology and human skills are the most important contributors to long-term economic growth' (1995: 23-48). EID also supports the development of emerging markets in the US nation's inner-city and improvised areas, both by taking advantage of under-served local markets and using under-utilised local labour pools.

13.3.3 Benefits to key demographic groups

In the USA, many public financing sources target: specific demographic groups or local characteristics (e.g. economically distressed areas); neighbourhoods or development projects that include a large number of businesses owned by women and/or minorities; areas that have been impacted by trade agreements. In addition, pursuant to the CRA, federally insured banks and thrift institutions are required to make capital available in low-income and moderate-income communities. Federal banking regulators conduct compliance audits to ensure, among other factors, that banks are participating in community development, not operating in a discriminatory manner, extending appropriate types of credit and generally meeting community credit needs. Therefore, it is wise to consider who the principal beneficiaries will be when one explores financing options.

13.3.4 Sustainable design

William McDonough and Michael Braungart, in their ground-breaking article 'The Next Industrial Revolution' (1998: 82-92), point out that, by design, 'many manufactured products are intended not to break down under natural conditions'. McDonough and Braungart (1998) argue that conflicts between industrial processes and the natural environment are design problems that can and should be solved through new ways of designing industrial production. Under the eco-industrial model, environmental considerations are incorporated into all aspects of product and process design, including the design of the facility. A number of public and non-profit institutions support sustainable design projects, either with funding or, more frequently, with technical assistance.

13.3.5 Open space and community development

EID by its nature has a strong industrial development component. As a result, the emphasis of EID financing efforts tends to be on job creation and economic development, largely because there are many financial incentives available for this purpose. Nevertheless, the community development aspects of EID projects, which improve local quality of life and preserve and protect the natural environment, are equally important aspects of EID projects. Just as some brownfield properties may better serve a community's needs by being cleaned up for re-use as a ballpark, soccer field, nature preserve or other green space, some EID projects include community development components. The Cape Charles, VA, Sustainable Technology Park, for example, includes wetland restoration, biking and walking trails and a community meeting space. Communities must be creative in identifying funding sources for these non-traditional development features. National, regional and local programmes that support parks and open space are all potential funding sources. In addition, communities should explore the potential to obtain funding from private foundations and corporate sponsors for this type of activity.

13.3.6 Competitive advantage

Lenders and capital providers are sometimes reluctant partners in EID, partly because of the uncertainty of financial risk associated with the EID model. In the best case, the process of assessing financial risk and developing mitigation strategies can cause project delays. In the worst case, a developer may not be able to secure financing. However, if one can eliminate or reduce this uncertainty by providing a greater understanding of the financial risks involved, capital providers may be more inclined to support the project.

This issue really boils down to the question: what is a better financial risk—EID or conventional industrial development? Is a company that is more sensitive to environmental issues a better risk than a company that ignores environmental considerations on a daily basis? Is a company with less capital tied down for operations and maintenance a better risk than a company with higher operating costs?

In the USA, if one were to ask Wall Street stock analysts to select their top stocks, they generally would provide a list of stocks based on the ability of those stocks to 'outperform the market' with less 'market risk'. As a general rule, this would include many large 'brand-name' firms (e.g. Wal-Mart Stores Inc., Pfizer Inc., Intel Corp. and Sony Corp.), but few small to mid-size companies that may have selected a highly competitive, niche, or 'high-risk' market. Nevertheless, even if a firm is in a higher risk category, it may conduct business in a way that gives it a long-term competitive advantage relative both to its current peers and to future rivals.

Sustainable business practices can give EID firms this competitive advantage in their industry and in the global economy. Eco-industrial practices help companies 'become more competitive by lowering their resource needs while reducing or eliminating their waste clean-up and disposal costs' (OSTP 1994: 2, 26, 54-44). In a recent *Washington Post* article on spirituality and values in the workplace, Ray Anderson of Interface Inc. discussed the technological and organisational retooling that he undertook to make his company's business practices sustainable. Anderson noted, 'the new emphasis on eliminating waste also saved the company $165 million from 1994 to 2000—a savings far greater than the $30 million cost of re-educating workers and re-engineering machinery' (Broadway 2001: A01).

Another case to be made in terms of assessing risk is that lenders tend to focus on financial and economic conditions, while ignoring underlying environmental and social factors that may affect that risk. In her paper, 'Sustainable Finance: Seeking Global Financial Security', Rosalie Gardiner (2002) notes that, 'Whilst these factors may be complex and inter-linked, they can be instrumental in raising risks in financial sectors in the short and long term.' Gardiner argues that environmental and social considerations are key components of financial security, noting that 'conservative estimates suggest that the cost of mitigating disasters related to climate change, i.e. floods, storms and droughts, will overtake global GNP by 2065', and that, '[although] educated, skilled, healthy and happy people are key drivers of economic growth and financial stability . . . 20% of the world's population continue to live in extreme poverty and even in the US some 20% currently live under the official poverty line' (Gardiner 2002).

As a community-based model with broad public benefits, EID takes into account all three of these variables—economic, environmental and social factors—and thereby creates projects that may indeed be smarter, less risky investments. Since sustainable practices make good business sense, it is possible to market eco-industrial projects to financiers as being of lower risk compared with conventional investments. In fact, it may be possible to argue for loans at a lower interest rate.

13.3.7 Socially responsible investing

No discussion of the financing of EID would be complete without mentioning the role of socially responsible investment (SRI). SRI integrates social goals and values into investment decision-making. Since EID is a community-based economic development strategy that responds to local social, environmental, economic and quality-of-life goals, these projects can often qualify as socially responsible investments.

A growing proportion of all investment portfolios in the USA are subject to some form of investment screen (e.g. in the tobacco and weapons industries and in terms of the company's performance relative to the environment, human rights, labour issues, animal welfare and so on). The amount of capital invested by community development banks on local development initiatives—including affordable housing and small-business ventures in the neediest urban and rural areas of the country—has increased dramatically. In fact, according to the Social Investment Forum's 2001 *Trends Report* (SIF 2001), more than US$2.3 trillion is invested today in the USA in a socially responsible manner through screened portfolios, shareholder advocacy and community investing. This accounts for roughly one in every eight dollars under professional management in the USA.

According to a report by Pax World Funds (PWF 2001) marking its thirtieth anniversary, in the past three decades assets of socially responsible funds grew five times faster than did those of other mutual funds. The report indicates that, from the relatively modest US$150 million invested in socially and environmentally responsible mutual funds in 1971, assets had grown to US$103 billion by mid-2001. This represents a growth rate of more than 68 times, compared with slightly more than 13.5 times for the assets of all other mutual funds (PWF 2001).

Eco-industrial firms are attractive investment candidates for SRI fund managers—many of whom are looking for new companies that meet their criteria. EID firms represent emerging industries, actively manage their waste-streams, have significantly lower liability risks and tend to be community-oriented—all traits that fit well with the tenets of SRI.

Furthermore, SRI may appeal to eco-industrial firms not only as a potential source of financing but also as an investment option for their employees' retirement and pension funds. SRI allows eco-industrial park businesses, as well as their employees, to make investment decisions that are consistent with their values and the tenets of EID.

As SRI consists of an increasing percentage of available institutional capital, it will provide a large percentage of the institutional investment necessary to support EID. Tightening the linkages between the EID community and the SRI community will encourage mutually supportive financial arrangements and present a win–win situation for all involved.

13.3.8 Trade

In a recent article from the Federal Reserve Bank of St Louis, Patricia S. Pollard states that 'During the past two decades, the value of world merchandise trade more than tripled, and for many countries trading opportunities with their neighbours have increased substantially' (2001: 1). Pollard notes that increases in trade have been characterised by a global movement towards regionalisation, with an increasing number of regional trade agreements, such as the North American Free Trade Agreement (NAFTA). In addition to regional agreements, Pollard indicates that other factors may have contributed to enhanced trade, including a general trend towards relaxed trade barriers among nations or regional income level increases.

EID projects bring into play similar factors to enhance efficient commerce. Just as international trade has been stimulated by regionalisation, international co-operation, reductions in barriers and increases in income levels, EID arrangements seek to capitalise on efficiencies in the exchange of goods within regional networks, to enhance co-operation among firms, to increase regulatory flexibility in the characterisation and use of 'wastes' and to provide living wages.

Recognising the global nature of commerce and potential economic benefits to their project, the founders of the Cape Charles, VA, Sustainable Technology Park secured designation as a foreign-trade zone. The Avtex Superfund revitalisation effort in Front Royal, VA, on the other side of the state, which is planning ecologically for its redevelopment, also is part of an inland free trade zone port linked to an international trade zone. These designations increase the park's appeal to companies with international markets and provides a number of other benefits, including: relief from inverted tariffs; duty exemption on re-exports; duty elimination on waste, scrap and yield loss; weekly entry savings; and duty deferral.[3] The integration of foreign-trade zones and other trade-related activities into the EID model not only can positively affect the financial bottom line locally but also can help to propagate and broaden the model in an international context.

13.4 Financing sources

No one source of public or private financing exists in the USA to support EID; as a result, existing eco-parks have had to pursue a broad range of financing sources to make projects work. EID requires conscious efforts to nurture a diverse mix of financing partners. EID entrepreneurs must work to secure public investment from a range of very specific stand-alone programmes, not EID-specific programmes. The key is a profoundly fundable project, not just appending an 'eco' prefix to a project that won't succeed.

The Cape Charles, VA, Sustainable Technology Park, for example, received over US$2.5 million from a variety of federal and state agencies, including the National Oceanic and Atmospheric Administration (NOAA), the Economic Development Administration (EDA), the US Department of Agriculture (USDA), the Environmental Protection Agency (EPA), the Fish and Wildlife Service, the US Department of Energy (DOE) and the Virginia Department of Transportation, among others. To supplement these resources, the local development authority, created as an ownership mechanism for the eco-park, issued a US$2.5 million general obligation bond, backed by Northampton County, which has an AAA rating and insurance.

The Intervale Food Center (formerly Riverside Eco-Park) in Burlington, VT, is another example of a community that has leveraged a diverse mix of public financing to advance its project. This project will create an agricultural–industrial business park on a brownfield site within the Intervale, a 700 acre floodplain approximately 1 mile from the city's

---

3   See the website of the Foreign-Trade Zone Resource Centre, at www.foreign-trade-zone.com.

central business district. As envisioned, the park will capture and exchange by-products, including some of the thermal energy available as a by-product from the nearby McNeil Biomass Electrical Generating Station. The proposed park will link low-grade heat from the generating station to a variety of biological and farming applications. According to Burlington mayor Peter Clavelle:

> the most intriguing of these applications may be biological technologies called living machines. These ecologically engineered systems combine fish-farming tanks, hydroponic produce and other greenhouse opportunities while, at the same time, purifying organic liquid wastes. This type of system opens new avenues for commercial urban food production (Clavelle 1997: 6).

US EPA brownfield funding has provided critical local capacity, including a brownfield co-ordinator, to support the development of this project. Other financiers include DOE (for the feasibility study), the Department of Housing and Urban Development (HUD; for community development block grants [CDBGs]), EDA (for infrastructure), Burlington Electric and the Intervale Foundation, which raised funding to acquire the land for the eco-park.

Making a project 'happen' involves strategic marketing to multiple sources of financing. The following sections are designed to assist local governments and community organisations in the USA in identifying revenue options available to them to finance EID activities. The public financing discussion (Section 13.5) may also be applicable to many international EID projects. In addition, the US public and private funding, technical assistance and support services (Sections 13.6–13.9) may be illustrative to international EID efforts. Although a wide range of financing sources is presented, not all available financing options are portrayed. Local jurisdictions must take into consideration their particular circumstances in determining which financing techniques, revenue options and other funding mechanisms—or combination of options—are most appropriate and likely to serve their project needs.

13.5 Public financing

Many EID projects have the potential to be economically viable, to help revitalise communities, to create good, living-wage jobs and to encourage business growth. Many projects, however, require some public-sector financial participation in order to attract the private-sector investment necessary to achieve this success. The challenge to the public sector is to provide the right tools to help make EID projects economically viable. In evaluating the appropriateness of public investment in support of EID, it is necessary to question the value of the activity. Will it have desirable distributional consequences? Will it create public benefits not fully taken into account by the private market? Will it enhance efficiency? Can it be done at a reasonable cost? It is on the basis of these and similar questions that public investments may be justified.

In this section I provide an overview of specific funding mechanisms that have played, or might play, a role in assisting states, localities and the private sector to finance EID successfully. The financing tools explored range from grant and technical assistance programmes to loans from government-sponsored enterprises and loan guarantees.

Seven revenue sources are explored as funding options for EID:[4]

- General revenues
- Taxes
- Fees
- Fines and penalties
- Bonds
- Grants
- Loans

13.5.1 *General revenues*

General revenues are resources in the general operating funds of local governments. Although general revenues are typically funded through the mechanisms discussed below, in some instances 'carry-over' revenues that have been set aside or accrued as a result of surpluses in previous years may be available.

13.5.1.1 Advantages

General revenues are a very flexible source of funding and generally are not tied to a particular activity or service. General revenues can be linked to new plans or visions for local or regional economic development.

13.5.1.2 Constraints

Competition for general revenues within communities is fierce, as these monies are used for a variety of civic purposes, including for policing and fire protection, libraries, parks and recreation, disaster and emergency responses, and homeless shelters and soup kitchens, to name a few. As a result, surpluses and set-asides are very rare. Nevertheless, tapping into general revenues may be an option for communities with a longer time-horizon, such as those in the planning and feasibility stages of EID. Projects with enough support may be able to garner statutorily imposed set-asides that would begin an accrual of funds to be used for implementation at a later date.

13.5.2 *Taxes*

Taxes are a levy against property, income or sales. Property taxes are by far the most common form of tax levied at the local level; however, local governments sometimes have authority to charge income, sales and use taxes through state-permitted surtax provisions.

---

4 The architecture of this section is adapted from the US EPA 1999a.

13.5.2.1 Advantages

Taxes typically have a broad revenue base: that is, they are levied on the value of property, money earned or a percentage of sales and therefore have the potential to generate significant revenue. They can discourage 'bads' in a city by increasing their costs while using the funds generated to help pay for more desirable benefits. In the US, for example, so-called 'sin' taxes on alcohol and tobacco have been used to fund local education and environmental protection initiatives.

13.5.2.2 Constraints

Legislative action is generally required to raise taxes and, depending on public sentiment, it may be difficult to impose new or to modify existing taxes. In addition, historically, taxes have not been dedicated sources of revenue; therefore, EID projects will in all likelihood be competing with any number of other tax-revenue-funded programmes and may be subject to the legislative whims of an annual or biannual budget process. Special accounts and trusts funds, although relatively uncommon, are accounting mechanisms used to dedicate revenues in some cases.

13.5.3 Fees

Fees are a charge for a particular activity or service. Fees generally establish a direct link between the provision of a service and the cost associated with providing that service.

13.5.3.1 Advantages:

Fees are frequently viewed as being more equitable than other sources of revenue because they are often linked directly to programme beneficiaries. For example, an 'impact fee' might be levied on the developer of a 'greenfield' to cover the cost of the new water and sewerage systems needed to service future users. Fees can sometimes be set administratively, negating the need for legislative approval or a voter referendum.

13.5.3.2 Constraints

Fees are typically levied on a particular service and therefore have a narrower revenue base than most taxes, limiting their revenue-raising potential.

13.5.4 Fines and penalties

Fines and penalties are imposed for violations of laws or regulations. They may be imposed for civil or criminal offences and may be charged administratively or judicially. Although fines and penalties may provide occasional revenue windfalls, they are not generally considered to be a reliable source of revenue.

13.5.4.1 Advantages
Fines and penalties can be used to create an incentive for a desired behaviour or action.

13.5.4.2 Constraints
The amount of revenue generated is unpredictable.

13.5.5 Bonds
Bonds are generally used to raise capital for significant community investments, including land acquisition and public facilities. A bond is a written promise to repay a loan by a specific date. Bonds are frequently structured over a period of 15 or 30 years, with interim payments of principal and interest. A US$2.5 million bond from Northampton County, VA, financed the first building, a 31,000 ft² flex industrial facility in the Cape Charles Sustainable Technology Park.

Tax increment financing (TIF) revenue bonds are a variation on conventional bonds that allow a percentage of tax revenue to repay a bond issue within a specific district. A special district consisting of the boundaries of an eco-industrial park (or other boundaries) is created. Two sets of tax records are maintained, reflecting pre-development and post-development property values for the district. The incremental change or difference in revenue earned on the property serves as the source of bond repayment.

13.5.5.1 Advantages
Bond financing may be particularly well suited to the financing needs of EIDs. Advantages are as follows: (1) the bond market typically offers capital at lower interest rates than the rates for commercial loans; (2) the source of repayment can be linked to user charges from the facility financed by the bond (revenue bond); (3) alternatively, repayment can be secured or guaranteed by the 'full faith and credit of the issuer', with repayment from specific taxes, fees or general revenues (general obligation bond); and (4) there is wide discretion in structuring bond issues to meet the needs of local communities. In addition, financing costs may be repaid through EID energy, production and waste disposal cost savings.

13.5.5.2 Constraints
In some localities, voter approval is required for bond issues. In other jurisdictions, laws prohibit public agencies from incurring long-term debts.

13.5.6 Grants
Grants are sums of money awarded to a local government or non-profit organisation. They do not require repayment.

13.5.6.1 Advantages

Local governments do not have to use their own resources; this may free resources for other activities or allow communities to undertake activities that they otherwise could not afford.

13.5.6.2 Constraints

Many grant programmes have strict eligibility and/or matching requirements. Furthermore, grant programmes may come with special conditions on the use of funds; these may or may not coincide with local priorities.

13.5.7 Loans

Loans are similar to bonds in that they must be repaid over an established time-frame, usually with interest. Frequently, the public sector will provide loans below the market rate through programmes designed to address capital gaps where private sources of financing are not otherwise available, affordable or accessible.

13.5.7.1 Advantages

Government-sponsored loan programmes may provide capital at below-market interest rates. Generally, loans may be acquired without voter approval.

13.5.7.2 Constraints

Criteria for subsidised loan programmes may be difficult to meet.

13.6 Federal sources of funding

A number of federal agencies and programmes provide public assistance for economic development, environmental protection and energy efficiency that can be targeted to support EID activities. Only by understanding the numerous federal entities can communities and the private sector fully avail themselves of financial resources that exist for EID. Some of the most widely used and/or the most promising sources of public financing in the USA include resources provided through:

- US Department of Commerce: Economic Development Administration
- US Department of Commerce: National Oceanic and Atmospheric Administration
- US Department of Housing and Urban Development
- US Environmental Protection Agency
- US Department of Energy

- US Department of Agriculture
- US Small Business Administration
- US Department of Health and Human Services
- National Endowment for the Arts

13.6.1 US Department of Commerce: Economic Development Administration[5]

The EDA's mission is to generate new jobs, help retain existing jobs and stimulate industrial, technological and commercial growth in economically distressed areas of the USA. EDA targets communities that are experiencing long-term economic decline affected by military base closures, impacted by sudden and severe economic dislocations, designated as empowerment zones, enterprise communities or brownfields or that are striving to achieve a par with more economically self-sufficient places.

To accomplish its mission, the EDA provides grants for local capacity-building and infrastructure development. It also capitalises locally administered revolving loan funds (RLFs) to support business development. Grants are available to state and local governments, to Indian tribes, to public and private non-profit organisations, to educational institutions and to regional economic development districts. On average, EDA grants cover 50% of project costs, although areas experiencing severe economic distress may be eligible for grants of up to 80% of total costs, and Indian tribes may receive up to 100% of total costs.

The EDA has been among the most active federal supporters of EID since 1996, when the President's Council on Sustainable Development (PCSD), in its first report, *Sustainable America: A New Consensus*, recommended that 'Federal and state agencies assist communities that want to create eco-industrial parks . . . [as] new models of industrial efficiency, co-operation and environmental responsibility' (PCSD 1996d: 104). Building on the work of the PCSD, EDA has provided support for planning, technical assistance, design and 'bricks-and-mortar' infrastructure for eco-industrial parks (PCSD 1996a, 1996c, 1997).

Four EDA programmes can be used to support EID:

- National and Local Technical Assistance
- Planning for States and Urban Areas
- Public Works and Economic Development
- Economic Adjustment

These and RLFs will be discussed below.

---

5 Information about EDA programmes is available at www.doc.gov/eda. A list of useful web addresses is provided at the end of this chapter.

13.6.1.1 National and Local Technical Assistance

EDA's National and Local Technical Assistance programmes provide grants to support problem-solving and innovative economic development techniques. In an EID context, these grants might be used to support market feasibility studies, business plans, site-specific development planning or other analyses necessary to fill knowledge or information gaps that inhibit development.

13.6.1.2 Planning for States and Urban Areas

EDA's Planning programme provides funds for local economic development capacity-building to assist distressed areas in developing their own locally based comprehensive economic development strategies (CEDS). CEDS guidelines require that communities engage in a long-term planning process that 'promotes economic development and opportunity, fosters effective transportation access, enhances and protects the environment and balances resources through sound management of development' (CFED 1999: 15). In addition to CEDS, other planning grants assist economic development planning and implementation activities, including economic analysis, definition of economic development goals, determination of project opportunities and formulation and implementation of local and multi-jurisdictional economic development districts. In an EID context, planning grants might be used to support initial project development, stakeholder involvement and visioning (e.g. community charrettes) as well as longer-term project implementation (e.g. strategic plans).

13.6.1.3 Public Works and Economic Development

EDA's Public Works programme awards grants for construction of public works and development facilities that aid economic self-sufficiency and global competitiveness. This includes the revitalisation, expansion and upgrading of all manner of physical infrastructure necessary to attract new industry, encourage business expansion, diversify local economies and encourage private-sector investment. In an EID context, public works grants can be used to support both conventional industrial park infrastructure (e.g. water, sewers and electric utilities; access roads; rail spurs; port improvements) and innovative infrastructure (e.g. energy capture and distribution; broadband fibre-optics; energy efficiency and renewable technologies; Living Machines™[6]).

The City of Dallas, TX, was awarded a US$1.5 million public works grant for the construction of a 50 acre eco-business park and a 20,000 ft² international environmental training and technology centre on a brownfield site adjacent to the McCommas Bluff landfill. As envisioned, the project will create a business incubator and an environmental technology transfer centre, with resources from three educational partners.

The Red Hills EcoPlex industrial park in Choctaw County, MS, obtained a US$1.5 million public works grant to construct infrastructure associated with facilitating heat exchanges tied to a 440 MW clean-coal power facility. The EcoPlex also plans industrial exchanges among a brick company (clay), a greenhouse company (steam) and a roofing tile company (ash). The project offers a great example of public and private funding sources coming together to finance complementary pieces of the development package.

6 See www.livingmachines.com.

In addition to EDA and State of Mississippi funds, Phillips Petroleum and Tractebel Power Inc. have invested over US$400 million to develop the project. Public funds were used to support the creation of additional public benefits (i.e. local jobs) where there was some financial risk that the private sector was unwilling to consider. Early tenants in the EcoPlex include two types of greenhouse, one supporting hydroponic gardening.

13.6.1.4 Economic Adjustment

EDA's Economic Adjustment programme helps areas experiencing economic dislocation—such as from industrial restructuring, defence base closures, natural disasters and depletion of natural resources—to develop and implement an adjustment strategy. EDA receives an appropriation for both regular economic adjustment and defence economic adjustment activities. Grants are available for strategic planning, project implementation and capitalisation of RLFs. In an EID context, economic adjustment grants might be used to support sustainable recovery strategies, to help provide the infrastructure to implement that strategy, and for the creation of a local capital pool (i.e. RLF) to support both immediate local lending capacity and long-term assistance through re-lending.

The Brownsville Economic Development Council secured a US$168,000 economic adjustment strategy grant to facilitate a series of interactive sessions between industry and the public sector to develop strategies and criteria for the Brownsville, TX, EIP. The grant allowed the Council to: compile existing research on eco-industrial parks; to work with Bechtel Corporation and the Texas Engineering Extension Service (TEEX) to develop a prototype computer model to facilitate product exchanges and to target industries necessary to implement the project; and develop technical specifications for the required eco-park infrastructure. The Brownsville EIP is unique in that tenants will not be physically located together in an industrial park setting. Instead, it will be a 'virtual park' of companies spread across Brownsville and Matamoros and linked by exchanges of materials.

The Port of Cape Charles Sustainable Technology Park used a US$400,000 economic adjustment implementation grant for the construction of infrastructure (i.e. roads, storm-water sewer lines, water mains, a pumping station and other assorted improvements) to allow for the development of phase 1 of the eco-park.

The Intervale Food Center and business incubator in Burlington, VT, obtained a US$1 million economic adjustment grant for the construction of a bio-shelter greenhouse and infrastructure for capturing and using waste heat from an electrical generating station through biomass gasification. The project is located within the Burlington Federal Enterprise Community.

In 1999, the Computer and Electronics Disposition Eco-Industrial Park (CEDEIP) in Austin, TX, used a US$200,000 economic adjustment strategy grant to perform feasibility and market analyses to develop a long-range business plan for an EIP in Austin. The park will focus on the re-use, recycling, remanufacturing, sale and ultimate disposition of electronic products. In 2001, CEDEIP obtained a second, US$320,000, economic adjustment grant to finalise the park's business plan and develop a master plan for implementation. The master plan will include: infrastructure design (water, waste-water, power, roads and so on); pre-engineering and environmental analysis of the proposed site; preliminary architectural work; site layout and planning; development of the park's management and organisational structure; and the development of bylaws, codes and covenants. This is an excellent example of phased EDA economic development support.

13.6.1.5 Revolving loan funds

EDA RLFs can provide 'gap financing' in instances where private financing is not otherwise available or available at competitive rates, and they can also leverage other private financing. Therefore, the benefits for EID projects are twofold: First, RLFs can provide financing for activities where funding would otherwise not be available. Second, private lenders may be more inclined to consider EID business investments if they are sharing the financial risk with a public loan pool. In addition, RLFs have the potential to provide long-term support for EID business development activities because, on repayment, principal and interest stay in the community for re-lending and to support continued economic development activity.

13.6.2 US Department of Commerce: National Oceanic and Atmospheric Administration[7]

The NOAA has a broad mission that includes weather forecasting, public natural disaster notification, protection of endangered ocean species and conducting scientific research to preserve the environment, including research on global climate change. A number of NOAA offices and programmes promote sustainable coastal development.

Like EDA, NOAA was an early supporter of EID. In fact, EDA and NOAA have signed a 'first-of-its-kind' Memorandum of Understanding (MOU) calling on these two bureaux from the US Department of Commerce (US DOC) to co-ordinate and jointly support their activities in relation to emerging national EID efforts.

13.6.2.1 Office of Ocean and Coastal Resource Management

NOAA's Office of Ocean and Coastal Resource Management (OCRM) provides technical and financial assistance to coastal and Great Lakes areas to develop and implement comprehensive coastal resource protection and management programmes. Technical assistance consists of dissemination of successful waterfront revitalisation and sustainable development efforts; financial assistance consists primarily of small grants for coastal revitalisation feasibility studies. In an EID context, OCRM grants might be used to support planning, engineering and design of EID projects located in coastal areas.

13.6.2.2 Office of Response and Restoration

NOAA's Office of Response and Restoration (ORR) provides technical assistance to assess and mitigate risks to coastal areas resulting from hazardous waste, oil and chemical spill impacts. In an EID context, ORR may be able to support assessment and mitigation plan development for EID development on brownfield sites in coastal areas.

---

7 Information about NOAA programmes is available at www.noaa.gov.

13.6.2.3 Office of Sustainable Development and Intergovernmental Affairs

In recent years, this office within NOAA has provided small grants for community workshops on a variety of sustainable development-related topics, including brownfield redevelopment and EID. Last year, for example, NOAA co-sponsored a workshop with the local chamber of commerce in New Bedford, MA. The workshop, titled 'The Bottom Line of Green is Black', focused on applied industrial ecology and EID.

13.6.3 US Department of Housing and Urban Development[8]

HUD's mission is twofold: (1) to provide housing assistance and affordable housing for low-income persons; and (2) to help economically disadvantaged communities to meet their development needs. HUD programmes are targeted to primarily benefit persons of low and moderate incomes.

Major HUD programmes with applicability to EID include:

- Community development block grants (CDBGs)
- Section 108 loan guarantees
- Economic Development Initiative (EDI) grants
- Brownfield Economic Development Initiative (BEDI) grants
- Empowerment Zones and Enterprise Communities
- Renewal Communities

13.6.3.1 Community development block grants

Each year the CDBG programme provides more than US$4.5 billion in federal formula grants to eligible (1) metropolitan cities and urban counties, known as 'entitlement communities'; and (2) states that administer awards of these grants to smaller communities and rural areas for use in revitalising neighbourhoods, expanding affordable-housing and economic opportunities and/or improving community facilities and services. Municipalities use CDBG funds to provide grants, loans, loan guarantees and technical assistance. To receive annual CDBG funding communities must develop and submit a consolidated plan that is based on citizen participation; in particular, involvement is required of residents from predominantly low-income and moderate-income neighbourhoods, from slum or blighted areas and from areas in which the grantee proposes to use CDBG funds. CDBG grantees have significant discretion in the allocation of funds, which may be awarded to public, private, non-profit and for-profit organisations. In an EID context, CDBGs might be used to support all types of activities, including community planning, deconstruction and materials re-use, construction of public facilities and improvements (i.e. infrastructure), activities relating to energy conservation and renewable energy resources and venture capital support for business development.

---

8 Information about HUD programmes is available at www.hud.gov.

13.6.3.2 Section 108 loan guarantees

HUD's Section 108 loan guarantee programme enables states and local governments (i.e. entitlement communities) that participate in the CDBG programme to borrow federally guaranteed funding for use in financing large economic development projects and other revitalisation activities. Security requirements for Section 108 guarantees require that grantees pledge future CDBG allocations as collateral. The Section 108 programme has objectives similar to those of the CDBG programme in that it requires that all activities benefit low-income and moderate-income persons, aid in the elimination or prevention of slums and blight or meet urgent community development needs. Section 108 loan guarantees are typically used for very large, sometimes community-wide, economic development initiatives. Eligible activities include: acquisition of real property; rehabilitation of publicly owned real property; construction of public facilities; related relocation, clearance and site improvements; payment of interest on the guaranteed loan and issuance costs of public offerings; and public works and other site improvements. In recent years, HUD has provided over US$1 billion annually in loan guarantee authority. In an EID context, Section 108 loan guarantees might be used to provide financing for large community development projects such as land acquisition, green building construction, eco-industrial park development and multi-modal transportation facilities.

13.6.3.3 Economic Development Initiative

HUD's EDI awards competitive grants to eligible communities participating in the CDBG programme in conjunction with a federal loan under the Section 108 loan guarantee programme. EDI grants can be used to pay for direct project costs or they can be held as loan reserves to enhance the feasibility of large economic development and revitalisation projects financed through Section 108. EDI provides communities with a way to reduce their CDBG risk by providing additional financial support for Section 108 projects. In an EID context, these grants might be used to help improve the financial feasibility of eco-industrial projects that have secured Section 108 funding.

13.6.3.4 The Brownfield Economic Development Initiative

The BEDI is a part of the EDI programme that provides funding specifically to finance the clean-up and redevelopment of environmentally contaminated industrial and commercial sites. As with EDI grants, BEDI grants must be used in conjunction with loans guaranteed under the Section 108 programme. These grants might be used to help improve the financial feasibility of EID projects that are located in brownfield-impacted areas and have successfully secured Section 108 funding.

13.6.3.5 Empowerment Zones and Enterprise Communities

EID projects located in federal empowerment zones and enterprise communities (EZ/EC) are eligible for several types of tax incentive (USDT 1999). These incentives fall into two broad categories: wage credits, and support for business expansion. EZ/EC wage credit programmes include an empowerment zone employment credit, Welfare-to-Work credit, a work-opportunity tax credit and an Indian employment tax credit. Business expansion programmes include enterprise zone facility bonds, an increased Section 179 deduction (a US tax code provision) and an environmental clean-up cost deduction for brownfields (HUD 1999). Since 1994, 85 urban areas and 38 rural areas have been competitively selected and designated as EZ/ECs.

13.6.3.6 Renewal Communities

In the near future, HUD will designate up to 40 Renewal Communities in areas of pervasive poverty, unemployment and general distress—at least 12 of which must be located in rural areas. This designation will provide special tax incentives for the period from January 2002 to December 2009. The tax incentives will include a 0% capital gains rate, a renewal community employment credit, a commercial revitalisation deduction, additional Section 179 expensing and an extension of the work-opportunity tax credit (HUD 2001). In an EID context, Renewal Community incentives might be used to leverage ongoing EID business and community development activities, thereby improving a project's overall chance of success.

In the case of mixed-use EID projects, a number of HUD housing-related programmes and initiatives may be applicable. These include: provision of subsidised public housing for low-income individuals and families; HOME Investment Partnership Program block grants that develop and support affordable housing for low-income residents; Federal Housing Administration mortgage and loan insurance that supports home ownership; and Section 8-subsidised rental certificates and vouchers for low-income households.

13.6.4 US Environmental Protection Agency [9]

US EPA's mission is to protect human health and the natural environment. To accomplish this mission, it implements environmental protection programmes across a variety of media, including air, water and hazardous waste as well as specific programme areas such as those involving underground storage tanks, use of pesticides and pollution prevention.

Significant US EPA programmes and initiatives relevant to EID include:

- Brownfields Economic Development Initiative
- Office of Policy, Planning and Evaluation
- Environmental Finance Programme

13.6.4.1 Brownfield Economic Redevelopment Initiative[10]

EPA defines brownfields as abandoned, idled or under-used industrial and commercial facilities where expansion or redevelopment is complicated by real or perceived environmental contamination. US EPA's Brownfield Economic Redevelopment Initiative (hereafter referred to as the Brownfield Initiative) is an effort to help communities revitalise such properties, both environmentally and economically, to mitigate potential health risks and to restore economic vitality to areas where brownfields exist.

9 Information about US EPA programmes is available at www.epa.gov.
10 Information about US EPA's Brownfield Initiative programmes is available at www.epa.gov/brownfields.

Several Brownfield Initiative grant programmes may support EID. These include:

- Brownfield assessment demonstration pilots
- Targeted brownfield assessment
- Brownfield Clean-up Revolving Loan Fund demonstration pilots
- Job training and development demonstration pilots
- Brownfield tax incentive

Brownfield assessment demonstration pilots
US EPA's brownfield assessment demonstration pilot grants (funded up to US$200,000 over two years) are designed to test assessment models and facilitate co-ordinated assessment and clean-up efforts at the federal, state, tribal and local levels. These funds are typically used to increase local capacity to address brownfield redevelopment issues. Assessment pilots may be used: to assess, identify, characterise and plan response activities; for public participation and community involvement activities; and for legal, planning and economic studies necessary to advance site assessment and clean-up, among other needs. An important component of the assessment pilot is planning for the long-term, sustainable re-use of brownfields. EID projects located in brownfield-impacted areas have successfully leveraged EPA assessment pilots, particularly to support sustainable redevelopment planning and community participation activities related to site re-use.

Targeted brownfield assessment
US EPA's targeted brownfield assessment programme is designed to help states, tribes and municipalities—especially those without EPA brownfield assessment demonstration pilots—to minimise the uncertainties of contamination often associated with brownfields. Under this programme, EPA provides funding and/or technical assistance for environmental assessments at brownfield sites throughout the country. A targeted assessment may encompass one or more of the following activities: a screening (phase 1) assessment, including a background and historical investigation and a preliminary site inspection; a full (phase 2) site assessment, including sampling activities to identify the types and concentrations of contaminants and the areas of contamination to be cleaned; and establishment of clean-up options and cost estimates based on future uses and redevelopment plans (US EPA 1998b).

In an EID context, a targeted assessment might be used to offset other costs associated with project development or implementation. In addition, as noted in Section 13.2.7, contamination, or a perception of contamination, can act as a significant barrier to lender and developer interest in an EID project. A targeted assessment might be used as a vehicle to put realistic bounds on the financial risk associated with a particular EID project.

Brownfield Clean-up Revolving Loan Fund demonstration pilots
US EPA's Brownfield Clean-up Revolving Loan Fund (BCRLF) demonstration pilots provide grants to capitalise RLFs for brownfield clean-up activities. BCRLF awards must be used only for environmental response activities and cannot be used for site redevelopment.

BCRLFs carry interest on loans (generally at low or below-market interest rates) that are captured through loan repayments and then relent for other brownfield clean-up activities. In an EID context, BCRLF pilots can provide badly needed funds for remediation in cases where brownfield clean-up is involved.

US EPA's Clean-Water State Revolving Fund (CWSRF) can also be used to finance the remediation of brownfields that affect surface water and groundwater quality. There are currently 51 CWSRFs—one in each state and in Puerto Rico—that provide low-interest or no-interest loans for water-quality projects, over terms as long as 20 years. Eligible uses include brownfield mitigation to prevent or correct water-quality problems, including abatement of polluted run-off, control of storm-water run-off, correction of groundwater contamination and remediation of petroleum contamination (US EPA 1998a). Anyone seeking to undertake EID projects that face brownfield remediation and/or water-quality issues should explore the applicability of this state-administered programme to their project financing needs.

Job training and development demonstration pilots
US EPA's job training and development demonstration pilots are designed to provide environmental employment training for residents in communities impacted by brownfields. This programme attempts to link meaningful training opportunities for economically disadvantaged populations with jobs created as a result of brownfield redevelopment activities. Job training pilots of up to US$200,000 are awarded over a two-year period. Among others, eligible entities include municipalities, colleges, universities, community job training organisations and non-profit training centres. EID projects have focused on the linkages between environmental protection, economic development and worker training. As noted in Section 13.2.6, the Green Institute in Minneapolis, MN, has implemented a very successful deconstruction worker training initiative as part of its EID activities. Brownfield job training pilots might be used to achieve similar results in communities affected by brownfields.

Brownfields tax incentive
Originally signed into law in August 1997, the Taxpayer Relief Act included a tax incentive to spur the clean-up and redevelopment of brownfields in distressed urban and rural areas. Under the brownfields tax incentive, environmental clean-up costs are fully deductible in the year in which they are incurred rather than having to be capitalised over time. The original tax incentive 'sunset' after three years but was modified and extended to 31 December 2003. Eligible properties must meet specific land-use and contamination requirements (US EPA 2001). In an EID context, the brownfield tax incentive is another financial tool that may be applicable to EID projects that are facing clean-up costs associated with the re-use of a brownfield site. Project developers should look for opportunities to utilise this and other tax incentives, such as the historic rehabilitation tax credits and EZ/EC tax advantages (see Section 13.6.3.5), as part of comprehensive financing packages for their EID projects.

13.6.4.2 Office of Policy, Planning and Evaluation

US EPA's former Office of Policy, Planning and Evaluation (OPPE) developed two prototype software models to support EID. Facility Synergy Tool (FaST) is a search-and-match database that identifies opportunities for materials, energy and water exchanges.[11] Designing Industrial Ecosystems Tool (DIET) is a decision optimisation model to identify the optimum mix of industries needed to achieve the goals of EID (cost savings, pollution reduction and job creation).[12] These tools can be used both in the design and development phase of EID to sell the project to communities and developers and in the financing phase to demonstrate cost savings to potential financiers.

13.6.4.3 Environmental Finance Programme

US EPA's Environmental Finance Programme helps communities identify creative approaches to fund environmental projects, including one-time capital investments and ongoing operation and maintenance (O&M) costs. The programme uses the expertise of the Environmental Financial Advisory Board (EFAB), university-based environmental finance centres (EFCs) and the Environmental Financing Information Network (EFIN) to research, provide technical assistance and disseminate information on innovative environmental financing techniques.[13]

Environmental Financial Advisory Board
EFAB is a federally chartered advisory committee consisting of a diverse group of independent financing experts from the public and private sector. It produces policy and technical reports on a wide range of environmental finance matters.

Environmental finance centres
The EFC network is a university-based programme that provides financial outreach services to the public and private sectors. The EFC network currently includes: the University of Southern Maine, Syracuse University, University of Maryland, University of North Carolina at Chapel Hill, University of Louisville, Cleveland State University, University of New Mexico, California State University at Hayward and Boise State University. Among other activities, the EFC network functions as a national clearinghouse for environmental finance-related information, provides workshops on financing mechanisms for local officials and conducts analyses on financing alternatives. The EFCs are an invaluable resource to EID communities and developers seeking solutions to the difficult questions raised by how to pay for the community's vision.

Environmental Financing Information Network[14]
EFIN is an electronic outreach service that provides access to environmental financing information and publications generated by US EPA offices (including from the programmes mentioned above) and by independent sources.

13.6.4.4 Office of Environmental Education

US EPA's Office of Environmental Education works to foster an environmentally conscious public. Its goal is to support informed environmental decision-making. To this

11   www.epa.gov/otaq/transp/modlmeth.pdf, page 4.
12   Ibid.
13   See US EPA's Environmental Finance Programme homepage, at www.epa.gov/efinpage.
14   Information about EFIN is available at www.epa.gov/efinpage/efin.htm.

end, US EPA's Environmental Education Grant programme awards grants 'that enhance the public's awareness, knowledge and skills to make informed decisions that affect environmental quality'.[15] Eligible applicants include state and tribal governments and non-profit organisations, particularly those with environmental and/or education related missions, such as colleges, universities and education associations. Approximately US$2–3 million is available for this programme each year; requests under US$25,000 are evaluated by US EPA's ten regional offices, and those above US$25,000 are evaluated by headquarter staff in Washington, DC. At least 25% of the total project costs must be paid for by non-federal matching funds. US EPA's environmental education priorities are articulated each year in the programme's annual grant solicitation notice. Past programme priorities have included regional capacity-building, models for communities and environmental justice issues. EID practitioners might use US EPA environmental education grants to develop a process to educate the public about the environmental ramifications of competing economic development proposals or methods and to disseminate a model for the types of environmental considerations that are part of the EID development framework.

13.6.5 US Department of Energy[16]

The DOE's mission is to foster a secure and reliable energy system that is environmentally and economically sustainable, to be a responsible steward of the nation's nuclear weapons, to clean up DOE facilities and to support continued US leadership in science and technology.

13.6.5.1 Office of Energy Efficiency and Renewable Energy

DOE's Office of Energy Efficiency and Renewable Energy (EERE) addresses the energy concerns related to five of the nation's principal sectors: industry, utilities, transportation, buildings and federal facilities. EERE provides technical assistance and/or small grants for research, development, demonstration and commercialisation of innovative energy-related technologies.

Several EERE offices and initiatives may support EID including:

- Office of Industrial Technologies
- Office of Power Technologies
- Office of Transportation Technologies
- Office of Building Technology, State and Community programmes
- Centre of Excellence for Sustainable Development
- Energy Efficiency and Renewable Energy Network
- Brightfields

---

15   Information about US EPA's Environmental Education Grant programme is available at www.epa.gov/enviroed/grants.html.
16   Information about DOE programmes is available at www.energy.gov.

Office of Industrial Technologies

The Office of Industrial Technologies (OIT) 'creates partnerships among industry, trade groups, government agencies and other organisations to research, develop and deliver advanced energy efficiency, renewable energy and pollution prevention technologies for industrial customers'.[17] OIT focuses on nine energy-intensive and waste-intensive industries: agriculture, aluminium, chemicals, forest products, glass, metalcasting, mining, petroleum and steel. OIT administers two financial assistance programmes that are relevant to EID efforts: the National Industrial Competitiveness through Energy, Environment and Economics (NICE3) programme and the Inventions and Innovations (I&I) programme.

National Industrial Competitiveness through Energy, Environment and Economics programme[18]

OIT administers the NICE3 grant programme jointly with US EPA. This programme is a public–private cost-share initiative that promotes energy efficiency, clean production and industrial competitiveness. State and industry partnerships are eligible to receive a one-time grant of up to US$525,000. NICE3 grants 'support technology development that can improve industrial cost-competitiveness, prevent pollution, conserve energy and reduce industrial waste'.

Inventions and Innovation programme[19]

OIT administers the I&I programme jointly with the US DOC. The I&I programme provides financial, technical and commercialisation assistance for innovative energy-saving inventions. It funds individual projects at two levels—up to US$40,000 or up to US$100,000, depending on the stage of development—for preliminary ideas and inventions related to saving energy. This programme targets specifically OIT's nine industrial sectors but also includes more general categories such as industry, power, transportation and buildings.

Industrial Assessment Centers[20]

OIT oversees the Industrial Assessment Center programme, which provides no-cost industrial assessments and energy audits to small and medium-size manufacturers to help them identify ways to improve productivity, reduce waste and save energy. Eligible firms are those with fewer than 500 employees, annual sales below US$100 million and energy bills of more than US$100,000 and less than US$2 million.

Office of Power Technologies[21]

The Office of Power Technologies (OPT) develops clean, competitive power technologies, including renewable energy (solar, wind, geothermal and biomass), energy storage, hydrogen and superconductors. The goal of this office is to use power technologies to lower energy costs, reduce greenhouse gas emissions and pollutants and to improve the reliability of service. OPT facilitates technology transfer between research and development (R&D) programmes and industrial end-users that may be beneficial to EID projects.

---

17 See the OIT website at www.oit.doe.gov.
18 NICE3 programme and application information is available at www.oit.doe.gov/nice3.
19 I&I programme and application information is available at www.oit.doe.gov/inventions.
20 Industrial Assessment Centre programme information is available at www.oit.doe.gov/iac.
21 Information about OPT is available at www.eren.doe.gov/power.

Office of Transportation Technologies [22]

The Office of Transportation Technologies (OTT) supports research in advanced power sources, alternative fuels, bio-based resources, fuel cells and hybrid vehicles. OTT also administers the Clean Cities programme, which facilitates voluntary public–private partnerships to support development and use of alternative-fuel vehicles (AFVs). Technical assistance provided through this programme may be of particular relevance to virtual eco-industrial parks that function based on regional by-product exchanges and that may have greater transportation needs than co-located businesses in an industrial park setting.

Office of Building Technology, State and Community programmes

The Office of Building Technology, State and Community (BTS) works in partnership with industry and government to make buildings more efficient and affordable. In addition to its research, technology-transfer and technical assistance roles, BTS awards small, targeted grants to states and communities to support their efforts to promote increased energy efficiency and the use of renewable energy resources.

State Energy Program [23]

BTS administers the State Energy Program (SEP), which 'provides funding to states to design and carry out their own energy efficiency and renewable energy programs'. The SEP provides funds to states, which then support local projects related to energy efficiency and renewable energy resources. There are two types of grants available to states through this programme: formula grants, and special project grants, awarded on a competitive basis. As one example of an EID that benefited from such assistance, the Cape Charles, VA, Sustainable Technology Park secured grant funds from the Virginia SEP to support its rooftop photovoltaic (PV) system, which currently produces over 50 kW of solar power.

Center of Excellence for Sustainable Development

The DOE's Office of Energy Efficiency and Renewable Energy is a significant resource on sustainable development and has sponsored the Center of Excellence for Sustainable Development to help communities address development concerns. The Center's website[24] is a valuable resource on all manner of topics related to sustainable development, including financing, land-use planning, transportation, green buildings and sustainable businesses.

Energy Efficiency and Renewable Energy Network

DOE's Energy Efficiency and Renewable Energy Network (EREN) is a resource for a variety of energy efficiency and renewable energy resources. Included within EREN is a web page[25] on energy efficiency and renewable energy financing resources, divided into the following sectors: homeowners, small businesses, industry, utilities, state and local programmes, federal buildings and international.

---

22 Information about OTT is available at www.ott.doe.gov.
23 SEP information is available at www.eren.doe.gov/buildings/state_energy/about.html.
24 www.sustainable.doe.gov
25 www.eren.doe.gov/financing

Brightfields[26]
DOE's Brightfields initiative supports incorporation of solar and other renewable energy technologies into the re-use of brownfields and vacant industrial properties. The goal of this programme is to improve air quality, to increase energy capacity and reliability, to foster local economic development and to support the sustainable re-use of brownfields. DOE provides training and technical assistance in financing, land-use planning and technology considerations to support local Brightfields development efforts. In addition, OPT (see above) recently announced small Brightfields grants (between US$30,000 and US$50,000 each) to three communities. This programme has considerable relevance to EIDs located on brownfield sites. Communities pursuing EID should avail themselves of the resources and technical assistance available through this innovative programme.

13.6.6 US Department of Agriculture[27]

Within the USDA, the Rural Development mission area is the most applicable to EID financing efforts. The USDA Rural Development mission is to improve the economy and quality of life in rural America. USDA financial programmes support such essential public facilities and services as water and sewerage systems, housing, health clinics, emergency service facilities and electric and telephone services. USDA promotes economic development by supporting loans to businesses through banks and community-managed lending pools. In addition, USDA provides technical assistance to communities undertaking community empowerment programmes, through the federal EZ/EC initiative (see Section 13.6.3.5).

The Cape Charles, VA, Sustainable Technology Park secured a grant of nearly US$1 million from USDA Rural Development to support infrastructure at the park.

USDA Rural Development programmes applicable to EID projects include:

- Rural Business Co-operative Service: business enterprise grants
- Rural Utilities Service: water and waste disposal system grants

Rural Business Co-operative Service: business enterprise grants
USDA's Rural Business Co-operative Service business enterprise grants promote industrial development, business expansion and job-creation activities in areas with a population under 50,000. Grants can be used for industrial development, business working capital and capitalisation of RLFs, among other activities. Typical projects include water and waste-water infrastructure, land acquisition, building rehabilitation, transportation facilities and energy and telecommunications utility extensions.

Rural Utilities Service: water and waste disposal system grants
The water and waste disposal system grants from USDA's Rural Utilities Service (RUS) provide assistance for rural water and waste disposal needs. Eligible applicants include local governments, non-profit organisations and Indian tribes. Grants may be used for

---

26 Information about Brightfields is available at the EREN website, at www.eren.doe.gov/brightfields.
27 Information about USDA programmes is available at www.usda.gov.

all manner of water and waste-water infrastructure as well as for solid waste facilities, but they must service areas with fewer than 10,000 people. In an EID context, RUS grants can be used to support water and waste-water infrastructure for development projects; in addition, the solid waste eligibility 'hook' might be exploited to fund innovative EID by-product exchanges. RUS grants can be leveraged with similar assistance from EDA and HUD, further strengthening EID proponents' financial packages.

13.6.7 US Small Business Administration[28]

The mission of the US Small Business Administration (SBA) is to aid, counsel, assist and protect the interests of entrepreneurs and small businesses. To accomplish this mission, SBA guarantees loans and provides business development assistance. SBA administers its programmes through district offices throughout the USA. SBA provides additional services through small business development centres (SBDCs), business information centres (BICs), tribal business information centres (TBICs), US export assistance centres (USEACs), women's business centres (WBCs) and one-stop capital shops (OSCSs) at various locations throughout the USA.

SBA programmes relevant to EID include:

- Section 7(a) loan guarantees
- Section 7(m) MicroLoans
- Certified Development Company (504) loans
- HUBZone empowerment contracting

13.6.7.1 Section 7(a) loan guarantees

SBA's Section 7(a) loan guarantee programme, the agency's largest programme, provides short-term and long-term loans to eligible small businesses that cannot obtain financing on reasonable terms through normal lending channels. The SBA provides financial assistance through its participating lenders in the form of loan guarantees, not through direct loans. The agency does not provide grants for business start-up or expansion. Loans guaranteed under the programme are available for most business purposes, including purchasing real estate, machinery, equipment and inventory, or for working capital. The loans cannot be used for speculative purposes. The SBA can guarantee a maximum of US$1 million under the Section 7(a) programme. The guaranty rate is 80% for loans up to and including US$150,000, and 75% for loans over US$150,000. Maturity is up to 10 years for working capital, and up to 25 years for fixed assets.

SBA's Section 7(a) loan guarantee programme may be used to assist small businesses engaged in EID arrangements, including supporting pollution-control and energy-efficiency technology. According to the DOE, 'approved technologies for SBA loans and loan guarantees are solar thermal and electric systems (photovoltaics), energy-efficient products and services, biofuels, industrial co-generation, hydroelectric power and wind

---

28   Information about SBA programmes is available at www.sba.gov.

energy'.²⁹ SBA Section 7(a) loan guarantees may also be used to market EID tenants' products and services.

13.6.7.2 Section 7(m) MicroLoans

SBA's Section 7(m) MicroLoan programme provides short-term loans of up to US$25,000 to small businesses for working capital or for the purchase of inventory, supplies, furniture, fixtures and machinery and/or equipment. Proceeds cannot be used to pay existing debts or to purchase real estate. Loans are made through SBA-approved non-profit groups, currently located in 46 states. These lenders also receive SBA grants to provide technical assistance to their borrowers.

13.6.7.3 Certified Development Company (Section 504) loans

SBA's Certified Development Company (CDC; Section 504) loan programme provides long-term, fixed-rate financing to small businesses to acquire real estate, machinery or equipment to expand or modernise or to undertake construction for business expansions or renovations. Typically, at least 10% of the loan proceeds is provided by the borrower, at least 50% by an unguaranteed bank loan and the remainder by an SBA-guaranteed debenture. The maximum SBA debenture is generally US$750,000 but may go as high as US$1 million. Loan maturities of 10 and 20 years are available. Section 504 debentures can be used in conjunction with private lending to finance major fixed assets for EID projects.

13.6.7.4 HUBZone Empowerment Contracting[30]

SBA's HUBZone Empowerment Contracting programme encourages economic development in historically under-utilised business zones—HUBZones—through the establishment of federal contract award preferences for small businesses located in such areas.

13.6.8 US Department of Health and Human Services[31]

The mission of the US Department of Health and Human Services (HHS) is to protect health and provide human services to those most in need.

Several HHS programmes and initiatives may support EID, including:

- Agency for Toxic Substances and Disease Registry
- Office of Community Services
- Indian Health Service

---

29 See the web page on SBA loan programmes for energy-related businesses, on the DOE website: www.eren.doe.gov/consumerinfo/refbriefs/l113.html.
30 Information on the HUBZone Empowerment Contracting programme is available at www.sba.gov/hubzone.
31 Information about HHS programmes is available at www.dhhs.gov.

13.6.8.1 Agency for Toxic Substances and Disease Registry [32]

The mission of the HHS Agency for Toxic Substances and Disease Registry (ATSDR) is to prevent exposure (and adverse human health effects and diminished quality of life associated with exposure) to hazardous substances from waste sites, unplanned releases and other sources of pollution present in the environment. ATSDR implements its mission through public health assessments of waste sites, health consultations concerning specific hazardous substances, health surveillance and registries, response to emergency releases of hazardous substances, applied research in support of public health assessments, information development and dissemination, and education and training concerning hazardous substances.

ATSDR provides small grants (up to approximately US$50,000) to public health departments to support environmentally neutral development (i.e. development that does not cause health hazards for communities). These grants have frequently been used to support brownfield redevelopment activities but are equally applicable to EID projects, particularly those that involve brownfields.

13.6.8.2 Office of Community Services [33]

The HHS Office of Community Services (OCS) works with state, local and community organisations to provide human and economic development services that 'ameliorate the causes and characteristics of poverty and otherwise assist persons in need'. The OCS Division of Community Demonstration Programmes (DCDP) administers a Rural Facilities programme that annually provides competitive discretionary grant funds to assist rural areas with water and waste-water facility upgrades. OCS also competitively awards discretionary grants for business, physical and commercial development purposes under its Urban and Rural Community Economic Development programme. Grants are targeted to assist low-income populations and are awarded to non-profit community development corporations and community action agencies.

13.6.8.3 Indian Health Service

The HHS Indian Health Service (IHS) provides federal health services to Native North American Indians and Alaska Natives. This includes culturally acceptable personal and public health services. The IHS Office of Environmental Health and Engineering (OEHE) accomplishes this mission by, among other things, providing financial and technical assistance for development and operation of water, waste-water and solid waste facilities. Those setting up EID projects on American Indian reservations or Alaska Native villages that promote a healthy environment should explore the applicability of OEHE programmes to their project financing needs.

---

32 Information about ATSDR is available at www.atsdr.cdc.gov.
33 Information about OCS discretionary grant programmes is available at www.acf.dhhs.gov/programs/ocs/01comply/ocspgm.htm.

13.6.9 National Endowment for the Arts[34]

The National Endowment for the Arts (NEA) has a new grant programme, called New Public Works, to support national design competitions for public works projects. The goal of the New Public Works initiative is to support excellence in design in the public realm. In 2002, NEA will award up to US$1.25 million to support as many as 20 design competitions. Awards are targeted to the disciplines of architecture, urban planning, industrial design and landscape architecture. In selecting awardees, interdisciplinary reviewers strongly consider evidence of community and political support for the projects as well as the financial ability to carry them out. Reviewers also look for projects that are of 'national significance' and provide replicable approaches for other communities. Eligible applicants include tax-exempt non-profit organisations, units of state or local government and federally recognised American Indian tribes. The emphasis of the New Public Works programme on innovative design may be ideally suited to EID project activities.

13.7 State funding sources

Like the federal government, many state governments offer innovative financing approaches to meet local development needs. Almost as varied as the number of states themselves, these approaches include grants, loans and tax incentives in a variety of different forms. Many states have also started to support EID-type activities. State-based initiatives to support environmental protection, economic development, technology commercialisation and energy efficiency, among other things, can be leveraged to support EID development concepts. A comprehensive overview of applicable state programmes is beyond the scope of this chapter, but in this section I provide a few anecdotal examples that demonstrate types of state-based financing available for EID.

According to the Environmental News Network, New York State is creating a business park to promote the development and commercialisation of clean energy technologies (ENN 2001). Companies that locate in the Saratoga Technology Energy Park (STEP) will receive development assistance from the University of Albany's (US DOE-designated) Clean Energy Incubator Program and the New York State Energy Research and Development Authority (NYSERDA). Assistance will be provided through 'in-kind' services as well as funding opportunities to help small businesses reduce the risks associated with new technology development.

In Virginia, a variety of state-based tax incentives are helping to increase the attractiveness of the Cape Charles Sustainable Technology Park for business location and expansion. These include: Virginia Enterprise Zone incentives, including income tax credits, real property improvement credits and grants to businesses that hire zone residents; Solar Photovoltaic Manufacturing Grants, for manufacturers of photovoltaic panels; Recycling Equipment Tax Credit, for the purchase of machinery and equipment for processing recyclable materials; and the Clean Fuel Vehicle Job Creation Tax Credit, for

---

34 For more information, see the web page for the NEA New Public Works programme, at www.nea.gov/artforms/design/npwgrants.html.

manufacturers of components for use in clean fuel vehicles.[35] Similar state financial incentive programmes may be applicable to EID projects in other states.

As another example of state assistance, the Phillips Eco-Enterprise Center in Minneapolis, MN, was able to secure a US$1.5 million grant from the Minnesota State Legislature to support its EID project. Structured as a forgivable loan, the state stipulated that the non-profit Green Institute own the project, maintain a predefined tenant focus and create local jobs. If these conditions are met, the state will forgive the loan in 20% increments. This 'equity stake' provided important leveraging to attract other project financiers, including local private lending institutions.

Financing to support EID activities has also taken shape in provincial authorities such as support for various projects at the Burnside Industrial Park System in Nova Scotia, Canada. In Italy and Austria, along with some other European countries, provincial authorities and businesses are exploring applications of industrial ecology.

13.8 Local sources of funding

As noted previously, a variety of local revenue sources exist that might support EID projects. Two examples of particular relevance to EID are tax increment financing (TIF) and lease rate incentives. Some communities have used TIF to support their development projects. TIF is achieved through local government assessment of property values. TIF has traditionally been used to encourage development in areas with low property values. By developing a property, developers increase land values and tax revenue. The incremental difference in tax revenue between the original assessment rate and the new, higher, assessed rate is used to finance development. One tool that might be used by communities using TIF is to set aside a portion of revenue as credit enhancement to issue bonds for development of future phases of eco-parks.

The Cape Charles Sustainable Technology Park established a 12% lease rate incentive for companies exceeding the minimum sustainability (i.e. social, environmental and economic) criteria established in the Park's governing covenants and restrictions. Park rates are competitive with other new industrial facilities in the region and currently stand at US$7.50 per square foot per year, before performance incentives. It is hoped that these incentives will indirectly raise more local revenue to support development activities—from fees, taxes, private investment, etc.—than would have been realised in their absence.

13.9 Private financing

To date, there has been a disconnect between the goals of EID project developers—who have been focused on community objectives, such as environmental and economic

---

35 Information on state financial incentives at Cape Charles Sustainable Technology Park is available at www.sustainablepark.com/incentives.html.

development objectives—and EID financiers, who are focused on the profitability of the undertaking, including projected return on assets and revenue flows. Nevertheless, few projects are undertaken completely outside the private financing realm, few are without market-driven financial objectives, and few are justified solely on their relevance to attaining some greater 'public good'. Private sources of capital including debentures and equity investments have become the funding mainstay of real estate project developers. Furthermore, although typically more expensive than municipal bonds and government-sponsored loan programmes, private-sector commercial loans are among the most flexible and largest sources of investment capital for public and private borrowers.

The private sector brings nearly infinite resources to the table and, when properly targeted, these resources can be used to achieve the dual goals of public benefits and private profits. In fact, the reality is that only the private sector has the financial breadth and ability to make EID a 'mainstream' development technique. Although the public financing techniques discussed above can help to improve EID project viability, the public sector alone cannot ensure adequate and affordable funding for EID projects. Broad-scale EID development can be achieved only if public resources can be used to leverage much larger private-sector investments.

In their paper, 'Strategies for Financing Eco-industrial Parks', Jonathan North and Suzanne Giannini-Spohn (1999) argued that private financiers have been wary of EID projects because: they are uncertain how many secondary activities they may need to underwrite; there is a lack of precedence on which to base financial rates of return; they are concerned about restrictions (e.g. EID codes and covenants) that may impede their ability to resell or transfer a property should they acquire a financial interest; and existing firms may regard an EID project as a competitor and politically not support its financing. As discussed above, the public sector can help to encourage greater private-sector involvement in EID by: providing incentives, such as tax credits, to improve a project's cash flow and increase its financial viability; reduce finance costs, such as with interest rate subsidies; and reduce lender risk, such as with loan guarantees or targeted financial assessment tools. What lenders do not account for is that there may be additional benefits to EID above and beyond conventional development including:

- Fixed administrative costs that can be spread over a large group of people and businesses
- Energy and location (i.e. materials transportation) efficiencies
- Shared infrastructure and operation and maintenance (O&M) costs
- Reduction or elimination of by-product disposal costs
- Tax incentives (e.g. for EC/EZs, brownfields, energy efficiency, historical preservation)

A number of private financial institutions have a track record in supporting innovative projects, including EID. The following discussion, although not comprehensive, provides examples of some of the institutions that may look favourably on EID projects.

13.9.1 ShoreBank Corporation[36]

Founded in Chicago's South Shore neighbourhood in 1973, ShoreBank was the first community development bank in the USA. ShoreBank's mission is to increase economic development opportunities in traditionally under-served rural and urban areas. Today, ShoreBank operates in Chicago, Cleveland, Detroit, the Upper Peninsula of Michigan and in the Pacific Northwest. ShoreBank's investments target economically disadvantaged neighbourhoods and minority businesses. Since 1973, ShoreBank has made nearly US$900 million of capital available for community revitalisation efforts, including loans for housing rehabilitation and small-business development. EID projects located in economically disadvantaged neighbourhoods within ShoreBank's service areas would be competitive candidates for financing.

13.9.2 Sustainable Jobs Fund[37]

The Sustainable Jobs Fund (SJF) is a community development venture capital fund and a US Department of Treasury-certified Community Development Financial Institution (CDFI) which strives to create quality jobs in the recycling, remanufacturing and environmental sectors for economically distressed regions in the eastern USA. SJF will invest up to US$1 million in equity or subordinated debt and targets high-growth firms that provide quality entry-level employment: 'SJF focuses on firms that provide unique products and services that are sustainable—meeting the social and environmental needs of the present without compromising the future.' SJF is an excellent source of business capital for EID projects.

13.9.3 The Ford Foundation

The Ford Foundation's Leadership for a Changing World programme seeks to recognise, strengthen and support leaders and to highlight the importance of leadership in improving lives and bringing about positive change in US communities.[38] Each year, the programme recognises 20 leaders and leadership groups not broadly known beyond their immediate community or field. Awardees will receive US$100,000 over two years to support their work and US$30,000 for additional learning activities that advance their efforts. Leadership areas of interest include, but are not limited to, economic and community development, human rights, the arts, education, human development, sexual and reproductive health, religion, media, and the environment. Eligible awardees are non-profit charitable organisations with 501(c)(3) status. As noted in this chapter, leadership is essential to ensuring the success of EID efforts (see e.g. Section 13.1). Those setting up EID projects should look to foundation resources, such as the Leadership for a Changing World programme, as flexible sources of funding to leverage their efforts.

---

36   Information about ShoreBank corporation is available at www.shorebankcorp.com/main/index.cfm.
37   Information on the SJF programme is available at www.sjfund.com/livesite/index.cfm.
38   Information on the Leadership for a Changing World programme is available at www.leadershipforchange.org. Information about the programme can also be found on the Ford Foundation website, at www.fordfound.org.

13.9.4 The Home Depot[39]

The Home Depot has long been recognised as a private company that embraces corporate social responsibility and strives to have a positive impact on the communities in which its stores are located. To that end, Home Depot accepts grant applications in four principal areas: affordable housing, at-risk youth, the environment, and disaster preparedness and relief. Environmental grants focus on: forestry and ecology; green building design; clean-up and recycling; and lead-poisoning prevention. Grantees may request materials, cash, volunteer support or some combination thereof for their projects. Grants are accepted from eligible non-profit organisations on a rolling basis. Many corporate foundations have been contributors to local US EID projects such as local foundations in Chattanooga, TN, and the Twin Cities in Minnesota.

13.9.5 Private non-profit organisations

Private non-profit organisations have the ability to raise funds from a variety of sources for targeted objectives. For example, the City of Burlington, VT, has successfully turned to private non-profit organisations to carry out a variety of tasks that would otherwise have been left to the resource-constrained and understaffed City government. This includes the Burlington Community Land Trust, which advocates and protects low-income housing, and the Intervale Foundation, which manages a job-producing urban agriculture operation and has purchased and held land for a variety of 'public' purposes, including the proposed Riverside Eco-Park. Both of these organisations derive their revenue from private and public sources. This public–private revenue-sharing model may be applicable to other EID projects.

13.9.6 Corporate sponsors

Some EID efforts have successfully secured significant donations and in-kind contributions from corporations (e.g. vendors of design and construction projects) that recognise EID projects as an opportunity to showcase their products and services. The Phillips Eco-Enterprise Center in Minneapolis, MN, received close to US$400,000 of in-kind goods and services, including: salvaged building materials; design, construction and legal services; and partial funding for the rooftop garden, active daylighting and the ground source heating and cooling systems.

13.10 Policy recommendations

Although a number of public and private institutions provide sources of funds, securing financing for EID projects still can be a challenge. An important goal of EID practitioners is to 'level the playing field' with regard to access to capital between EID projects and

---

39 Information on Home Depot grants is available at www.homedepot.com.

conventional industrial projects. The following discussion offers policy recommendations to facilitate financing for this forward-looking, useful development model.

13.10.1 Efficient mortgage financing

The notion of a location-efficient mortgage (LEM) has been discussed in the context of controlling sprawl.[40] A LEM is a Smart Growth tool that makes loans more readily available to homeowners living close to public transportation. A LEM allows homebuyers to qualify for larger mortgages by taking into account the savings that they will realise by using public transport as opposed to cars. LEMs decrease traffic congestion and pollution, increase affordable housing opportunities and encourage use of public transportation. Financing has been subsidised by several groups, including Fannie Mae, the nation's largest private-sector provider of capital for low-income, moderate-income and middle-income home mortgages.[41]

Although this model may have some applicability to EID in and of itself, a variation on this model may provide an even more powerful financial incentive for EID—that is, production-efficient mortgages, lease rates and so on. Under this concept, eco-industrial entrepreneurs would borrow money at reduced or incentivised rates based on the fact that they are able to demonstrate lower operating costs and therefore a greater ability to repay compared with a conventional business. Similarly, local governments could provide lease rate incentives for going 'beyond compliance' when implementing environmental improvements, such as those being used in the Cape Charles, VA, Sustainable Technology Park.

Recommendation

> Generate a list of comparable options ('comparables') (analogous to the real estate comparables that homebuyers use when purchasing property) from a diverse mix of representative EID projects to assess and demonstrate the benefits of this development strategy relative to conventional industrial development.

13.10.2 Tax incentives for technology development and commercialisation

As noted previously, development and diffusion of advanced environmental technology are key aspects of the EID strategy (Section 13.3.2). Environmental technologies allow companies to pursue no-waste or low-waste industrial processes, as they gradually move along a continuum from waste management, to recycling, to pollution prevention and, ultimately, to industrial ecology (i.e. in-stream recovery and re-use of materials). In the energy arena, technologies help facilitate low-emission or no-emission energy production that may be more efficient and reliable than more conventional polluting energy sources and help avert global climate change. Coupled with other incentives that have

---

40 For more information about LEMs, see the website of the Center for Neighborhood Technology, at www.cnt.org, and the website of the Institute for Location Efficiency, at www.locationefficiency.com.
41 For more information, see the website for Fannie Mae at www.fanniemae.com.

been used to support EID, such as the Brownfields Redevelopment Tax Incentive and EZ/EC tax credits (see Section 13.6), new federal tax credits for environmental and energy technologies would provide strong support for this rapidly emerging, environmentally benign, development model.

As noted previously, the US DOE's EERE (see Section 13.6.5.1) has supported the development, demonstration and deployment of advanced energy technologies in the EID context. Examples include: the biomass power facility at the centre of the Intervale Food Center in Burlington, VT; the solar cells and wind turbines being manufactured and used at the Cape Charles, VA, Sustainable Technology Park; and the geothermal heating and cooling system being used at the Phillips Eco-Enterprise Center building in Minneapolis, MN.

Recommendation
> Establish a federal tax incentive to support the development and commercialisation of advanced environmental and energy technologies.

13.10.3 *Procurement*

Federal, state and local governments—through their acquisition of goods and services—have an enormous ability to influence markets. In the past, this influence has been used to promote specific economic and social objectives. Government purchasing has been used to support minority-owned businesses, to foster small-business development and to promote economic growth in low-income areas.

Only recently, environmental considerations have taken a more prominent role in procurement decision-making. For example, government policies targeted at use of recycled content paper, energy efficiency (e.g. Energy Star® appliances),[42] water conservation and use of renewable resources help sustain 'environmentally sound' industries. As the next logical step in this progression, government can use its purchasing power to promote firms within the EID sector—firms that have voluntarily elected to operate in an environmentally benign and socially and economically progressive manner.

The growing interest in procurement policies has not been limited to the public sector. A recent report by the US EPA's Environmentally Preferable Purchasing (EPP) programme profiles the growth in environmentally preferable purchasing by several private-sector companies. In its report on EPP, US EPA (1999a: 1) notes:

> Although environmental purchasing is a new concept for many companies, others are beginning to solve some of the challenges encountered when incorporating environmental considerations into purchasing decisions. Through a variety of environmental and cost-saving initiatives—design for environment, greening the supply chain, full-cost accounting, zero-waste initiatives, ISO 14000 certification, environmental accounting and others—private-sector companies are identifying, manufacturing and purchasing 'green' products and services.

42 www.energystar.gov

The profiled companies have discovered that, in addition to traditional measures used when making purchasing decisions (e.g. cost, availability, performance), use of alternative environmental considerations may also promote the company's financial objectives.

Eco-industrial firms should compete for multi-year government contracts—a source of future revenue that would clearly appeal to prospective investors. The US Department of Defense is currently exploring the possibility of targeting its extensive procurement needs to assist small and disadvantaged businesses located on American Indian reservations that are exploring EID activities.

Recommendation
> Develop explicit federal, state and local government procurement policies and procedures that, whenever possible, encourage the procurement of goods and services from firms that have elected to operate within the fabric of EID arrangements.

13.10.4 Eco-industrial development revolving loan funds

Although a number of EID projects have successfully secured financing, EID is still at a disadvantage when compared with conventional development. Federal and state organisations, along with foundations, could capitalise a national EID RLF for both capital investment and business development. The fund could be administered by a national non-profit organisation in conjunction with a national lender that has a background in environmental or corporate social responsibility lending, such as the ShoreBank Corporation.

As an alternative, regional, local, or even project-specific RLFs could be created to support EID. Even small amounts of seed funding could grow quickly if some portion of all documented energy and other cost savings were captured by the RLF and then lent, and re-lent, for continual upgrades, retrofits and expansions. If necessary, this type of RLF could be initiated even without any start-up capital.

Recommendation
> Develop a partnership between interested foundations and government organisations to set up an EID RLF.

13.10.5 Eco-industrial development loan guarantee programme

As discussed previously (Section 13.9), one of the primary barriers to securing financing for EID is the inability of private-sector lenders to accurately and adequately assess financial risk. Lenders evaluate projects based on market risk, return on investment, available security and so on, and have even developed software packages to assist this process (WEI 2001). However, EID factors such as waste management and energy cost savings are not accounted for in these software models. Standardised, meaningful, financial risk assessment tools (e.g. metrics and 'comparables') are needed for EID projects. EID project financial transparency and independent assessments of completed projects will be important in convincing private-sector capital providers of the value of the model.

One way to address this barrier is to develop comparables as well as new metrics and models to assist lenders in developing a more accurate accounting of project benefits. Another way would be simply to eliminate the risk by offering EID loan guarantees. The SBA offers loan guarantees for a large number of different programme areas (see Section 13.6.7), and EID loan guarantee programmes could be modelled on these existing examples.

Recommendation

Develop and implement a federal EID-specific loan guarantee programme.

13.11 Conclusions

As this chapter demonstrates, EID has become a financially feasible alternative to traditional industrial practices. Past economic development strategies focused on ways to take advantage of industrial techniques developed as part of the late-19th-century Industrial Revolution. Today's economic development should promote sustainable strategies as part of the 21st-century eco-industrial revolution.

A vast array of programmes, tools and resources are already available in the USA to help finance EID efforts, and, with a few policy reforms, existing federal, state and local financing programmes can be given an EID 'spin' to further advance this emerging industrial development technique and put it on a level playing field with conventional development.

The challenge for community organisations, public officials and businesses alike is to educate themselves on how to best take advantage of these opportunities. This requires strong leadership, innovative partnerships and, above all, the ability and willingness to question traditional development conventions.

Web addresses

- Cape Charles, VA, Sustainable Technology Park, State financial incentives: www.sustainablepark.com/incentives.html
- Center for Neighborhood Technology: www.cnt.org
- Fannie Mae: www.fanniemae.com
- Ford Foundation: www.fordfound.org
- Leadership for a Changing World programme: www.leadershipforchange.org
- Foreign-Trade Zone Resource Centre: www.foreign-trade-zone.com
- Home Depot: www.homedepot.com
- Institute for Location Efficiency: www.locationefficiency.com

- National Endowments for the Arts, New Public Works programme: www.nea.gov/artforms/design/npwgrants.html
- ShoreBank Corporation: www.shorebankcorp.com/main/index.cfm
- Sustainable Jobs Fund: www.sjfund.com/livesite/index.cfm
- US Department of Agriculture: www.usda.gov
- US Department of Commerce
    - Economic Development Administration: www.doc.gov/eda
    - National Oceanic and Atmospheric Administration: www.noaa.gov
- US Department of Energy: www.energy.gov
    - Brightfields: www.eren.doe.gov/brightfields
    - Center of Excellence for Sustainable Development: www.sustainable.doe.gov
    - Energy Efficiency and Renewable Energy Network, financing: www.eren.doe.gov/financing.
    - Industrial Assessment Center: www.oit.doe.gov/iac
    - Inventions and Innovation: www.oit.doe.gov/inventions
    - National Industrial Competitiveness through Energy, Environment and Economics: www.oit.doe.gov/nice3
    - Office of Industrial Technologies: www.oit.doe.gov
    - Office of Power Technologies: www.eren.doe.gov/power
    - Office of Transportation Technologies: www.ott.doe.gov
    - State Energy Program: www.eren.doe.gov/buildings/state_energy/about.html
- US Department of Health and Human Services: www.dhhs.gov
    - Agency for Toxic Substances and Disease Registry: www.atsdr.cdc.gov
    - Office of Community Services: www.acf.dhhs.gov/programs/ocs
    - Office of Community Services discretionary grant programmes: www.acf.dhhs.gov/programs/ocs/01comply/ocspgm.htm
- US Department of Housing and Urban Development: www.hud.gov
- US Environmental Protection Agency: www.epa.gov
    - Brownfield Initiative Programs: www.epa.gov/brownfields
    - Environmental Education Grant Program: www.epa.gov/enviroed/grants.html
    - Environmental Finance Program: www.epa.gov/efinpage
    - Environmental Financing Information Network: www.epa.gov/efinpage/efin.htm
- US Small Business Administration (SBA): www.sba.gov
    - HUBZone Empowerment Contracting Program: www.sba.gov/hubzone
    - SBA loan programmes for energy-related businesses: www.eren.doe.gov/consumerinfo/refbriefs/l113.html

14
REAL ESTATE AND ECO-INDUSTRIAL DEVELOPMENT
The creation of value

Edward Cohen-Rosenthal
Work and Environment Initiative,
Cornell University, USA

Mark Smith
Pario Research, USA

In real estate it is a tired but true statement that the most important considerations for a buyer are 'location, location and location'. Behind this statement is a greater truth for the marketing and sales of industrial or commercial real estate that bears discussion. Location is a placeholder for the concept of **value**. The location helps create value for the buyer either for operations, for image or for other reasons. The goal of the developer is to enhance the value of property for the potential buyer in order to achieve higher returns. This leads to a win–win situation for the buyer and seller.

Eco-industrial development (EID) is an emerging approach in economic development circles. A score of communities are experimenting with new eco-industrial parks (EIPs), retrofitting existing industrial areas and/or creating virtual connections within a region. The President's Council for Sustainable Development (PCSD) has tagged these experiments as a wave of the future. One way of understanding this phenomenon is as a new dictum: 'value, value and value' as the best way to the best sales.

In this chapter we discuss three ways to ramp up the value of industrial and commercial siting that not only provide benefits to the immediate participants in the deal but also have advantageous community and environmental outcomes.

The three areas are:

- Increased value through targeted marketing for location that meets market and material requirements

- Increased value through green buildings that preserve operating capital and increase productivity and property valuation

- Increased value through the creation of business networks that reduce operating costs, decrease inventory and space requirements and lead to market advantages

14.1 Finding and nesting the right kinds of investment: the value of place

Over the past five years, the Work and Environment Initiative (WEI) has used the model outlined in Figure 14.1 as the primary way to locate and identify potential business opportunity. It is based on the fact that business success is and always will be a factor of where an organisation sits within the larger ecology of its markets, access to materials and labour, technology and delivery systems. When all of these stars are lined up right then a company is very successful. When they are not, problems result that can require all sorts of adjustments to the deal and can lead to outright failure. This is a fancy way of saying what most brokers know: the client checks to see where there are roads and rail links around a location, whether there are suppliers and contractors who can reach the facility, whether there is an available labour force and so on.

We contend that location based *primarily* on tax incentives or other local 'baubles' is a signal of business trouble ahead. It is the business drivers that should come first. These being equal, then other items can be factored into the decision. We do understand that speculation and personal preferences do form a basis for many decisions, but a community cannot plan around these peccadilloes or possibilities when developing a strategy.

From a business perspective, the essential driver behind all successful location decisions is bottom-line profitability. Although various location factors may have different strengths, what counts is how they all tote up towards a successful bottom line. Embedding a business in a favourable and supportive environment leads to market success and sustainability.

Our approach says locations will be most successful in generating value when their marketing approach draws on the unique characteristics of an area and the unique advantages it can offer a prospect. Although the selling of shopping centres that are 'cookie-cutter' versions of each other has been a major source of real estate earnings in the USA and elsewhere, the numbers of failed locations and abandoned sites, and the competition among them has led to decreased value for some investors. The ability to target and differentiate in the market allows developers to focus more intensively and to attract higher-calibre clients.

From our perspective, markets are the primary driver of success for EID. It does not matter how cheap land is in Uzbekistan, if you cannot get your goods to the buyer in a reliable and affordable manner then it does not make a difference. Market access is a function of the amount of local demand for the product or service and the ability to access transportation routes that make local and/or broader markets accessible. In eco-industrial siting, the first ring of prospects are those with potential for just-in-time (JIT) delivery and strong customer–supplier relationships. It is preferable from a customer and

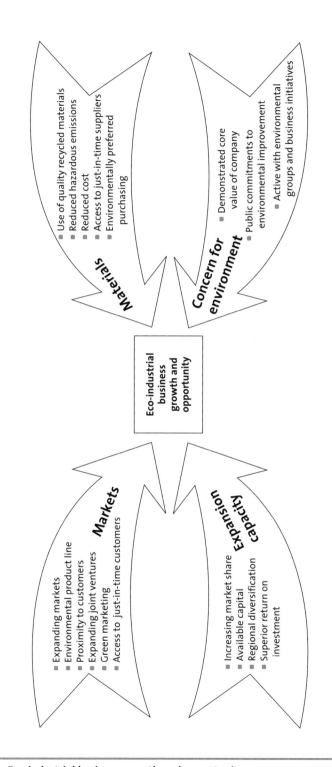

Figure 14.1 *Eco-industrial business growth and opportunity*

environmental viewpoint if there is a significant concentration locally. This reduces psychological distance to customers and lowers the environmental impact caused by longer supply lines.

Understanding the unique store of materials available in an area is also important. These materials are found as local natural resources and local production output, the industrial, commercial and residential waste-stream, and in dormant materials left in abandoned buildings and properties. These materials can be used as local input resources either in their current state, by being further refined (e.g. aluminium) or by breaking them down into constituent materials or parts that can be usefully reintroduced into a productive process. Ideally, these are requisite components of another process—either in a chemical state or as a product. There is a wide range of other possibilities, ranging from the remanufacturing of appliances, to producing local aggregate for roadfill, to links of energy that cascade from one place to another, to deconstruction of buildings and re-use of their fixtures and materials. To the degree that locally harvested resources are used, the local economy is improved. To the degree wastes that would have been externalised are productively recaptured, the local environment receives benefits.

From our perspective, all four of the factors outlined above—markets, expansion capacity, materials and concern for environment—can be sufficient reason for recruitment into an EID project. The more of the factors that have a strong connection in a locality, the stronger the possibility of a premium-value location there. Hence, market access alone could be the primary driver independent of environmental awareness. Yet those with a focus on the environmental market may find that location in an EIP would be a core concern based on values of the company or market image. Similarly, materials access could be the predominant driver. Every industrial real estate broker also understands that cash seeking a home or ambitions for market share growth can be the key factor in location decisions.

14.2 Increasing the value of real estate investment: the value of property

From the perspective of the real estate investor and user, an eco-aware approach to the property and built environment offers substantial opportunities to increase value. The real estate investor and user can realise greater value in many forms:

- Lower costs in operating buildings
- Higher real estate valuation because of higher net income
- Higher productivity of money invested in real estate assets
- Higher productivity from the workforce
- Lower costs associated with reduced workforce turnover
- Positive company image associated with ecodesign

Compared with traditional ways of doing business, many eco-business real estate opportunities are not yet well defined or documented. This is a new way of thinking for most people in real estate. In the past, the topic of the environment has generally meant increased cost and time—instead of cost savings, increased value and an increase in human health and productivity.

Often, eco-business opportunities suggest changes in the way business relationships are conducted. In most cases, boundaries that inhibit communication or economic linkages need to be spanned both physically and conceptually.

For the real estate manager, resource efficiency in building and business park operations is an important area for profitability. Energy efficiency is where many developers and users are rethinking the way they do business and are finding substantial cost savings. Maximisation of energy efficiency requires spanning specific aspects often disconnected in producing and operating real estate. Design, finance, building programming, occupant behaviour and lease structure all play an important part in how energy efficiency actually brings value to commercial real estate. The design of energy infrastructure to allow waste heat to be cascaded, the use of cooling towers, the use of co-generation, the creation of a link to solar fields or geothermal sources and/or the use of energy storage capacity to provide cost advantages are all possible ways to conceive energy interconnections. In buildings, climate-responsive design should be incorporated to reduce the need for resources to heat and cool buildings. Properly implemented, an energy-efficient building can also create a more pleasant and healthier physical space for workers.

The programming of buildings and their relationship to surrounding businesses is another prime opportunity to use EID to improve profitability. Eco-business principles encourage the optimal use of space relative to production programming needs and the economics of real estate. These two factors suggest that businesses in proximity to each other work together to optimise space utilisation, material efficiency and other inputs such as labour, transportation and information. Building and landscape design are important in realising savings. Experienced design teams can create savings by combining opportunities for climate, technology, materials, landscape and building programming. Eco-design and programming can change relationships between the buildings in a park and between buildings and the sites in which they are located. For an industrial or office park, eco-design and programming suggests resource sharing, which can be facilitated by proximity of shipping, receiving and storage facilities and by traffic flows and parking relationships. Equally important for the developer is the lowering of the cost for construction and maintenance of infrastructure through more compact design, innovative approaches and shared use.

In most cases, a well-designed eco-industrial approach saves money. To realise the savings, real estate investors and users need to share not only the processes of design and building but also the distribution of the gains. For example, energy-efficient design will save money for the party responsible for the energy bills, usually the occupant. To motivate the developer or investor to save somebody else's money, some portion of the savings should go to the developer. To accomplish this sharing, parties that have previously had different motivations enter a new dialogue. Leases can be restructured to distribute a portion of the energy savings to the developer or owner, resulting in higher net income to the developer or owner. When a developer builds an energy-efficient building to be sold, a higher building value can be calculated based on the higher net income allowed by energy savings. Building shells can be energy producers—thus recognising advantages from conservation and generation of energy. Neither of these allowances is part of mainstream real estate practice.

Change through green design and programming not only affects costs and revenues but also their timing. For example, there may be higher design fees incurred early in the building process to achieve a green building design, or construction costs may be higher, to be later offset by higher net operating income. For a company that builds and owns its own facilities, these changes reconcile favourably. However, when a developer of speculative space or a developer wishing to sell the completed building realises these changes in cash-flow timing, boundary-spanning discussions and new business relationships are required.

Labour and material efficiency can be enhanced by the sharing or pooling of resources used in company operations as well as by sharing building and park operation and maintenance. Overhead costs are also reduced. Capital can be conserved further, as many resources can be purchased only as needed instead of being incurred as a constant cost.

Several important changes create an environment where resources are shared. First, a philosophical commitment to eco-industrial principles should be a core business value as should willingness to make necessary changes. A company needs to be open to doing business differently. Second, the physical design and business programming of a business park changes. This is reflected in the site plan and architectural and engineering specifications.

Some eco-business benefits are less tangible, but they contribute important gains to business operations. For example, green buildings generally lead to a more efficient workplace because of better lighting, ventilation and a more functional design and programming. Less absenteeism and workforce turnover also results from these efficiencies. The superior working environment produced by green building design, as well as the company culture created by adopting eco-business principles, attracts and retains employees. These factors lead to a company's workforce being more productive, increasing return on labour.

When buildings and company operations reduce costs, an increased amount of capital is available for other key corporate priorities. The increased profitability available from eco-business principles in building design and programming and in company operations comes at a time when in business every aspect is being examined to increase shareholder value and to focus the company on its core competences.

For long-term value creation and later competitive advantage over less-well-designed buildings, eco-industrial approaches can yield higher resale returns after initial costs may already have been amortised, because the building owner can realise and offer buyers lower fixed operating costs. Further association with other higher-value firms should also increase the long-term value of the location and the property asset. Developers move beyond selling space to providing functionality, flexibility and service to the client. This added value can be translated into higher profits.

14.3 Archetype for the Quantum Connection Eco-Park™

Why do commercial and industrial concerns invest in more space than they need? Traditional approaches to real estate brokerage argue for urging the purchase of more space rather than less, within reason. Why is this? When return is linked solely to square footage or land purchased then the seller pushes for more from the buyer. When the metric is monetary return per square foot of functionality and value, a more intensive use of space is proposed. Also, business will often acquire more space than currently needed, in anticipation of growth. We have given thought to the primary reasons for current practice, which include demands for:

- Parking
- Inventory and storage
- Transportation access
- Amenities for food, meeting space, childcare, health facilities and so on
- Possible growth
- Buffer against future costs of designing, getting approval for and building for future space needs.

This results in built environments that require more construction than might be needed, thus adding to the initial cost of occupancy. It also costs more to condition and maintain that extra space.

We have instead developed the Quantum Connection Eco-Park™ prototype to address these issues. Figure 14.2 provides a schematic of the overall approach, which has enormous flexibility in terms of planning of structures, use of materials and access approaches. As a concept it reduces the need for asphalt by at least 66% over traditional design and provides a minimum of a 50% increase in leasable space.

By using consolidated inventory and transportation management at the primary entrance to handle large lorries and storage, the various participants in the park take much of their inventory off their books by having it handled by a company dedicated to this service whose job it is to find more cost-effective ways to ship, handle and store material. Smaller vehicles do most of the work for JIT deliveries in the park.

In a version of New Urbanist residential development applied to commercial and industrial applications, 'clusters' are formed in a more compact land-use pattern of various companies, based on their real needs. A common facility, the Business Connection Center, serves as the home for meetings, advanced communication, networking of businesses and employee amenities. This provides a staging area for interaction of the businesses in the park. They can use the space for common functions or locate shared services in this structure.

To deal with flexibility needs, a Quantum Connection Flex Building is part of the core infrastructure of the park. This multi-use facility serves several roles. It can be an incubator for start-up companies, a staging area for companies moving into the park, a

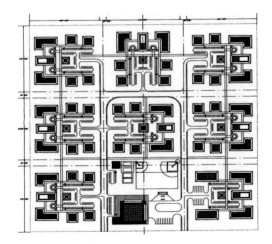

Figure 14.2 **The Quantum Connection Eco- Park™ prototype: schematic**
© 2002 Edward Cohen-Rosenthal and Mark Rodham Smith

location for overflow or seasonal needs or for research and mockup development, or for other uses, as required by tenants.

The builder constructs the spine of the Quantum Connection Eco-Park™ with parking, the Business Connection Center and the Quantum Connection Flex Building as well as the ring road. Potential tenants can begin in the flex building while construction is taking place on their new facilities. This conserves capital for the developer that might be sunk into speculative structures with assumptions about client needs and allows cash flow to be managed in line with known needs. An artist's impression is provided in Figure 14.3.

The net result for the occupant is more parking for each location, better eating, health, daycare, meeting and communications facilities, greater flexibility for growth and lower construction and maintenance costs. The environmental result is reduced air pollution, noise, greenhouse gas emissions and water run-off problems and a better physical environment.

As we said in Section 14.2, the developer needs to share in the benefits and increased value that result from this kind of design, hence the square-foot price should be adjusted to accommodate the increased value, thus delivering higher return per square foot to the developer and lower costs to the tenant at the same time. Instead of just selling space or collecting rent, the developer, through partnership with tenants, realises higher value.

14.4 Co-location and industrial symbiosis: increasing bottom-line performance

There are several ways the operation of an integrated eco-industrial park or connection helps the client realise long-term savings. Industrial ecology seeks to find ways in which

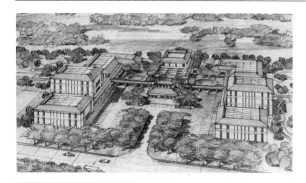

Figure 14.3 **The Quantum Connection Eco-Park™ prototype: artist's impression**
© 2002 Edward Cohen-Rosenthal and Mark Rodham Smith

by-products can become feedstock for other linked entities. In Kalundborg, Denmark, over the past 20 years an industrial symbiosis has developed that shares energy, water and other materials in ways that loop the materials back into productive use. Kalundborg is discussed in more detail in Chapter 16.

When waste energy is channelled and used by another user, such as the steam energy from Kalundborg, then energy costs can be reduced while improving air quality. Energy cascading is a possibility. When the gypsum that would need to be carted away from one location is used as part of the plasterboard factory nearby, both parties realise savings. In a number of situations in the USA such as Brownsville, TX, and Research Triangle, NC, these possible interconnections have been extensively mapped. With the data in hand, clusters of companies can look for profitable exchanges. In short, the effort is to do everything possible to reduce overall costs by looking for desirable exchanges of materials and energy.

Although these kinds of exchanges are more likely to occur in larger manufacturing situations, interaction that leads to business and environmental benefits are possible along a whole range of issues. Cornell's WEI has developed a model for potential integration teams within eco-industrial networks, where the participants seek to find mutual advantage through common efforts in the areas identified. In Chapter 11, methods that lead to savings in a whole variety of areas are described, including materials, energy, transportation, information systems, human resources, production technologies and in areas giving rise to other costs of doing business; in addition, these methods lead to increased revenue from joint venture and marketing opportunities. For potential buyers, there may already be glaring opportunities for cost reductions in the current configuration of an eco-industrial network—but their own capabilities and needs will also be factored into the potential mix and can serve as a magnet for recruitment as well. What they get in an eco-industrial network is a systematic scan to look for these opportunities to reduce costs and increase revenues, whereas traditional real estate is often unaware of these possibilities or does not build them beyond serendipity to factor into the locational decision.

Use of the network as a means to increase overall organisational success leads to higher value to the buyer. The precise set of benefits will vary from location to location and be influenced by the imagination and willingness to form partnerships of those involved. Unlike a large corporation that can mandate certain exchanges between subsidiaries, an eco-industrial network is based on voluntary interconnections that make business sense because of cost, market opportunities, common interests, quality, reliability and so on. These interconnections are based on criteria set by the parties involved and by specific agreements.

In short, the eco-industrial network creates its own value-added connections. Participants form those that make sense, demur from those that do not and create specific processes between themselves that mine, extract and use productively what is possible.

14.5 Conclusions

This bundle of benefits of eco-industrial parks and networks have real business bottom-line results for developers and tenants, but there is also another set of benefits that is less easy to quantify but that is very much a part of any real estate broker's decision process. By taking an eco-industrial approach to development and communicating benefits effectively, the time it takes to win approval from the community and various permitting agencies can be dramatically reduced—turning NIMBY ('not in my back yard') into enthusiasm for well-designed, community and environmentally conscious development. This has real financial implications in terms of time to market and cash requirements but also makes life easier for the developer and the potential tenants.

Value creation provides a means for all parties to benefit from EID. The broker is marketing high-value properties. The tenant sees increased value as the location choice becomes more valued and valuable through reduced operating expenses and enhanced image and through the higher value of the built structures and of the property itself. The community sees higher and better use of its land, with less pollution and higher tax valuations. What caps off the value chain is that we can profitably find ways to shift our development approach to a much more sustainable and life-enhancing way of doing business.

Appendix. Quantum Connections and Planned Property Densification: summary description

Quantum Connections is a plan for developing eco-industrial (and eco-business) parks. The authors of this chapter conceived this plan. Quantum Connections integrates five key characteristics:

1. Ecology and business integrated into site planning, with the unique tool of 'Planned Densification'*
2. Sustainable building design
3. Sustainable park management
4. Sustainable, continuous improvement in company operations
5. Sustainable relationships (such as by-product exchanges and resource sharing) and links to the local regional economy

A major goal with Quantum Connections is cost reduction and higher overall asset values. A growing number of companies recognise that the strength of the communities in which they locate has a direct and important impact on their total costs. These companies seek and support culture for its ability to improve academic skills, workforce skills, reduce crime and numerous social ills, and connect people for stronger community. With Quantum Connections, business can participate in the exploration and production of indigenous culture, and use the arts and culture as a prime venue for community and family activities. There is profound new research showing how this occurs, and this information can be beneficially used in revitalisation and connecting commerce and culture. Functional issues such as jobs-to-housing balance, affordability and commute times are important.

Regarding the market for green buildings and business parks, Pario is observing a remarkable integration of sustainability into mainstream business and manufacturing. A key indicator of the sustainable progress within business is the broad coverage of sustainability in 'The TFM Show 2003',† the largest manufacturing convention in the United States. There are over ten education modules dedicated to sustainable topics. In essence, mainstream businesses that occupy the buildings provided by the construction industry are becoming green. Ironically, the real estate development industry that provides businesses their space is falling behind the curve of innovation and adoption. In their operations, the businesses that occupy the space we build are surpassing the real estate industry's sustainable business performance as represented by the product we provide to them—as well as by our own processes and procedures.

---

* Somewhat related to these characteristics, Pario categorises 'performance levels' as: level 1, green buildings; level 2, site ecology preservation; level 3, green business requirements and green park management (which could include distributed solutions for resources such as energy and water); level 4, by-product exchanges (BPX) and flows, resource sharing and links to regional economy. Levels 3 and 4 incorporate key eco-industrial principles. These represent a progression of sustainable accomplishments from easier and more isolated to more difficult and integrated.

† See The TFM Show at Construct America 2003 at www.tfmshow.com/about_tfmshow.htm and the sustainable education modules at www.tfmshow.com/matrix.pdf. For five years, Pario has been anticipating that the mainstream manufacturing industry will adopt sustainable business methods aggressively. The reason is that, without calling it sustainability, manufacturers are uncompromisingly trying to reduce their costs and, to do so, reduce material use, disposal, and embrace technology and process change. Essentially, business is on its own track towards sustainability. However, the real estate and construction industry that builds their facilities is falling behind in sustainable implementation.

Planned Densification/Quantum Connections

Quantum Connections site-planning approach includes planning to add capacity to buildings, parking and other site functions, as needed. This 'planned densification' solves problems on a regional and property-specific level.

1. Issue

A productive way to reduce sprawl and increase productivity of land assets is to confront specific process barriers to increased density from a property development perspective. Then, see how those solutions can be translated into policy.

2. Problem structure

Higher-density development that reduces sprawl is often difficult to accomplish, especially in new edge development and initial revitalisation. The reason is the relationship between the market life-cycle and the property life-cycle. As a region develops, or as a neighbourhood redevelops, market demand initially supports lower-density, lower-value construction. Households and firms that move to edge locations expect lower prices, and regional real estate markets are usually structured in this manner. The same occurs with initial revitalisation—lower density and lower value is part of the economic equation as revitalisation begins and, indeed, is often the very catalyst for it. Because this economic circumstance changes, however, a problem exists. As edge areas grow and become more central, and as revitalising areas realise stronger demand, the highest and best use for properties changes, from both a regional planning perspective and a property-owner perspective. The problem is that the real estate asset, once designed, built, financed and occupied, has an expected life that in many cases substantially exceeds changing market demand.

Developers can generally only design and build for current or short-term market conditions. With most revitalisation and edge development, the relatively low cost expectation by users establishes a limit on revenue expectations and therefore construction cost and type.

Most developers would like to build at higher density, and most landowners would rather operate under higher density assumptions.

Because a market for higher density often occurs after initial construction, which is usually low-density, we should try to accommodate this demand and opportunity for increased financial and social productivity from land assets.

Usually, as growth occurs, a property in a well-serviced location can best function in the region by providing more density, increasing the functionality of associated civic infrastructure already in place—such as roads, utilities and transit systems. For example, the properties in a neighbourhood around a premium location such as a transit stop had best be allowed to densify in order to maximise the utility of the transit stop, and the return on its civic investment. Also, accommodating growth within an existing infill property and neighbourhood reduces additional edge growth requirements. But the inability to effectively create infill development fuels urban sprawl—costly both financially and socially. Effectiveness can be increased with better and more flexible zoning, design, ownership and finance structure—that is, anticipating and facilitating densification.

So, two things occur. Either a lower-density product gets built, casting in place a fixed asset that is soon functionally suboptimal, or a developer/owner holds property until the market supports more expensive construction (this pattern often 'freezes' development and evolution of the community). Either way, land often becomes out of sync with the community's needs. A more productive solution is to have the land and building evolve more closely with the market.

3. Mechanism for improvement

A major opportunity for better regional development is to plan property development in a flexible, evolutionary manner that allows for a property to naturally evolve with market economics. Planned densification aims to make increases in density a natural part of the property life-cycle, accommodating the market life-cycle changes around it. Typically, property life-cycle determinants are not matching market life-cycle circumstances. Planned densification suggests that, for appropriate properties and neighbourhoods, initial zoning and design anticipate additional square footage and function be added, as well as possible use changes, as the market evolves. There are many mechanisms to do this, including modular structures, structures that will accept additional square footage (such as additional storeys) and site areas targeted for densification (open space to be built upon, maybe with a roof garden, a parking lot that turns into a parking structure or building with a parking structure)—these and many other design possibilities that allow more capacity and function to be added.

Normally, site and building redevelopment is expensive, requires time-consuming and costly entitlement, and thus does not occur as efficiently as possible for the municipality or for the property-owner. When entire neighbourhoods are faced with this constraint, the neighbourhood's ability to change and utility to the region is reduced, and undesirable sprawl is encouraged.

Quantum Connections 'Planned Densification' suggests that selected areas be realised for their future 'highest and best' role in regional growth, and that infrastructure, zoning, property design and ownership allow for the property function to evolve with the market—meeting both short- and long-term needs effectively. Thus, a more resource-efficient, sustainable community will emerge. Asset values will be higher for the owners and municipal assets will have a higher return on investment.

From a financial perspective, construction costs may increase slightly, because of both inflation and additional costs of 'preparing' for additional density in initial construction. However, land value will increase to help finance the marginal costs, and the owner will gain value in the short and long term with mortgage finance leverage. Further, if 'external' costs can be captured—say, only a very small share of the 'edge subsidy' that is given to support sprawl—the costs may be mitigated. Essentially, the marginal funding that may be required for this change initiative is available if process participants can connect true cost and benefit. A public–private partnership should be able to accomplish this connection.

With a typical industrial park developing in an edge location or otherwise low-value market, initial development may be at a low FAR and a low-cost building targeted for tenants that expect low prices. As the area matures with more businesses, support retail, and residential communities in close proximity, the use of the industrial park can be intensified for everybody's benefit. Floor area ratios can be increased two to four times or more, which will bring a corresponding increase in the value of property, a strong motivator for investors and developers.

As with many other sustainable design ideas, there are some institutional barriers to address. One might be the suitability of traditional financing and terms for investments with which key cost and income relationships change. Another issue is the potential for increased first cost for some of the designs that provide the desired flexibility. Many flexible, modular and deconstructable building concepts are emerging that may address this issue.

Because planned densification should provide very substantial benefits to property-owners, municipal and regional stakeholders, there should be a group of people and organisations empowered to craft these specific and relatively minor solutions.

Pario feels that our industry needs to move towards more modular construction. The flexibility will increase utility and value. One of Pario's clients, BioSpace Development Company, is using modular construction for biotech labs.‡ And Workstage has the right approach with a modular chassis to which interior and exterior systems are attached.**

Dell makes my computers; they say 'made to order'. They show me pictures representing my computer being assembled part by part, to my specs. I can take it apart myself, install new components as I need new functionality or when something needs repair. Such modular construction instills flexibility, and ability to increase capacity as needed.

---

‡ See www.BioSpacedevelopment.com.
** www.workstage.com

When discussing this with a world-famous architect, he said: 'You know the problem with this, don't you? Architects will look at it and say "that's not how we do it".' I don't suggest that this change will occur with one project; rather, it is something our industry needs to work toward. However, a large community developer and the associated municipality could probably realise tremendous value gain by putting the ideas in place on a community scale, starting today.

4. Benefits

Flexibility is a primary benefit
With Quantum Connections, Planned Densification allows eco-industrial park planning areas to develop initially at densities that work with 'emerging' market and industrial park economics. Then, as the area matures, and as firms need to expand (which often occurs quite soon), there is ability above or adjacent to buildings for added capacity. With this functionality, firms can grow place, the project can mature with the market, and the urban form can mature with the needs of the region in a manner serving the region's best interest.

Increased value is an important incentive
The property-owner will realise a property value increase with the added density. Densification is triggered when the market will support the cost of the additional square footage. This is a common occurrence; however, because of the way we design and build space, and with traditional zoning, the ability to increase the asset's utility is restricted.

The owner is motivated for the change
Because of increased flexibility and increased value, the owner has the motivation to investigate the site-specific Planned Densification solution.

The municipality is motivated for the change
Just as with the owner, the municipality should realise greater value and greater return on investment for infrastructure, and is therefore a stakeholder in change. Municipal and government responsibilities, such as better air quality and better environmentally related health conditions, should result from reduced sprawl and therefore be a motivator for government support. Once Planned Densification is understood, resources can be allocated to supporting its adoption. Indeed, some of the subsidisation for sprawl can be much more productive with Planned Densification.

Pre-empt NIMBY-ism
By planning for density, people will know it is coming and have less logical ground for resisting it. Density increases will be part of a community plan and not a zone 'change' process. Planned infrastructure will also mitigate some of their key arguments that there is not sufficient infrastructure for additional development.

Reduce sprawl and increase regional land asset productivity
By understanding and accommodating the market, finance and stakeholder interests of development, Planned Densification can reduce the amount of greenfield land required for new development, and thus reduce sprawl. Reduced sprawl will save resources and lower costs, and increase productivity of scarce urban land assets while providing more opportunities for open space and preservation.

5. Actions

a. Identify stakeholders
Assess local circumstances, challenges, and solutions posed by Planned Densification. Determine who will benefit and enlist them as stakeholders/supporters.

b. Realise the cost of the problem
Determine who pays the cost of sprawl, and where externalities create opportunity for additional stakeholder roles.

c. Hence a reason for, and resources for, change
Local initiative leadership can use these and other relevant issues as reasons to create a compelling argument for change.

d. Determine mechanisms

- Zoning. Allow Planned Densification capacity and easy approvals.
- Design. Use the tools identified above.
- Marketing. Because of higher values, owners and municipalities can be enlisted.
- Finance/ownership. Create flexibility to capture the large value increases with property densification.

e. Implement
Grow programme and leverage skill investment gained from initial phase.

Comments

- Size may matter. There may be more flexibility with larger parcels and a larger number of parcels to use as tools.
- There are benefits from addressing on all size levels. All increased density will help.
- Urban and suburban relevance. Our metropolitan areas are realising record suburban growth while increasing inner-city revitalisation. As more development occurs with old patterns, opportunity is lost.
- Implications for style and materials

15
EVALUATING THE SUCCESS OF ECO-INDUSTRIAL DEVELOPMENT

Marian R. Chertow
Yale University, USA

At the end of the day, a critical question for any new practice, policy or world-view is: how do you measure success? This has been a confusing question in environmental studies, where villains can become heroes and heroes villains according to choices in system boundaries and human understanding of ecological conditions. DDT (dichlorodiphenyl-trichloroethane), for example, is clearly harmful to flora and fauna, and yet some African nations have decided to continue its use as a deterrent for a human scourge: the spread of malaria. Nuclear power, anathema to environmental advocates, is being touted as worth a second look, in part because of the lack of greenhouse-gas emissions in this form of power generation. The analysis paralysis around even small issues—the ponderous debate summarised by the words 'paper or plastic?'—illustrates the dearth of simple answers and the uncertainty of intuition.

Evaluation of eco-industrial developments (EIDs) has an even more complex cast. Beyond the conventional concerns of project economics and functionality, EIDs include environmental and community considerations that must also be analysed. To be successful, project benefits must outweigh the costs for each firm as well as for the development as a whole. An added responsibility for eco-industrial developers is to show that the newer form of organisation they envision is as good as or preferable to traditional ways of doing business. Tolerance for innovation comes and goes, so there is a great burden on proponents to show that, put under the analytic microscope, both in the planning stage and then in implementation, there are sound reasons and quantifiable benefits to balance the accompanying risks.

In many ways, the story of the industrial district of Kalundborg, Denmark—the most famous example of an industrial ecosystem (see Chapter 16)—is the perfect metaphor for those attempting to measure the success of EID. Until some high-school students did a science project in 1989 in which they made a scale model of all the pipelines and connections in their small community, the environmental aspects of the project went largely unnoticed.[1] Like the 2,000-year-old man who discovers women with the unforgettable Mel Brooks intonation, 'There's ladies here!', recognition of Kalundborg's environmental attributes was actually a revelation of what already existed rather than a new frontier. A critical measurement task is to know what we already have, even prior to taking on more complex challenges.

Since the coming to light of Kalundborg, there have been many more conscious attempts to create and develop projects incorporating the types of co-operation and exchange that justify the colourful name coined by the power plant manager in Kalundborg: 'industrial symbiosis'.[2] In the USA, the President's Council on Sustainable Development (PCSD) began an 'eco-industrial park project' using industrial symbiosis and other sustainability concepts in 1994. The PCSD defined an eco-industrial park as:

> A community of businesses that co-operate with each other and with the local community to efficiently share resources (information, materials, water, energy, infrastructure and natural habitat), leading to economic gains, gains in environmental quality and equitable enhancement of human resources for the business and local community (PCSD 1996c).

Parsing this definition reveals many objectives of these EIDs, including:

- Efficient sharing of resources
- Economic gains
- Gains in environmental quality
- Equitable enhancement of human resources for the business community
- Equitable enhancement of human resources for the local community

Indeed, EID has been proposed as the answer to many problems, including job creation, urban redevelopment, brownfield reclamation, community identity, revitalisation of agricultural land and adequate energy supply. This is clearly overdrawn. Such expectations could be the death knell of these projects, even before they have had the chance to become established. Neither does experience with one project fully inform the experience with another: just as each natural ecosystem is distinct, these projects are quite different from one another, except that all involve long time-frames, significant capital investment with varying risk profiles and multiple parties with numerous objectives in diverse cultural settings.

Still, it is reasonable to develop criteria for evaluating the success of eco-industrial projects that go beyond the way conventional industrial or commercial projects would be measured such as by jobs created, profits earned, trainees trained and so forth. Several baseline questions, tied to the objectives defined above, frame issues pertaining to the success of EID. Five objectives and accompanying framing questions are outlined below.

1 Jørgen Christensen, Kalundborg Industrial Development Council, personal communication, 1998.
2 Modifying an expression from biological symbiosis, the power station's Valdemar Christensen of Kalundborg originally defined industrial symbiosis as 'a co-operation between different industries by which the presence of each . . . increases the viability of the other(s), and by which the demands [of] society for resource savings and environmental protection are considered' (see Engberg 1993).

- Objective 1: efficient sharing of resources
    - How does co-location play a role in the success of a project? If there are not synergies based on some measure of geographic proximity, the project is unsuccessful as an EID, even if it is economically successful.
    - What symbiotic exchanges take place? Although every project need not have large quantities of physical waste being shared as in Kalundborg, exchanges are a good indicator of the level of co-operation being achieved.
- Objective 2: economic gains
    - Is the development commercially viable or does it require outside subsidy? Often the public sector serves a catalytic role in EID, but after a defined time a project will not be sustainable if it is significantly dependent on such subsidies.
    - Is the EID project structure more or less costly than conventional methods? Analysis concerning costs must be performed but should also include monetisation of environmental benefits.
- Objective 3: gains in environmental quality
    - In what ways is environmental performance enhanced collectively as well as for individual firms and organisations? Addressing the question of performance involves operational effectiveness of the parts and the whole.
    - What are the costs and benefits to the ecological community? New development can be restorative or degrading to the broader ecosystem beyond the question of improved environmental performance for operating entities.
- Objective 4: equitable enhancement of human resources for the business community
    - What strategic advantage accrues to individual companies to justify their participation? It is unreasonable to expect companies to choose EID only because they may wish to be 'green'. A clearer understanding of possible strategic advantages such as risk reduction, employee engagement or reduced waste or energy costs should be evaluated.
- Objective 5: equitable enhancement of human resources for the local community
    - Is there a specific community involvement programme based on awareness of place? A successful EID is not a 'cookie-cutter' project imported from outside but must include a nuanced view of community factors.

15.1 Breaking down the evaluation problem

Evaluation is the art of systematically comparing project goals and objectives with the work accomplished. The most important question is: 'What was achieved?' When goals and objectives are clearly articulated, the evaluator can ask how closely this achievement matched its stated ambition. When they are not, the evaluator must provide a set of reasonable benchmarks. Evaluation research requires both adherence to some sort of scientific method as well as pragmatic responsiveness to context (on the evaluation of social programmes, see Rossi and Freeman 1993).

There are many kinds of EIDs, as described throughout this text, as well as 'green development' more broadly.[3] To facilitate evaluation, we can begin with green developments with fewer parties to the transaction, such as the establishment of a single environmentally conscious building—a 'green building'—or the instigation of a single material exchange—known as 'green twinning'. There are still economic, environmental and community benefits to be measured, but the task is greatly simplified.

15.1.1 Green buildings

Many analyses have been done of single-standing green buildings—buildings planned from the start to incorporate environmental design throughout the building's life-cycle to achieve results such as energy conservation, water re-use and reduction of toxic substances. These buildings fulfil their programmatic functions while demonstrating that other objectives, including cost reduction and aesthetic appeal, can also be met through environmental design.

To illustrate the process of matching goals with actual results, let us begin with five principles of environmental architecture as defined by Thomas Fisher in 1992 and summarised below.[4] Such architecture should:

- Create a healthful interior environment, particularly with respect to toxic emissions
- Ensure energy efficiency
- Use ecologically benign materials
- Consider the environmental form, addressing the form and plan of the design to the site, region and climate
- Engender good design, described as 'well-built, easy to use and beautiful'

It is easier to develop measures for some of these principles than it is for others. The renovation and restoration of the Audubon House in New York City, the 'green' headquarters of the Audubon Society, clearly met many of these goals. With respect to a healthful interior, the building uses no chlorofluorocarbons (CFCs), which contribute to ozone depletion, in its cooling or insulation, the fresh air ration per person greatly exceeds standards and the combination of materials chosen and a highly efficient ventilation system address the principle of a healthy indoor environment. With respect to

---

3 In the Rocky Mountain Institute's *Green Development: Integrating Ecology and Real Estate* (RMI 1998: 4) the authors point out that there is no single face to this kind of enterprise, since 'for one project, the most visible "green" feature might be energy performance; for another, restoration of prairie ecosystems; for yet another, the fostering of community cohesion and reduced dependence on the automobile'.

4 See the web page of the Building Energy Efficiency Research Project at the Department of Architecture, University of Hong Kong, http://arch.hku.hk/research/BEER/sustain.htm.

energy efficiency, the building is self-reported to use 62% less energy than a 'conventional New York City code-compliant office building'.[5] Ecologically benign materials were selected; for example, conference tables were made from certified wood, and numerous recycled materials were used such as dry walls, and tiles made of 60% post-industrial recycled light bulbs. Whether these environmental inputs actually achieve healthful outcomes over time is a subject requiring further research.

The Adam Joseph Lewis Center for Environmental Studies at Oberlin College, Oberlin, OH, addresses environmental form and good design by orienting the building along an east–west axis to optimise passive solar performance. It provides daylighting for all the interior spaces, a practice that has been correlated with greater productivity and lower absenteeism of employees (see e.g. Loftness *et al.* 1998). With regard to Fisher's general notion of environmental form, some green buildings are better integrated into a particular landscape than are others, but, in general, even quantifiably 'good design' at the building level is not enough to change the feel of an entire neighbourhood.

15.1.2 Green twinning

The term 'green twinning' has evolved to describe a single material or energy exchange. Industrial symbiosis, in contrast, involves several such twinnings. A common, measurable example of green twinning is co-generation, where an industrial operation creates heat, and the heat is captured in some form and used for another industrial use. An example of green twinning of materials comes from Chaparral Steel and its related cement company, Texas Industries. Working with the Business Council for Sustainable Development of the Gulf of Mexico, a newly patented process has been developed to add slag from the steel plant to the raw materials in the cement mix.

Examining this material exchange, we see that many of the PCSD objectives are realised. Measurable results include a 10% increase in cement production as well as a 10% decrease in energy consumption. The value of the slag increased 20 times over the previous market price offered by road contractors, and landfill costs have dropped significantly. Moreover, the twinning has led to additional instances of by-product re-use, including baghouse dust and automobile shredder residue. Community, in this case, is primarily confined to the work environments of these companies. The new process has engendered positive feelings of being innovative and environmentally aware as well as cost conscious. Much less waste is delivered to local landfills, which leaves a scarce resource, landfill capacity, available to others in the broader community (see Forward and Mangan 1999).

15.2 Evaluation of three eco-industrial developments

At the level of a single building or resource exchange as described above, evaluation objectives were clearly framed, and economic, environmental and community costs and benefits could be analysed. Moving up in organisational complexity allows examination

---

5   See the website of the Audubon Society, at www.audubon.org/nas/ah/index.html.

of three EID projects that provide space for multiple organisations, all very different, and each originating in a different sector. The Phillips Eco-Enterprise Center (PEEC) in Minneapolis, MN, is run by the Green Institute, a non-governmental organisation (NGO). The Sustainable Technology Park in Cape Charles, VA, is run by a quasi-governmental industrial development authority. The Londonderry Ecological Industrial Park in Londonderry, NH, was sold to and is operated by a private office park developer. These projects generally fall within the PCSD definition of eco-industrial park. Each project has seen some level of success, based both on conventional criteria as well as on expanded EID criteria elaborated above. The purpose of considering these projects is to identify concretely what the successes have been and to point to areas where the analysis becomes more ambiguous.

15.2.1 Phillips Eco-Enterprise Center

As described in Chapter 17, the PEEC is a green office building and light industrial facility on four acres in a disadvantaged neighbourhood in Minneapolis. The Green Institute, the NGO that developed the facility, is described as 'an entrepreneurial environmental organisation creating jobs, improving the quality of life and enhancing the urban environment in inner-city Minneapolis'.[6] The group working on the PEEC identified three specific overall objectives, concerning energy conservation, use of recovered materials and a healthy indoor work environment. Director Michael Krause reports that the project has met or exceeded these goals and that it has paid in total about 10% more upfront than would have been the case for conventional building construction. The group expects to recoup this expense over time in reduced utility costs.

With regard to the community, PEEC has brought jobs and investment to an area that needed them, and the Green Institute, a community group itself, places a high value on further involvement with neighbourhood development. This project does not currently emphasise symbiotic exchange of materials and energy, but, as an 'eco-enterprise centre', has subscribed tenants, most of whom are involved with environmental business and activities. With respect to evaluation based on prescribed environmental, economic and community criteria, the project has clearly been successful with what it has attempted to do. However, as Krause wisely comments in this volume (page 277), 'Only time will tell whether the PEEC can have a broader and more systematic effect on surrounding land uses, providing a catalyst for an evolved system of industrial ecology.'

15.2.2 The Sustainable Technology Park in Cape Charles, Virginia

This project represents increasing complexity, with several buildings, a 200 acre site and substantial public infrastructure development at a site beside the Chesapeake Bay. In an effort to bring jobs to a poor, rural region of Virginia, the community created a joint industrial development authority, a quasi-public organisation with a board appointed by the Northampton County government and participating towns. A master planning process created Northampton's Sustainable Development Action Strategy. As in the Green Institute case described in Section 15.2.1, the strategy identified specific goals

---

6   See the Green Institute's website, at www.greeninstitute.org.

relating to the economy (job creation and a strong and lasting economic base), environmental protection of natural, cultural and historic assets, and equity for current and future generations. To date, the project has had an extensive community involvement process and, with respect to the environment, has emphasised habitat protection and wetland construction in sensitive areas. The business aspects are a work-in-progress as the buildings are being developed and as more of the master plan is implemented over time. There has not yet been a materials and energy exchange—more effort has been focused on the potential for water re-use.[7]

15.2.3 The Eco-Industrial Park in Londonderry, New Hampshire

This project is more industrial in nature than are the previous examples and is anchored by a 720 MW combined-cycle natural gas power plant. The 100 acre site was assembled by town staff and then leased to a private developer. According to the vision statement for the project, its intent is to develop 'systems and processes which minimise the impact of industry and business on the environment, improve the economic performance of the member companies and strengthen the local economy'.[8] Specific covenants that the developer must follow call for many types of environmental protection, including the requirement that all tenants develop an environmental management system (see Chapter 19). Development plans of prospective tenants are required to be reviewed by environmental design consultants. Even with an extensive community planning and design process, siting of the power plant proved controversial and divided the community, although it was eventually approved.

Although more time must pass to determine how the project will fare, several tenants have located in the park. One challenge will be to determine if the power plant can operate synergistically with tenants still to be identified by using its waste heat, carbon dioxide and other power plant by-products. It is already set to use millions of gallons of municipal waste-water being pumped in for cooling. Geographically, the location of the park in a busy commercial area close to Manchester Airport is a positive factor in attracting and retaining tenants, a locational benefit not shared by many other EIDs.

15.2.4 Applying the framing questions to two eco-industrial parks

Further analysis is offered of the Cape Charles and Londonderry projects reviewed here as a means of testing the framing questions listed in the opening text of this chapter. As EID projects go beyond conventional commercial and industrial development, the evaluation includes whether each project, at least conceptually, provides amenities different from those of traditional office parks. For example, a traditional office park may provide shared business services or van transportation, or even organised community involvement, so these items would not be further considered as an aspect unique to EID. The Green Institute PEEC, as discussed in Section 15.2.1, is closer to the green building

---

7 www.sustainablepark.com
8 www.londonderry.org/page.asp?Page_Id=528

model housing an environmental business centre, so is excluded from this comparison of two self-described 'eco-industrial parks'.[9]

15.2.4.1 The role of co-location
Co-location has enabled the Cape Charles Park to install photovoltaic cells and other green features in the first building. Other amenities parallel traditional parks. Londonderry has also emphasised green design and has used covenants and governance agreements to ensure that co-location will result in increased attention to environmental and community issues.

15.2.4.2 Symbiotic exchanges
These are not currently part of the Cape Charles Sustainable Technology Park, although there has been work put into re-use of water in connection with a cement plant. The Londonderry Ecological Industrial Park, as noted in Section 15.2.3, is using municipal waste-water for cooling but is not currently using other by-products. It has been reported that two remaining lots in the park will be targeted for firms that can use the synergies of the power plant (Lowitt 2001).

15.2.4.3 Commercial viability and subsidy of project
The Cape Charles Park was financed along the lines of a public-sector model of industrial development, with additional loan and grant money. The level of this support seems to be in the same range as that for comparable public developments. The Cape Charles Park has the additional goal of bringing economic development to a disadvantaged area, and the financial structure is in line with these objectives. As Londonderry was turned over to a private developer, the developer has not received substantial financial assistance, although the developer, Sustainable Design and Development was reported to have bought the property at a subsidised price. The developer has undertaken infrastructure improvements and some increased costs to achieve the design standards negotiated. The power plant developer, AES Granite Ridge, offered reduced rates for steam and hot water to nearby tenants and price breaks to businesses in Londonderry. The politically charged atmosphere surrounding the plant's approval and the town's reaction to it has led to uncertainty about the full extent and nature of this offer.

15.2.4.4 Costliness of the project structure of eco-industrial development compared with that of conventional development
The Cape Charles and Londonderry parks face price competition from conventional development and therefore have had to price their space at competitive rates for new construction and to market their amenities alongside those of other developers. Cape Charles has numerous incentives available because of its location in the Virginia Enterprise Zone and its classification as a Foreign-Trade Zone. Sustainable Design and

---

9 The Londonderry park is called an 'eco-industrial park', and the Cape Charles website describes its development as 'America's premier eco-industrial park' (www.sustainablepark.com).

Development, the Londonderry developer, is reportedly pricing the two remaining lots in the Londonderry park at above-market rates to capitalise on the synergies produced by the AES Granite Ridge Plant.

15.2.4.5 Enhancement of environmental performance

Both projects have used covenants and guidelines to support environmental design and performance. Cape Charles has emphasised energy conservation and choice of materials and requires companies to fill out a 'sustainability matrix'. The basic performance requirements for the Londonderry park require each company to participate in an environmental management system, to seek synergies and to file an annual ecological report.

15.2.4.6 Costs and benefits to the ecological community

The Cape Charles project has put significant emphasis on restoring the natural systems of the coastal watershed and improving the ecological community as well as on social and economic goals. Clearly, even a clean power plant offers some disturbance to the ecological community, but great emphasis has been placed in Londonderry on minimising its environmental footprint. Much land has been reserved for interpretive trails throughout the Londonderry park.

15.2.4.7 Strategic advantage to individual companies

If a company decides to locate in an eco-industrial park, clearly it is choosing this location over competing choices. Several environmentally related firms have chosen the Sustainable Technology Park in Cape Charles. It is certainly arguable that the AES power plant in Londonderry would have faced too much community opposition to have gained acceptance in another location that did not already foster environmental and community values. A German firm recently approached Sustainable Design and Development on its own and sought to locate within the eco-industrial park because its corporate philosophy was in alignment with the park's goals.

15.2.4.8 Community involvement programmes based on sense of place

Both eco-industrial parks relied on extensive community involvement, design charrettes and public-sector participation in the development of the projects. Londonderry has a Community Stewardship Board to ensure the public accountability of tenants. Both projects have had extensive local, state and national scrutiny, given their pioneering roles in EID.

15.3 Analysing the results

The examples described in this chapter show the three EIDs to include the goals of conventional development, with additional 'eco-industrial' requirements as well. The projects tend to involve more complex governance structures and voluntary limitations on the market through, for example, deed covenants concerning design or type of businesses that are acceptable. These restrictions do place an added burden on these projects, often with tangible costs and less tangible benefits.

Each of the goals identified earlier—efficient sharing of resources, economic gains, gains in environmental quality and equitable enhancement of human resources for the business and local community—is clearly a component of the three projects reviewed. In particular, each project has involved much cross-sectoral co-operation, and these examples illustrate that EID projects often come with extensive community involvement processes as well. These processes are characterised by attempts to identify missions and goals that, in turn, make the framework for evaluation concrete and transparent. Unlike the green building and green twinning evaluations, however, multi-party projects, ultimately, have fewer variables under the control of proponents, which increases the uncertainty of meeting objectives. Still, it is important to consider vision and goals as a starting point that can then be tailored and specified according to what is appropriate for each project.

The new century does not give us a large number of EIDs to consider when we seek to establish criteria for success. The high mortality rate of proposed EIDs suggest that completion alone is an enormous hurdle. Few of approximately 20 projects discussed at the seminal Cape Charles meeting of the PCSD on October 1996, for example, are still under active consideration. The framing questions identified here were applied to two eco-industrial parks. EIDs, more broadly, take many other forms in addition to industrial parks for which the framing questions must also be considered (for identification of other material exchange types, see Chertow 2000).

Recognising two evaluation questions—whether the EID has met its own goals determined during the planning process and, ultimately, what was achieved overall—returns us to the underlying questions of this chapter. Generally, the three EIDs discussed here laid out goals and objectives by which success could be measured. As the projects have worn on, initial objectives have been modified but kept in sight. The broader questions suggested here regarding co-location, symbiotic exchanges, commercial viability, strategic advantages for the businesses, community involvement, cost and environmental footprint can only be fully evaluated over time. Whether a failed or marginal EID project would have succeeded as a conventional project involves a level of modelling not yet available to researchers.

Several approaches would facilitate EID evaluation. Basic data such as energy and materials use by individual organisations would help to establish an information base that could be very helpful in the search for synergistic tenants. Adoption of design standards would also provide a benchmark for measurement. Means of recognising and valuing communal benefits should be developed, as well as measures of the internal, private, benefits achieved by individual companies and concerns choosing to participate in EID. In the final analysis, the question must be asked whether, in the long run, the benefits of EID are retained or whether they fall victim to various market pressures. In 20 years' time we will have many more examples to study to provide insights to guide sustainable economic development.

Part 4
ECO-INDUSTRIAL COMPENDIUM OF CASES

16
THE INDUSTRIAL SYMBIOSIS IN KALUNDBORG, DENMARK
An approach to cleaner industrial production

Noel Brings Jacobsen
The Symbiosis Institute, Denmark

16.1 Highlights

Symbiosis: in nature, this refers to the association of dissimilar organisms to their mutual advantage. In short, it is helpful interdependence. Five process industrial plants and the municipality of Kalundborg have combined forces in a unique environmental collaboration using nature's symbiosis model. Calling it 'industrial symbiosis', the collaboration is based on the idea that one company's waste product is another company's raw material.

Kalundborg's industrial symbiosis came into existence over a period of 25 years (Table 16.1), evolving until its present state which sees partnerships in 19 projects (Fig. 16.1). Today, these 19 different projects concern recycling of water, transfer of energy and recycling of waste products between the six independent symbiosis partners.

The original incentive for the industrial symbiosis was profitability: a desire by the companies involved to make sensible investments in order to save costs. All the exchange projects were negotiated commercially between the partners, most often between two partners at a time. The industrial symbiosis, then, did not develop as a well-planned network according to a decision of joint management or with the academic knowledge of environmental network theories. Rather, the industrial symbiosis developed as a number of single projects, initially quite independent of each other. It developed as a simple attempt to exert good management practice and to improve environmental performance. This is, in fact, the strength of the Kalundborg approach: business leaders have done the 'right thing' for the environment in the pursuit of rational business interests. The industrial symbiosis has shown that environmental concern and business can go hand in hand.

16. THE INDUSTRIAL SYMBIOSIS IN KALUNDBORG, DENMARK *Jacobsen* 271

Year	Symbiosis development
1962	Surface water from Lake Tissø to Statoil
1972	Excess gas from Statoil to Gyproc
1973	Surface water from Lake Tissø to Asnæs power plant
1976	Biomass from Novo Nordisk/Novozymes to local farmers
1979	Fly ash from Asnæs power plant to cement industry
1980	Cooling water from Asnæs power plant to fish farm
1981	Excess heat from Asnæs power plant to district heating system
1982	Steam from Asnæs power plant to Statoil and Novo Nordisk/Novozymes
1987	Surface water from Lake Tissø to Novo Nordisk/Novozymes
1987	Cooling water from Statoil to Asnæs power plant
1989	Yeast slurry from Novo Nordisk to local farmers
1990	Liquid sulphur from Statoil to fertiliser industry
1991	Waste-water from Statoil to Asnæs power plant
1992	Excess gas from Statoil to Asnæs power plant
1993	Gypsum from Asnæs power plant to Gyproc
1995	Subsequent treatment of Novo waste-water at the public waste water plant
1995	Re-use basin to waste-water from Asnæs power plant and Statoil
1998	Sludge from public water treatment plant to Biotechnical Soil Cleaning/SOILREM

Table 16.1 **The historical sequence of industrial symbiosis in Kalundborg, Denmark**

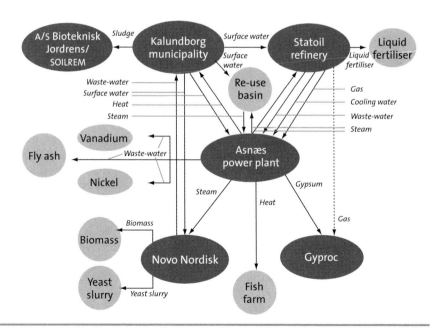

Figure 16.1 **The industrial symbiosis in Kalundborg, Denmark, 2001**

16.2 Facts

Kalundborg is a city of approximately 20,000 inhabitants, about 100 km west of the Danish capital of Copenhagen. The business community of the city is characterised by four large process industries, namely Asnæs power plant (a coal and Orimulsion-fired power station), Statoil (an oil refinery), Novo Nordisk/Novozymes (a pharmaceutical and biotechnology company), Gyproc (a plasterboard manufacturing company) and a number of smaller companies, among these A/S Bioteknisk Jordrens/SOILREM (a soil remediation company). Together with the municipality of Kalundborg, these five companies form the industrial symbiosis, a collaboration that has gradually grown as new, remunerative projects became feasible over the years.

At the heart of the industrial symbiosis is the Asnæs power plant, the largest power plant in Denmark. It is fired by coal and Orimulsion, has a maximum capacity of 1,372 MW and produces about half of the electricity used in eastern Denmark. It employs about 250 people. Since 1980, Asnæs Power Station has supplied district heating to the city of Kalundborg and process steam to both Statoil's refinery and Novo Nordisk. By combining the production of electricity and heat, the fuel is better utilised and heat can be produced more cheaply than by the purchasers of the heating themselves.

Only part of the cooling water from Asnæs power plant goes into the Kalundborg Fjord. The rest is utilised by a fish farm in a trout-rearing programme that produces 200 tons of fish a year. Sludge from the fish farm is used as manure to enrich the fields of neighbouring farms.

Asnæs power plant uses a lot of water to produce electricity and steam. Previously, only groundwater was used, a resource that is becoming increasingly scarce in the region. Today, however, this has been replaced by surface water from a local lake, Lake Tissø, and large quantities of cooling and waste-water from the Statoil refinery. The waste-water is stored in a 200,000 m^3 re-use buffer.

Finally, the Asnæs power plant has an electrostatic filter system to collect fly ash from coal combustion. The fly ash is re-used in the cement industry; the fly ash from the combustion of Orimulsion contains large amounts of nickel and vanadium, which is extracted for industrial use.

The Statoil refinery is owned by the Norwegian concern, Statoil Ltd. The refinery employs 300 people and treats annually about 5 million tons of crude oil. The Statoil refinery is the largest refinery in Denmark. Like Asnæs power plant, the Statoil refinery uses large quantities of surface water, which is taken in from Lake Tissø, thereby saving the scarce groundwater resources. Formerly, the refinery's waste-water was discharged directly into Kalundborg Fjord. Now, the waste-water and cooling water is piped and carried off to Asnæs power plant, and a part of the recycled water is returned to Statoil as steam.

Like all oil refineries, the Statoil refinery has an eternal flare that is an integral part of the refinery's security system. Formerly, this excess gas from the flare was delivered to Gyproc and Asnæs power plant in the place of coal and oil, respectively. Today the gas delivery to Gyproc has changed to natural gas, and the delivery of excess gas to the power station is temporarily suspended. However, from a historical point of view, the gas exchanges have been very important for the co-operative development of the industrial symbiosis.

Statoil has also installed a desulphurisation plant. The sulphur is removed from the flue gas in a process from which ammonium thiosulphate is a major by-product. This by-product is used in the production of liquid fertiliser.

The desulphurisation plant at Asnæs power plant produces about 200,000 tons of gypsum per year. Some of the gypsum is sold to the plasterboard factory, Gyproc, where it is used in place of natural gypsum. Gyproc has 180 employees and produces plasterboard for the building and construction industries. Gyproc is a part of British Plasterboard, one of the world's largest gypsum groups serving markets in more than 45 countries.

As mentioned, Gyproc buys gypsum from Asnæs power plant. The 'industrial gypsum' from Asnæs power plant is more homogenous than is natural gypsum and is therefore more appropriate for the production of plasterboard. At the same time, a natural resource is saved, natural gypsum, formerly imported from Spain and transported all the way to Denmark.

Process steam from Asnæs power plant and surface water from Lake Tissø are used by Novo Nordisk/Novozymes in its pharmaceutical and enzyme production. Novo Nordisk/Novozymes is a leading global producer of insulin and industrial enzymes and has a total of 18,000 employees, about 2,000 of whom work in Kalundborg. Enzyme production is nourished by the fermentation of starch products such as potato flour and cornflour. This results in large quantities of biomass, which is pumped via pipelines to farmers in Western Zealand, where it is spread on fields as a replacement for commercial fertilisers. Novo Nordisk's insulin production also yields a fermentation cream—yeast slurry—which is used as pig fodder.

A/S Bioteknisk Jordrens/SOILREM and its affiliated divisions specialise in remediating soil contaminated by chemicals and heavy metals. The pollutants are decomposed by natural bacteria and fungi. Sludge from the public water-treatment plant is utilised at A/S Bioteknisk Jordrens as a nutrient in the bioremediation process.

The municipality of Kalundborg is also very much part of the industrial symbiosis. The municipality supplies surface water from Lake Tissø to industries and sludge to A/S Bioteknisk Jordrens and buys district heating from Asnæs power plant.

The industrial symbiosis has developed over the years as individual projects became feasible, and recent years have seen an increase in the frequency of new projects. The present status of the total symbiotic system is shown in Figure 16.1.

16.3 Reflections

It is important to note that all agreements between the various symbiosis partners are based on financially sound principles. As the products exchanged are low-economy residual products, the resulting profits therefore partly lie in the price difference between the residual products and the raw materials and partly in the price difference between the industrial symbiosis activity and the least expensive alternative. However, an accurate assessment of the profits attained in the industrial symbiosis project is extremely difficult to make for the parties involved because the individual projects overlap financially and because the actual price of the alternatives will remain unknown until the companies are forced to use them. An estimate of the industrial symbiosis economy, however, shows that the total investments in all 19 projects together amount to around US$75 million. The present annual savings are over US$15 million. The total savings accumulated throughout the years is likely to be around US$160 million, and the average repayment time of the individual projects is about five years. It is therefore beneficial to bear the industrial symbiosis idea in mind when new industrial areas are being established and new industries are being sited. Examining the possibility of creating new symbioses and collaboration could benefit the companies involved as well as the environment.

The environmental strain on an area with several large industries has so far, in traditional industrial development, been equal to the sum of the strain of the individual industries. This means that the more industries there are gathered in one place, the greater the strain on the environment. However, in an industrial symbiosis, this linear correlation is invalidated as residual products from one company are used as raw material by another company. In comparison, residual products of a company working independently would have to be deposited or discharged, and the other company would have to purchase its raw material elsewhere or excavate it from nature.

The experiences from the Kalundborg industrial symbiosis show that an optimum composition of companies that are different and that are mutually of the right size can have the effect of reducing the total environmental strain on the area in which they are located. For example—in the case of Kalundborg—almost 2 million m³ of groundwater, 1 million m³ of surface water and 200,000 tons of natural gypsum are saved every year, which underlines the environmental potential of industrial symbiotic activities.

16.4 Lessons learned

The experiences from Kalundborg show that a number of preconditions must be fulfilled before an industrial symbiosis can be established. The most important precondition is that a relationship must be in place among the top management of the companies involved. The top management must not only know each other but also be well informed about one another's production processes and have the necessary confidence in and extend an openness to one another. The Kalundborg industrial symbiosis is basically built on openness, close co-operation and mutual trust between the partners.

Another precondition is that the involved companies' products and by-products be different. There should be no elements of competition that might prevent the type of close collaboration that industrial symbiosis requires, and the fact that two identical industries have nothing to trade underlines the necessity of diversity in the symbiotic network.

A third precondition in the case of the Kalundborg industrial symbiosis is based on the fact that a substantial volume of piped material is involved, therefore making it necessary for the companies to be placed geographically close to each other. Long pipelines are costly, and the greater the length the greater the energy losses. In other words, a short distance—physically and mentally—has been a central promoting factor for the development of the Kalundborg symbiosis.

Although the industrial symbiosis results in direct financial and environmental advantages, there are other benefits from this form of collaboration. The fact that the symbiotic industries have become well known for caring about the environment has created tremendous publicity and goodwill in the environmentally conscious public. This has also helped boost the involvement of employees in the project and has instilled them with pride in their companies. The industrial symbiosis has therefore also translated into employee loyalty and low labour turnover, therefore heightening the survival prospects of the companies in markets with strong competition.

17
THE GREEN INSTITUTE PHILLIPS ECO-ENTERPRISE CENTER IN MINNEAPOLIS, MINNESOTA

Michael Krause
The Green Institute, USA

Corey Brinkema
Trillium Planning & Development, USA

The Phillips Eco-Enterprise Center (PEEC; see Fig. 17.1) looks out of place as you approach it along busy Hiawatha Avenue in Minneapolis, MN. The modern office and light industrial facility is unique in the midst of a heavy industrial area and is adjacent to an ageing 1960s-style shopping centre that has obviously seen better days. From a distance, there are only a few hints that this is a model green building, named in the Earth Day Top Ten in the year 2000 by the American Institute of Architects and winner of numerous other awards, including the international Bremen Partnership Award and the Energy Star Award of the US Environmental Protection Agency (EPA).

Figure 17.1 *The Phillips Eco-Enterprise Center, Minneapolis, MN*

So, what is a green office and manufacturing facility doing in one of the poorest neighbourhoods in Minnesota? How does it relate to the industrial and commercial businesses around it and to the residential community at its doorstep? The answers lie in an environmental justice battle that began more than 20 years ago and with neighbourhood activists who founded the Green Institute and never gave up on their dream of a 'model green building for green businesses'. In fact, the PEEC is an example of how low-income communities can work together to clean up the environment and control their economic future. Only time will tell whether the PEEC can have a broader and more systemic effect on surrounding land uses, providing a catalyst for an evolved system of industrial ecology.

17.1 Environmental justice origins

The PEEC is named for the Phillips neighbourhood, an area of concentrated poverty and unemployment south of downtown Minneapolis that was designated a federal empowerment zone in January 1999. The neighbourhood itself is named for the 19th-century progressive social commentator Wendell Phillips. Despite the proud origins of its name, Phillips is one of the poorest communities in Minnesota, with about 40% of its households earning less than US$10,000 a year. More than half of the neighbourhood residents are members of minority groups, including a large American Indian population and many recent immigrants from East Africa and Mexico.

More than 20 years ago, the property on which the PEEC sits was chosen by local government officials as the site for a large rubbish transfer station. Five small businesses and 28 houses were cleared from the site in Phillips, and the residential zoning was changed to industrial use. As they at first watched what various government bodies were preparing to do to them, neighbourhood residents asked why a rubbish transfer station is the only kind of 'economic development' that poor neighbourhoods and communities of colour ever seem to get. The battle for environmental justice was joined, and residents fought for more than 12 years to block the transfer station project.

In the early 1990s, when it looked as if Phillips residents would finally prevail, a group of environmentally oriented neighbourhood activists founded the Green Institute. They were mostly self-taught environmentalists, having worked on lead paint issues and community gardening. One of the activists, Annie Young, became a member of the Minneapolis Park Board and began a more in-depth study of sustainability issues. As a result of their pointed critiques of the transfer station over the years, residents had also learned a great deal about solid waste issues. Rather than wait for another undesirable use to be proposed for the transfer station site, residents created their own non-profit development organisation to gain control of the site and develop an industrial facility that would bring clean, good-paying jobs into the neighbourhood. They were also determined to turn the growing stream of solid waste into a source of materials for nurturing other community-based ventures.

County officials who had been pushing the transfer station 'turned on a dime' and provided funding for a business plan for an alternative to disposal of construction and demolition debris. In 1995, the Green Institute opened the doors to the ReUse Center, a retail store that sells salvaged and re-usable building materials. The programme added active deconstruction crews, and by 2002 these enterprises were selling more than US$1 million a year in salvaged building materials that would otherwise end up passing through transfer stations and be buried in landfills.

17.2 From grass-roots non-profit organisation to real estate developer

Years of coalition-building and political action by neighbourhood residents gave the Green Institute enough clout to win development rights from Minneapolis City Council to four acres of the site once planned for the transfer station. However, development rights alone do not a project make. The Institute had to wrestle with the many facets of real estate development, from assembling a capable development team to preparing a solid market-based business plan, recruiting companies and securing the financing to construct and operate a facility.

While undertaking a feasibility study, the Green Institute gauged the market opportunity for a sector-focused business centre targeting environmentally oriented firms and designed according to the principles these firms espoused. The development team identified over 500 businesses in the Twin Cities area that operated within the environmental, energy-efficiency and alternative-energy industries. The firms were primarily a mix of small and medium-sized enterprises (SMEs) but also included a few major corporations such as the Home and Building Control division of the Honeywell Corporation. Some of these firms were successfully competing in rapidly growing markets, and many of the firms were accustomed to on-the-job training of their employees, a factor that was helpful with an inner-city workforce that lacked education and job skills. These businesses also had a commitment to effective environmental management, including energy and resource efficiency as well as community involvement.

The feasibility analyses favoured targeting this segment of so-called green businesses. However, in order to successfully attract these firms into a low-income, inner-city neighbourhood it was clear that the facility would have to uniquely incorporate the ethics that these businesses went out and fought for every day. Hence, the already strong interest in incorporating sustainable design into the project was reinforced.

The tenant recruitment process began in earnest in late 1997, even with only the schematic design phase of the architectural process completed and little in the way of committed financing. The strategic situation of the property was a promising attribute, but the stigma related to typical inner-city issues became a problem for some prospective tenants. This being said, only two deal failures were attributed to safety concerns. At the same time, successful deals were being completed with forward-looking companies who were attracted by the uniqueness of the conceptual plans. The first signed tenant, Impact Energy Controls Corporation, agreed to a lease contract with an easy-to-assent precondition that their advanced energy management system be installed in the building at their manufacturing cost. The Institute also succeeded in attracting a manufacturer of energy-efficiency testing equipment, the Energy Conservatory. Before breaking ground, the Institute was able to secure four leases, totalling 40% of the available space.

Clearly, though, the most daunting task faced by the organisation was the financing of the facility. For logistical reasons, and to best make the case that such projects can indeed be profitably structured by the greater development community, the Green Institute chose to finance the bulk of the project with market-based debt instruments. Late in 1997, the Green Institute began courting commercial banks to provide a construction loan and permanent mortgage financing. Three local banks agreed in principle to consider the project. The Institute eventually selected Bremer Bank of St Paul because of the terms offered and the bank's demonstrated enthusiasm for the project. With a target debt limit of US$3–4 million (75% of appraised value) according to the bank's terms, the Institute found itself requiring from US$1–2 million in equity and/or secondary debt financing. For this equity share, the Institute twice sought and twice failed to secure grants from the federal Department of Housing and Urban Development (HUD). Then, in early 1998, the Institute decided to test its development concept with the Minnesota State legislature.

The neighbourhood was fortunate to be represented in the legislature by two women who were also fierce champions on behalf of the Green Institute. During the 1998 session, State Senator Linda Berglin and State Representative Karen Clark manoeuvred legislation that gave the Green Institute a US$1.5 million loan from the State of Minnesota. The loan will be forgiven after ten years as long as the Green Institute continues to own the facility and at least 67 jobs are created. The sum of the state and commercial bank funding and some prior grants amounted to approximately 90% of the total project costs. By late autumn 1998, the Institute secured a creative combination of new small grants, additional secondary debt and loan guarantees and in-kind contributions of building equipment and materials to finally close on the state and bank financing packages and was able to break ground.

17.3 Implementing sustainable design

An inclusive and participatory design process for the facility was led by LHB Engineers and Architects, a Minneapolis firm that had limited experience with industrial development but was a leader in indoor air quality and energy efficiency. One of the firm's strengths, in the eyes of the Green Institute, was a design process that was very open to client direction and goal-setting. Working with the Institute, the neighbourhood, architecture students from the University of Minnesota and a computer modelling effort funded by the local utility, the PEEC design focused on three primary objectives: (1) creating a healthy indoor work environment with superior air quality, abundant natural light, access to green space and a comfortable climate; (2) reducing operating energy use by 50% over a new code-compliant building; and (3) maximising the productivity of resources assembled for construction, including using salvaged and recycled-content materials, sourcing materials locally and re-using and recycling construction site waste.

When completed in October 1999, the PEEC had incorporated more than 25 different sustainable design strategies and nearly 100 product selections that helped it meet most of the original design goals (for a list of the major design features, see Box 17.1, or visit the Green Institute's website at www.greeninstitute.org). On the energy side, the office

The following are used in the Phillips Eco-Enterprise Center:

- Occupant health
  - Active daylighting systems
  - Indirect artificial lighting
  - Low-emission coatings
  - Multiple-zone climate control
  - Air-quality sensors and controls
- Energy efficiency
  - Geo-exchange heat-pump system
  - Air-to-air energy-recovery system
  - Energy-management system
  - Ground-source hot-water system
  - Low-density, high-efficiency lamps
- Efficient construction
  - Local sourcing of more than 90% of all materials
  - Salvaged brick façade
  - Salvaged steel superstructure
  - Salvaged wood and millwork
  - Salvaged fixtures and flooring
  - Fly ash substitution in concrete panels
  - Linoleum flooring
  - Recycled glass tiles
  - Recyclable carpet
  - Millwork from agricultural by-products
  - Polished concrete block interior
  - 79% construction waste re-use or recycling
- Other areas
  - 100% on-site storm-water retention
  - Green roof, with native landscaping
  - Restoration of half an acre of prairie

Box 17.1 *The Phillips Eco-Enterprise Center, Minneapolis, MN: major sustainable design features*

component of the facility (22,000 ft^2) is conditioned by a geo-exchange system that captures the energy potential of the soils underlying the site to heat and cool office space. A total of 18 heat pumps and 124 wells are piped together to transfer thermal energy between the building and the earth, providing a 35% improvement in energy efficiency compared with the highest-efficiency natural-gas furnace system. A heat pump also supplies hot water in the building, including water for showers that are designed to encourage bicycle commuting. An active daylighting system uses insulated skylights and mirrors on tripods that track the sun to concentrate daylight into the interior spaces of the industrial building. This system is so effective that no artificial lighting is needed in the industrial portion of the facility on all but the cloudiest days. An integrated system of sensors throughout the building measures temperature, motion, light and carbon dioxide levels. This information is used by a computerised energy-management system to provide heat, light and ventilation only where it is needed in the building. Low-density, high-efficiency lighting systems and energy-recovery ventilation also contribute to energy use that is approximately 40% less than in other new buildings.

One of the most unique features in the PEEC is difficult to discern with the eyes alone. More than 50 tons of steel roof joists and 22,000 bricks were re-used in construction. There are also re-used doors, windows, carpet and workstations in office areas and salvaged sinks in the bathrooms. In choosing new materials, the development team evaluated life-cycle implications, including the material's durability and the impacts of its maintenance over the life of the building. The design process also considered the amount of energy embodied in various building materials. The re-use of the steel roof joists, for example, saved 110 million BTUs of energy that would have been used to manufacture new steel joists. Materials with minimal transportation distances from producer to job site were given preference. Life-cycle analyses also led to the use of flooring made from linseed-oil-burnished block that eliminated the need for sheet-rock, precast concrete panels with fly ash content and pigmentation that eliminated the need for painting, and recyclable carpet tiles. Zero-emission paints and adhesives were also specified both for indoor air quality control and overall toxicity and resource impacts.

The PEEC development incorporated xeriscaping (water-conserving landscaping with use of plants that are suited to Minnesota's variable climate) or native species planting strategies that eliminate the need for artificial irrigation and save over half a million gallons of municipal water a year. In addition, all the site's storm-water, some 1.5 million gallons a year, is allowed to percolate naturally on-site. The storm-water is contained largely by a shallow depression constructed in the midst of a half-acre restored prairie and a quarter-acre 'bounce zone' wetland. Supplemental storm-water management is accomplished by parking lot islands with bio-filtration strips and a 4,000 ft^2 green roof.

To the surprise of the Green Institute and the development team, among the facility's most unique and discussed features are its operable windows. Many of the seasoned contractors working on the construction project had never before installed operable windows in a commercial setting. And the tenants and their employees cherish them, noting that they have the same ability as in their homes to manually access fresh air.

Some of these energy-efficiency and environmental upgrades were less expensive than conventional approaches, but the net impact of all upgrades added about 3% to the cost of the building's construction. That added cost will be repaid in less than five years from secured rental premiums. The energy and water savings alone realised by tenants nearly offsets their rental premiums. However, it is the anticipated employee productivity gains attributed to the availability of natural light, comfortable temperatures, high-quality indoor air and access to green space that are the real attraction financially. Although challenging to measure, even a 1% improvement in employee productivity brings the overall financial return on investment of the sustainable design measures to nearly 40% and the simple payback period to less than three years.

17.4 Where we go from here

Within one year of opening, the PEEC was 75% occupied and cash-flow positive. As of September 2001, two years after first opening its doors, the PEEC is fully occupied. Of the 18 tenants, there are a handful of small non-profit organisations dedicated to environmental issues, and the balance of occupants are for-profit businesses, with only two that

Figure 17.2 **Solar panels on roofing**

are not in green business fields. The PEEC has brought nearly 150 jobs to the Phillips neighbourhood and adds US$5 million to the local tax base. The Green Institute occupies 1,200 ft² in the facility and provides recruitment, training and business development support to tenants. A key relationship continues to be with the University of Minnesota, with the Green Institute linking student interns with PEEC tenants through a programme funded by the Kauffman Center for Entrepreneurial Leadership in Kansas City. The Green Institute is also broadening its employment and training initiatives with an EPA-funded project to train low-skilled workers for entry-level positions in site testing and brownfield remediation.

Having established its bona fides as a developer, the Green Institute is turning its attention to the area surrounding the PEEC and is looking for opportunities to incorporate eco-industrial development strategies. One track is to continually look for ways to improve the operation of the building, including internal waste exchanges and abatement strategies and the future addition of wind and solar energy systems to the building.

A second and far more comprehensive task is to expand the eco-industrial approach to emerging projects within a six-block area around the PEEC. These efforts will be aided by the construction, beginning in 2001, of the Twin Cities' first light-rail transport (LRT) system adjacent to the PEEC in the Hiawatha Avenue corridor. LRT represents a transforming transportation technology that is stimulating land-use planning and a rethinking of the current area around the PEEC toward more mixed-use, transport-oriented development.

Among the projects is an effort by three industrial businesses—a foundry, a bituminous and paving company and a large roofing contractor—to relocate to a larger, brownfield site in the area. These firms will work with the Green Institute to design a co-ordinated site with certain shared facilities and opportunities for energy and waste exchanges.

The ageing shopping centre adjacent to the PEEC will include a major station for the future LRT system, making the shopping centre a priority for transport-oriented redevelopment. The Green Institute is working with other community groups and developers on a plan for higher-density commercial and residential development on the site that would retain the existing local businesses in the shopping centre. The transport station site is also an opportunity to showcase sustainable design and construction techniques and energy efficiency and use of renewables in mixed-income housing.

The site that would be vacated by the three industrial businesses that are relocating nearby is directly across the street from the PEEC. It has redevelopment potential for medium-density housing; it is close to the future LRT station and would be adjacent to a major bicycle and pedestrian trail system. Part of the site includes a large rubbish incinerator built in 1939 that has been dormant since the 1960s. Recently, the Green Institute has teamed up with a local electricity generation co-operative and energy consultants to study the feasibility of retrofitting the incinerator facility into a combined heat and power plant to burn clean waste wood and other fuels. The plant could potentially generate 5–15 MW of electrical power and also produce hot water that, supplemented by some of the waste heat from industrial businesses, could be the backbone of a district heating system for new and existing developments in the vicinity of the PEEC.

An existing asset in the area is Pioneer Cemetery—the first cemetery in Minneapolis, which is full of history and provides 26 acres of beautiful green space. The cemetery has been largely walled off from the community but there are plans to open it up and give it a higher profile in residents' consciousness about the area. Another asset is the Midtown Greenway, a below-grade, abandoned railway corridor that is being rebuilt as a bicycle and pedestrian trail, with future potential for transport services. A new US$22 million YWCA (Young Women's Christian Association) had its grand opening in September 2000. Two major educational institutions, South High School, the public school magnet for science and technology, and PPL Edison School, a charter school, are also within the area.

The transformation of this corner of the Phillips neighbourhood from an area of brownfields, older polluting industries and an ageing auto-oriented shopping centre will not occur overnight. However, the PEEC has seeded a different approach to development that could result in significantly reducing the environmental footprint of industry while retaining its job base. This approach will bring up to 1,250 housing units into an area that had none and will preserve existing retail businesses. This transformation will create mixed-use, transport-oriented areas where residents of Phillips can live, work, shop and find recreation.

A key to the transformation will be to build and maintain communication links and strong working relationships among the many stakeholders. With its PEEC project setting the standard and with its understanding and interest in eco-industrial and sustainable development, the Green Institute will play a major role in the development of its eco-community.

17.5 What we learned

17.5.1 Planning and marketing

- A thorough feasibility analysis is essential to ascertain the attractiveness of a proposed eco-industrial development plan and to determine whether or not to proceed and what programme elements best address market needs. Initial findings of the PEEC feasibility study steered the project away from traditional notions of eco-industrial development (e.g. material and energy exchange foundations) and toward a sector-focused business centre. Later findings largely discounted a planned business incubator component in favour of targeting more established, creditworthy businesses. The business incubator is still part of the long-term programme but has been delayed as a building fund is being built and space 'turns over'.

- If presented with the option, many environmental industry firms will seriously consider locating in a sustainably designed commercial facility, even if holding costs are slightly higher. Non-profit organisations, particularly those that are environmentally oriented, are even more likely to consider such facilities.
- The PEEC successfully achieved rents that exceeded market comparables by 5–10%, a premium that is largely attributed to sustainable design measures and the nature of the environmental industry cluster.

17.5.2 Financing

- Creativity in deal structuring and funding sources is essential. Without a development portfolio for leverage or a long history of building projects, the Green Institute was motivated to assemble development funding from numerous and diverse sources. Though a commercial bank and the State of Minnesota provided 80% of the necessary funding, the total development package included equity funding from no less than 18 different sources and debt financing and guarantees from four sources.
- Targeted public-sector and/or private foundation financing can be pivotal for projects that have significant barriers to entry and/or lack of equity but a high chance of success once these limitations are addressed. Predevelopment 'soft' costs, being the most challenging to fund in the open market, are an attractive avenue for such funding vehicles.
- The PEEC project could have been successfully developed by a determined private for-profit developer with an equity investment and/or second mortgage of approximately 30% of the project's hard costs, or US$1,200,000. First-mortgage financing in the region is typically 75% of projected appraised value. The PEEC's cash flow at full absorption could conceivably cover a debt service equal to total project costs (US$5,200,000) at 1999 market interest rates of 8.5%.

17.5.3 Design and construction

- Comprehensive sustainable design can be accomplished in a commercial–industrial setting at significantly less than the traditional 10% premium expected by industry players. A post-opening evaluation of the PEEC project found that the sum of sustainable design measures cost the project 3% more than conventional code-compliant development. Many of the individual sustainable design elements of the PEEC had zero net impact construction costs or even reduced first costs.

- Simple payback analyses based on first costs and various resource savings are insufficient to evaluate the investment decisions for sustainable design. An analysis of the internal rate of return (IRR; see Table 17.1) is an important and enlightening exercise as are the addition of other inputs and potential impacts. These further analyses require other fixed and variable inputs such as opportunity cost of capital, projections of employee productivity effects and allocation of benefits to stakeholders. The same post-opening analysis mentioned above found that the PEEC is realising a 4.3 year payback for the Green Institute as developer, an immediate return to tenant businesses (considering they are paying above-market rents) and a 2.5 year payback for the combined savings for the developer and tenants. IRRs are 23% for the Green Institute and 39% for the developer–tenant combination.

- Because sustainable design remains an anomaly among design and construction professionals, careful encouragement and enforcement of plans and objectives by the developer is critical throughout the development process. The Green Institute found itself in the challenging position of serving as the 'sustainability police' during the construction effort to ensure adherence to and, if possible, improvement on building specifications.

17.5.4 Operation and long-term impact

- With little promotion the PEEC has become a major educational destination, entertaining several visits a week from individuals and organisations curious about sustainable design and eco-industrial development. More than 1,000 visitors have toured the facility since its opening.

- The PEEC has influenced development plans for properties surrounding the facility. The project has caused its neighbourhood, the City of Minneapolis and local businesses and developers to consider how to improve on conventional design and development. The construction of the LRT line has cast a spotlight on this area, and many development stakeholders are calling for the incorporation of sustainable design into projects along the rail corridor.

- Green buildings are an exercise in continuous improvement and are not complete after the construction and commissioning crews are gone. The energy efficiency and occupant health features of the facility require attention to ensure maximum productivity and to identify other opportunities for overall building improvement.

- By its nature, eco-industrial development is also an exercise in continuous improvement. Though the successes achieved thus far have been considerable, the PEEC continues to be a promising work in progress. The combined effort to

Feature, and financial benefit	Owner capital cost differential (US$)[a]	Annual return[b]		Payback time (years)	IRR (%)[c]
		TENANT	OWNER		
Sum of occupant health features: minimum 1% worker productivity improvement	143,000	43,000	N.A.	3.3	29
Ground-source heat pump: 35% more efficient than boiler or furnace; not subject to gas price volatility	48,000	6,500	N.A.	7.4	12
Efficient lighting and controls: reduced electrical load	6,000	700	N.A.	8.8	10
Air-to-air energy-recovery system: 79% energy recovery from exhaust air	10,000	3,500	N.A.	2.9	35
Energy-management system: co-ordinated energy allocation reduces the HVAC and lighting load	36,000	4,000	N.A.	9.0	9
Use of salvaged material installations: provides low-cost to no-cost substitutes for virgin materials	−20,000	0	N.A.	0.0[d]	N.A.
Native landscaping: eliminates need for irrigation infrastructure and reduces operating costs	−55,000	3,500	N.A.	0.0[d]	N.A.
Active skylights (energy):[e] reduced lamp installations and lighting load	90,000	5,000	N.A.	18.0	1
Lease premiums: rate premiums are chargeable as a result of design and networking amenities	169,000	39,700	39,700	4.3	23
Net total	169,000	26,500	39,700	2.5	39

HVAC = heating, ventilation and air conditioning
N.A. = not applicable
a The incremental increase (positive values) or decrease (negative values) in construction costs from current code-required building practices
b Estimate of financial benefit to building owner and tenant businesses arising from improved worker output, lower utility bills, prevented facility maintenance and rent premiums
c Internal rate of return; this is the implied return based on the capital cost differential, annual return and a 20-year holding period
d Immediate
e Skylight costs are also included in occupant health features; this calculation evaluate this feature based on energy savings alone

Table 17.1 *Payback and return on investment for selected sustainable design features*

get to this point was greater than ever imagined, and the Green Institute cannot assume that future endeavours will be any less challenging. This being said, our accomplishments and the public response to the project have also exceeded expectations, and it is clear that the PEEC will have a significant impact on how others conduct economic and real estate development.

18
CAPE CHARLES SUSTAINABLE TECHNOLOGY PARK
The eco-industrial development strategy of Northampton County, Virginia

Timothy Hayes
County of Northampton, Virginia, USA

Economic development and environmental protection are often viewed as competing interests at best and mutually exclusive at worst. Clean air and water, protection of wildlife and resource conservation are often pitted against jobs and business expansion. Proposals for new industry may be met with cries of 'not in my back yard' because of fears of pollution and negative community impacts.

To overcome this seeming incompatibility, Northampton County, VA, has opened the first phase of a new kind of industrial facility in which waste-streams are cycled into revenue streams and in which industrial processes are based on the design of natural systems. This ecological industrial park is the cornerstone of an innovative county strategy whereby economic development is protecting valuable environmental assets and whereby environmental protection is fostering development of a sustainable economy.

18.1 The land between two waters

Venture across the Chesapeake Bay from the mainland of Virginia to the southern tip of the slender finger of land known as Virginia's Eastern Shore and you will find Northampton County. With the Chesapeake on the west and the Atlantic on the east, Native Americans referred to this Virginia treasure as 'the land between two waters'.

Northampton County is a place rich in natural and historic resources, including miles of pristine beaches, a string of preserved barrier islands, thriving marshes and tidal creeks, fish and shellfish, birds and wildlife, open land and clean water, small towns and historic villages, woodlands and farms. Recognising the global importance of this ecosystem, the United Nations designated much of Northampton County and the surrounding region as the Virginia Coast World Biosphere Reserve (UNESCO 2003: USA, #31).

In sharp contrast to the county's natural and historic wealth, many of its people live in severe economic poverty. The closure of nearly all of the area's seafood and agricultural processing plants during the 1980s resulted in the loss of more than 1,500 jobs—more than 25% of the county's workforce. By the early 1990s, 28% of the population was living in poverty, while 10% of the county's homes lacked plumbing and 12% lacked adequate sanitary facilities.[1]

Having declined economically to become one of the poorest communities in Virginia and in the nation, the county began to wrestle with difficult, seemingly contradictory questions. Should the environment be sacrificed for economic development? Should a stagnant economy, lack of jobs and poverty be accepted as the price of protecting a unique globally significant environment? Should the county try to strike a balance between economic development and environmental protection that would provide for 'manageable' levels of both poverty and pollution? None of these scenarios was acceptable.

18.2 Triple bottom line

Rather than compromising the local economy for the environment, or vice versa, the county instead decided to pursue a strategy that would simultaneously maximise both the economy and the environment for the benefit of the entire community now and for future generations. Northampton County Administrator, Thomas Harris, summed up the challenge: 'We must do business today in a way that won't put us out of business tomorrow.'

The community and its governing body came to understand their role as trustees of a valuable portfolio of natural, cultural and human assets. Consequently, they began to explore ways to both *invest* and *protect* these assets in order to build a strong and lasting economy and to preserve one of the last truly exceptional unspoiled environments on America's Atlantic coast. From now on, success would be measured in terms of a triple bottom line: economy, environment and equity.

18.3 Sustainable development action strategy

In 1993 the county formally began its sustainable development initiative with a mission to 'build a strong and lasting economy by capitalising on and protecting Northampton's rich natural, cultural and human assets'.[2] The commitment was to develop in a manner

1 United States Census, 1990.
2 Northampton County Sustainable Development Action Strategy.

that would simultaneously benefit business, the environment and all the county's people both now and in the future.

Through a partnership with the National Oceanic Atmospheric Administration (NOAA) and the Virginia Coastal Programme (VCP), the county hired the nation's first local director of sustainable development (this author). With funding from the NOAA, VCP and other partners, Mr Hayes organised Northampton's Department of Sustainable Development and led the county in formulating a vision for its future and a strategy to achieve that vision.

This Northampton County Sustainable Development Action Strategy was developed through an intensive, collaborative process involving community workshops, task forces, meetings and events. The county involved a broad cross-section of its diverse citizenry in the strategy development process, leveraging their combined experience to identify industries with realistic, significant, immediate and ongoing potential for development. As a result of this process, Northampton decided to target the following industry sectors: agriculture, seafood and aquaculture, heritage tourism, research and education, arts and crafts, local products, and sustainable technologies. The community also identified the vital natural, historic and community assets that would need to be preserved and capitalised on to successfully develop and sustain these targeted industries. The final step in the process was the formulation of an action plan for implementing the strategy. In June 1994 the Board of Supervisors adopted its Sustainable Development Action Strategy as the county's official development policy and immediately began its implementation.

One of the keys of the Sustainable Development Action Strategy was to leverage private investment in the county in specific ways that would achieve the integral goals of building an economic base and protecting the natural and cultural assets that support that base. It was clear that severely limited public financial resources would have to be carefully focused to achieve the greatest possible return for each dollar invested. It was also clear that Northampton's strategy would need to be phased in over time using major impact projects as long-term building blocks. To launch implementation of its sustainability strategy most effectively, county officials had to determine which course of immediate action would yield the highest economic and environmental return in the shortest period of time. The decision was to create an eco-industrial park to be known as Sustainable Technology Park and to be located in the town of Cape Charles.

18.4 A world-class eco-industrial park

As it began planning infrastructure for business development, Northampton set its sights on building a world-class facility commensurate with its place and purpose. Rather than developing an industrial park that would be one of the last of the old ways of doing business, an industrial park here would have to be developed as one of the first of a new industrial revolution. It must be an *eco*-industrial park; 'eco' both for the county's ecology and for its economy. The goal was to provide an 'environment for excellence' that would attract and grow companies that would share the county's high business, environmental and human equity standards. The vision was for a 'green' industrial park—green as in the colour of a healthy environment and green as the colour of money (the US dollar) and a healthy economy.

In late 1994, Northampton County and the Town of Cape Charles passed a joint Resolution of Commitment pledging vital local resources to develop the eco-park. Immediately they hosted a launch event and planning workshop that brought together a diverse team representing virtually everything and everyone necessary to make the eco-park a reality. Along with about 150 members of the local community, the planning team included designers, architects and engineers led by the University of Virginia School of Architecture; federal, state and local government regulatory and support agencies; the owners of the future eco-park site; public and private potential funders; non-governmental organisations; potential corporate tenants; and representatives of the President's Council on Sustainable Development. This gathering led to the formation of the Eco-Industrial Roundtable led by Cornell University's Ed Cohen-Rosenthal, and later to the incorporation of the Eco-Industrial Development Council (EIDC).

Out of this event came a strong consensus vision for a park that would help build a strong and diversified economic base by attracting and incubating new companies and by retaining and expanding existing companies. These companies would provide quality jobs with competitive wages, benefits, and opportunities for training and advancement.

The park would preserve natural and cultural resources, protect habitat and water quality and strive to eliminate waste and pollution. It would showcase green technology companies and maximise efficient use of resources through industrial symbiosis and eco-industrial networks whereby the by-products of one industrial process would serve as raw material for another within the park or the region.

Figure 18.1 **Sustainable technologies: Northampton County's Atlantis Energy Systems manufactures products that integrate solar electricity generation into building materials such as roofing, siding and window glass, making buildings producers—not just consumers—of energy**

Photo: Tim Hayes

18.5 Funding

Building on the momentum created by the public launch event, the Northampton Department of Sustainable Development raised more than US$6.6 million of local, state and federal funding to develop the park's first phase (Table 18.1). A tangible demonstration of the community's commitment to its eco-park, this funding included a US$2.5 million bond issue approved by local voters. To date, this public investment has leveraged another US$8 million from private companies locating in the first phase of the park.

The public funding covered everything necessary to open the eco-park facility, including master planning, community involvement, covenants and sustainability standards, environmental assessments, land-purchase, engineering, permits and approvals, infrastructure construction, multi-tenant green building construction, a solar electricity sys-

Funding body	Amount (US$)
Land and infrastructure funding	
Northampton County	600,000
National Oceanic and Atmospheric Administration, Virginia Department of Environmental Quality, Virginia Coastal Programme	173,000
Virginia Department of Transportation	665,200
Virginia Governor Opportunity Fund	170,000
US Department of Agriculture	490,000
US Department of Commerce	500,000
Total infrastructure funding	2,598,200
Funding for sustainable technology incubator	
National Oceanic and Atmospheric Administration, Virginia Department of Environmental Quality, Virginia Coastal Programme	10,000
Northampton County	1,900,000
Energy Recovery Inc.	196,320
US Department of Agriculture	500,000
US Department of Energy, Virginia Alliance for Solar Electricity	146,422
Total funding for sustainable technology incubator	2,752,742
Ecological infrastructure funding	
National Oceanic and Atmospheric Administration, Virginia Department of Environmental Quality, Virginia Coastal Programme	156,600
US Fish and Wildlife Service, Virginia Department of Conservation and Recreation	346,000
US Environmental Protection Agency, The Nature Conservancy	75,000
Northampton County	200,000
Total ecological infrastructure funding	757,600

Table 18.1 *Cape Charles Sustainable Technology Park, Phase 1: public funding distribution*

tem, purchase of natural areas, construction of lakes and wetlands, the creation of trails and amenities, marketing, legal costs and the leasing of initial corporate tenants.

Private investment provided for tenant improvements for office, research and development (R&D) and manufacturing space as well as partial funding for the park's solar energy system. In addition, US$7.8 million was committed to development of a 9 MW wind farm in the park that will produce more electricity than used by the entire county. The park will export this wind power to utilities within the mid-Atlantic US energy grid, serving customers in Pennsylvania, Maryland, New Jersey and Washington, DC, with clean, renewable energy.

18.6 Master plan

Early in the eco-industrial development process, a master plan for the park was created through an intensive three-day community design workshop known by architects as a 'charrette'. The plan carefully integrated the park with the historic town of Cape Charles and the natural landscape adjacent to Chesapeake Bay. The site centred on redevelopment of former industrial land surrounding the town's harbour. The specifics of the plan included roads, utilities, sewers, water and storm-water management and wetland tertiary treatment for water recycling. Fully half of the site was reserved for 'ecological infrastructure', including creation of the Cape Charles Coastal Habitat Natural Area Preserve along the shore of the Chesapeake Bay. This preserve is the focal point of a network of natural and created wetlands, woodlands and scrub/shrub wildlife habitat especially valuable for the millions of neotropical migratory songbirds that concentrate there each autumn. As an ecological industrial park, facilities plans were based on the design of natural systems. Electricity would be generated from sunlight and wind. A water re-use and recovery system was planned to recycle water for industrial use. Porous paving was designed into parking areas to reduce storm-water run-off, which is collected and filtered by constructed wetlands.

18.7 Management structure

To most effectively realise its eco-park, the community formed its own development company, the Sustainable Technology Park Authority. This body was incorporated as a political subdivision of the Commonwealth of Virginia with the mission to 'develop, manage, and market the Cape Charles Sustainable Technology Park as a world-class eco-industrial park providing all necessary facilities, services and surroundings to support an unsurpassed environment for business and industrial excellence'.[3] Its seven-member board of directors is appointed by the governing bodies of Northampton County and each of the towns within the County are to help facilitate an eco-industrial network that will

3   Sustainable Technology Park Business Plan.

Figure 18.2 *Synergy of industry and nature: construction of Sustainable Technology Park's multi-tenant eco-industrial incubator building included excavation and planting of a new lake and wetland*

Photo: Tim Hayes

transcend the park's formal boundaries and engage the region's farming, fishing, manufacturing and other industries.

18.8 Land assembly

Slightly more than 200 acres of land was acquired in five transactions between 1995 and 1998, including 30 acres adjacent to Chesapeake Bay to be permanently preserved as the Coastal Habitat Natural Area Preserve. Contrasting this pristine natural land, plans for the eco-park also included 45 acres of brownfields centring on abandoned industrial sites surrounding Cape Charles harbour and a former town dump. Recognising the opportunity the Cape Charles eco-park afforded as a model for sustainable re-use of brownfields, the US Environmental Protection Agency (EPA) funded environmental assessments that documented the lack of hazardous materials or pollution problems associated with the site. This gave a green light for acquisition, financing and construction and exemplified the eco-industrial principles of re-use by cycling abandoned land into valuable real estate.

18.9 Infrastructure

Detailed engineering designs were created based on the vision and plans of the original community charrette. Construction documents were subsequently prepared for the first phase of construction of roads, utilities and other infrastructure to serve the park. Local, state and federal permits and approvals were obtained through a streamlined process, as representatives of most of the agencies involved had participated in the community design charrette. Ground was broken on 17 October 1996 during a conference of the international Eco-Industrial Roundtable held in Cape Charles. The first phase of construction of roads and utilities was completed in 1998.

18.10 Sustainable technology incubator

Completed in January 2000, the showpiece of the first phase of Sustainable Technology Park is a 31,000 ft² multi-tenant manufacturing and office building. The building meets the 'silver' rating of US Green Building Council's Leadership in Energy and Environmental Design (LEED) standards as a 'green' building. It features an integrated solar photovoltaic roof system that converts sunlight into 42,000 watts of electricity. With the building operating at full capacity, this will meet up to half of the building's total annual electrical demand.

Other sustainability features include skylights for natural daylighting, a common meeting and conference centre, enhanced insulation, interior environmental sensing, carbon monoxide sensors and alarms, low-energy lighting, low-water-use fixtures, porous parking lot paving and native non-irrigated landscaping. The building, which was constructed primarily from local materials, has an enhanced structural strength and a longer lifespan than typical designs. Interior space can be divided flexibly to accommodate up to eight companies for the purposes of manufacturing, R&D, office space and other uses.

Figure 18.3 *Clean and renewable: the rooftop of Sustainable Technology Park's first building generates 42 kW of electricity, powering eco-industrial tenants and selling the excess to the regional energy grid*

Photo: Tim Hayes

The building's features are designed not only to reduce energy and resource demands but also to reduce operating costs and to increase occupant productivity and health. As such, the building is expected to be superior economically as well as environmentally.

18.11 Ecological infrastructure

In addition to the more traditional infrastructure of roads, utilities and buildings, Northampton County and its partners integrated key natural resources into Sustainable Technology Park. A local–state–federal funding package of more than US$757,000 was used to fund the park's 'ecological infrastructure' (Table 18.1). Included are a natural preserve, beaches and dunes, a critical migratory bird habitat, constructed wetlands and ponds, a system of trails and boardwalks, more than 4,000 new trees and a total of 90 acres of protected natural area.

This ecological infrastructure created a win–win–win synergy between the economic development, environmental protection and community improvement objectives of the park. The natural amenities have enhanced economic development efforts by helping to attract the corporate tenants the county has targeted. Without the projected financial income produced by the corporate tenants of the technology park, it would not have been possible to fund protection of the natural areas or construction of the trails, wetlands and ponds. The community, meanwhile, has gained a new natural area park, which has proved to be popular with joggers, birdwatchers and families alike.

18.12 Target market

Northampton combined a basic definition of technology—'the practical use of science and engineering'—with the United Nations definition of sustainable development: 'development that meets the needs of the present without compromising the ability of future generations to meet their own needs' (WCED 1987). The result became the county's working definition of **sustainable technology**: 'the practical use of science and engineering to meet the needs of the present without compromising the ability of future generations to meet their own needs'.

Within this broad definition, the county targeted: companies working with renewable energy, including solar, wind and bio-fuels; companies working on clean-water technologies involving purification, desalination, and water re-use and recycling; companies in fish and shellfish aquaculture; and companies involved in adding value to new and traditional local agricultural products. In 1998 Northampton County contracted Ed Cohen-Rosenthal and the Cornell University Center for the Environment, which focused on these industry areas and compiled a database of more than 3,000 companies targeted as priority prospects for expansion or relocation to Sustainable Technology Park. These companies were seen as possible initial building blocks for an industrial ecosystem within the park and surrounding county and region.

To help attract these companies, a matrix of sustainability criteria was built into the legal codes and covenants for the eco-park. These include incentives in the form of rent reductions for implementation of industrial ecology within or among companies that cycle by-products and would-be waste into marketable products.

18.13 Results

The sustainable technology strategy has already attracted several diverse companies to Northampton County and its eco-park. The Norwegian-based company, Hauge Technologies, originally leased space in the park to develop and manufacture 'pressure exchanger' equipment that significantly reduces the cost and energy demand of converting sea water to fresh water. This technology promises to make safe drinking water more affordable and available to more people worldwide. ProVento America Inc. is a subsidiary of a German wind energy developer that expects to grow to 50 employees within five years. Initially, ProVento has leased sites for six wind turbines that will produce 9 MW of electricity to be sold to the national power grid.

The newest company located in Sustainable Technology Park is producing a pharmaceutical from the blood of horseshoe crabs which are taken briefly from the Chesapeake Bay adjacent to the Park and returned unharmed. Wako Chemicals USA, a subsidiary of Wako Pure Chemical Industries of Osaka, Japan, uses the horseshoe crab as the source of a protein 'limulus amoebocyte lysate' (LAL), a reagent used by the pharmaceutical industry for detection of certain types of toxins prior to the release of pharmaceuticals into the market (see Fig. 18.4).

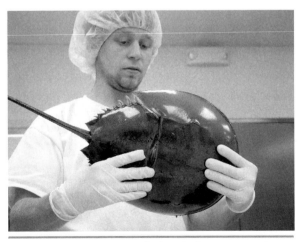

Figure 18.4 *Tapping nature's chemical treasure: Wako Chemicals USA produces a high-value, life-saving pharmaceutical from the blood of horseshoe crabs which are returned to the water unharmed*

Photo: Tim Hayes

The park has also formed an eco-industrial partnership with a neighbouring company, Bayshore Concrete Products Corporation, the county's largest manufacturer with more than 400 employees. Through an innovative exchange, the company has dedicated land adjacent to the Chesapeake Bay for expansion of the park's natural area while the park provided land for expansion of the company's operations. Also, plans are under way for re-use of water from the park and the Cape Charles municipal treatment plant for Bayshore's concrete products manufacturing.

In addition to the companies on-site in the park, Northampton's sustainable development efforts have attracted several companies that have located throughout the county and its towns. One of these companies, Atlantis Energy Systems, is the successor of a Swiss firm that produces 'architecturally integrated photovoltaics'. These are commercial and residential roofing, siding, windows and skylights that generate electricity from the sun as well as serving as components of a building's exterior. Scientific and Environmental Associates is a consulting firm that provides coastal resource management services to government and corporate clients the world over. The company moved its headquarters to Northampton because of its commitment to sustainable development and technologies. Delisheries is a home-grown company using space in the park's incubator to produce gourmet cookie (biscuit) mixes.

Within a year of the park opening, these companies have combined to generate more than 50 new jobs, a significant impact given Northampton's rural economy. Over the next two years, they are expected to create an additional 50 jobs and to bring US$15 million in direct real estate and equipment investment to the county. Cape Charles Wind Farm alone is valued at US$7.8 million and will generate US$120,000 annually in business personal property, machinery and tools taxes.

Annual debt service for the park is approximately US$200,000, with annual operations and maintenance costing about US$100,000. As soon as the first building is fully leased it will generate an annual rental income of US$180,000. Considering annual tax revenues of US$120,000, the park's direct revenues and expenditures are balanced for its first phase. Such financial sustainability will remain a key measure of the park's success as development continues.

A total of 800,000 ft^2 of building area is planned for future phases of the park, which will provide space for 200 or more companies and 1,200–1,500 jobs. The park's first incubator building contains 31,000 ft^2 of space to accommodate its first four to six companies and as many as 100 employees. Depending on market conditions, build-out of the park will occur over the next two or three decades, providing a long-term public framework for private investment.

18.14 Toward a sustainable future

Building on the success of the Cape Charles eco-industrial park, Northampton is already aggressively moving forward with the next phase of its sustainable development strategy. One key project is the reclamation of the county sanitary landfill as a 'seaside ecological farm'. The goal is to turn trash into treasure by transforming an economic and environmental challenge into a valuable community asset. The county landfill will reach capacity within two years, at which time it will have to be closed and capped. It forms the highest point in the county and affords breathtaking views of the Atlantic Ocean, coastal wilderness marshlands and a string of barrier islands. Conceptual development plans include a network of hiking trails, a high-point observation platform and the restoration of ponds, wetlands, forest and other bird habitats.

The seaside ecological farm project has already been selected as a Brownfield Redevelopment Showcase by the US EPA, which has committed US$400,000 to support the project. In addition, the National Oceanic and Atmospheric Administration has committed US$10,000 to fund partner workshops for project planning. The county has also entered into a Rebuild America partnership with the US Department of Energy to develop the renewable energy components.

The county is investigating the potential of a renewable energy farm that would harvest methane from the landfill, bio-fuels from soybeans grown on the county's farms and wind energy from the strong coastal winds. The six wind-energy turbines planned for the landfill property will produce 7.8 MW of clean renewable electricity to be exported to the northeastern US power grid. Early financial projections indicate that tax revenue from these turbines will offset a significant portion of the debt service required to fund the multi-million-dollar landfill closure costs.

Northampton's strategic plan also focuses on ways to further support the county's growing aquaculture industry, which involves cultivating fish and shellfish for harvest. The most profitable species being grown in Northampton's aquaculture farms is the famous Cherrystone Clam, which has grown from less than a US$1 million dollar industry in 1990 to an industry worth more than US$20 million in 2001. Other fish and shellfish species are also being developed for commercial aquaculture farming, and, by integrating aquaculture with agriculture, the county hopes to create additional eco-industrial synergies whereby soy–diesel by-products serve as fish food, and where fish farm effluent provides nutrient-rich agricultural fertiliser. Virginia Tech, The Nature Conservancy and Cornell University's Work and Environment Initiative have provided initial support for this work.

18.15 Conclusions

Northampton's Sustainable Technology Park is living proof that economic development and environmental protection are not mutually exclusive and that these goals can be realised simultaneously through eco-industrial development to enhance a community's overall quality of life. Although the county has a long way to go to completely overcome its severe poverty and to rebuild a healthy economic base, its experience demonstrates that applying eco-industrial principles to integrate asset development and protection is a powerful strategy for success. In Northampton's experience, a comprehensive, action-oriented community-based plan provided an effective framework for launching an eco-industrial park. A collaborative eco-industrial round-table process was the foundation for building an extensive local, state and federal partnership for initial project financing and for leveraging long-term private investment. Eco-industrial development is a cornerstone of Northampton's development strategy thus far and is expected to be relied on as the county continues its course toward a sustainable future.

SUSTAINABLE LONDONDERRY

Peter C. Lowitt, AICP
Devens Enterprise Commission, USA

Londonderry, NH, was seeking an economic development strategy to establish itself as a viable location for industrial development, when the concept of eco-industrial development fell into its lap. Stonyfield Farms Yoghurt (hereafter referred to as Stonyfield) was approached by a company that wanted to locate a plastic recycling operation on the vacant land next to it to utilise Stonyfield's grey water to rinse the plastics. The town had acquired this land from a previous developer a few years earlier for failure to pay taxes. Stonyfield Farms is an environmentally conscious, socially responsible firm intrigued by the opportunities that industrial ecology presented to it and to its host community. Londonderry saw this as an opportunity to attract companies similar to Stonyfield to the community and to create a win–win situation for industry and the environment. The town and Stonyfield decided to explore the opportunity to create an ecological industrial park on the 100 acres of town-owned land adjacent to Stonyfield. Thus was born the sustainable economic development strategy that has come to be called 'Sustainable Londonderry'. In a presentation to the American Planning Association, I stated that:

> Sustainable Londonderry is the story of a community with three Superfund sites, five orchards, two interstate exits and a regional airport and how the theme of sustainable development weaves these disparate elements together. The end result includes Londonderry Ecological Industrial Park (LEIP), the municipality participating in ISO 14000 training and efforts to preserve the community's original sustainable economic development base, its agriculture (Lowitt 1998: 1).

19.1 Londonderry, New Hampshire

Londonderry is a suburbanising community of 22,000 people. It is located in southern New Hampshire, just 40 miles north of Boston, MA. Chartered in 1722, Londonderry grew from a sleepy agricultural community of 2,000 people in 1960 to its present size when Interstate 93 was built in the early 1960s. During the past three decades Londonderry's agricultural heritage has come under assault as one of the fastest-growing communities in New Hampshire. Rapid uncontrolled growth has brought its share of problems to this region. Shopping centres sprawl where apple orchards once grew. Town infrastructure such as schools and police and fire departments have been striving to 'catch up', yet operations are not properly scaled to population size. Today, the community continues to grapple with growth-related issues.

19.2 Lessons learned

Londonderry had three Superfund sites—Tinkham dump site, Auburn Road Site and the Radio Beacon site. With no sales or income taxes in New Hampshire, property taxes have had to bear the brunt of paying for cleaning up these sites. Citizens have spent upwards of US$13 million to correct these problems. Having experienced the environmental and economic impact of the negative sides of development, residents are now mobilising to preserve the remaining agricultural heritage and to promote appropriate and well-planned development. A number of initiatives resulting from these lessons, especially Sustainable Londonderry, were recognised by the State of New Hampshire with the awarding of the first-ever Governor's Municipal Pollution Prevention Award in 1998 (Lowitt 1998).

19.3 Londonderry Ecological Industrial Park: a work in progress

Today, the Londonderry Ecological Industrial Park continues to establish successful precedents for communities interested in eco-industrial parks (EIPs). It is the first community to partner with and eventually sell the park to a private-sector developer (Sustainable Design and Development LLP of Bedford, NH). Londonderry established a model for park management and for a community stewardship board. Equally exciting is that the park has synergies built into it. For example, by attracting AES to develop a 720 MW combined-cycle natural gas power plant at the park using treated waste-water pumped from the City of Manchester's Waste Water Treatment Facility, it has shown that one really can turn a 'waste-stream' into a revenue stream. The park also illustrates that the lots within a park can be marketed at a premium price, and one unintended consequence has been the establishment of green building design principles for power plants. These successes have not come without controversy.

The Conservation Law Foundation (CLF), a New England regional environmental advocacy group, brought AES to the Londonderry Ecological Industrial Park. CLF saw that combined-cycle gas-fired plants would have a beneficial impact on air quality throughout the region and recognised AES's commitment to social responsibility as a suitable match for the Londonderry Ecological Industrial Park. Unfortunately, the last power plant sited in New Hampshire was the controversial Seabrook Nuclear Power Plant. Seabrook visually dominates the southern New Hampshire coast and created a poor perception of power plants for the entire state. In 1999, the New Hampshire State Facilities Siting Council granted a building permit to the AES Londonderry plant. Neighbours appealed against the decision to the New Hampshire Supreme Court, which upheld the Facilities Siting Council's decision to permit the plant. It is currently under construction.

One of the requirements for locating within the Londonderry Ecological Industrial Park is that the development go through a review by the community's green building or ecological design consultants. In the process of addressing these requirements, AES hired the artist Michael Singer to pull together a design team to create a green design for the Londonderry Plant (see Fig. 19.1).

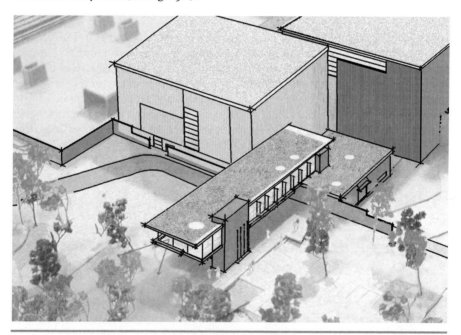

Figure 19.1 **The AES Granite Ridge plant. The design has a green roof, landscaped berm and buffer shielding the transmission area from public view and a visitor centre adjacent to the pond (on the right-hand side of this figure). The balance of AES's property is conservation land.**
Image: Michael Singer and Blackbird Architects

19.3.1 A model of sustainable development

EIPs offer a new paradigm for an antiquated system. There is mounting concern that the traditional linear system of industrial processes whereby resource-based inputs are mined, processed and dumped in concentrated form into the air, water and soil threatens future generations. The linear model is based on the assumption that resources are limitless and that 'waste' is an inevitable and acceptable consequence. Although there is great disagreement as to the Earth's carrying capacity and resource base, most will agree that neither is limitless. Applying the awareness of finite resources and the concern for providing for the needs of future generations has led to a new paradigm of a cyclical industrial system whereby industrial processes model natural ecosystems. Resource efficiency, elimination of the concept of waste by turning it into food for another part of the system and functioning on current solar gain are integral principles of industrial ecology.

Environmental and cost benefits to industrial systems operating under this new cyclical ecological approach include: reduction in virgin material input, reduction in pollution (output), increased systemic energy efficiency, reduced systemic energy use, reduction in volume and cost of waste and an increase in the amount and type of process outputs that have market value (Gertler 1995). Put another way, industrial systems operating under the EIP concept will result in the reduction and elimination of air, water and soil pollution and of solid and hazardous waste and in decreased mining of the Earth's resources as well as in cost savings (Lowitt 1998).

19.3.2 The vision

The Londonderry Ecological Industrial Park evolved out of a deep understanding of the principles of sustainability. The vision statement (1996) is as follows:

> The Eco-Park recognises as its primary function developing systems and processes which minimize the impact of industry and business on the environment, improve the economic performance of the member companies and strengthen the local economy. Through modeling the Park's industrial systems on natural ecosystems, decreased environmental impact will be realised.[1]

The vision statement identifies six key principles: sharing a common mission through long-term partnerships, accountability, striving for continuous improvement and innovation, land stewardship, serving the local community, and serving one another.

19.3.3 Covenants

The town, in consultation with Stonyfield, appointed an EIP advisory board and made a conscious decision to utilise covenants as an enforcement tool for ensuring that the park is developed as an ecological industrial park. The use of existing tools and existing zoning was designed to make the park easily replicable for use as a model. Covenants are a commonly used vehicle for privately regulating business and industrial parks. An EIP adds an ecological element to this traditional business-sector tool. The covenants and governance agreements require that all EIP tenants develop an environmental management system, track their resource use, set environmental performance goals, perform third-party ecological audits and report progress to the Community Stewardship Board (a self-management board to be composed of park tenants, the developer, town, citizen and environmental representatives). Practices incorporated into the governance agree-

1 www.londonderry.org/page.asp?Page_Id=528

ments affect all media, including standards for decreasing business impact on air, water and soil, ecological auditing for continuous environmental improvement, environmental goal-setting, input and output management and sharing, inter-firm collaboration, energy efficiency, water conservation, product stewardship, environmental reporting, stakeholder accountability, facility design and material use, design for environment and restorative activities.

The park management structure enforces adherence to the legally binding covenants to ensure continued compliance in fulfilling the mission of the EIP and protecting the site from environmental degradation. The advisory board worked closely with Sustainable Design and Development (the park's current private-sector owner) to develop the governance structures. They are the result of a dynamic process of 'give and take' between local government and environmental constituencies and private-sector development interests. At one point, an environmental law firm was brought in to mediate the discussion and move the parties concerned closer to agreement. The result is a balanced governance structure that reflects compromises by both sides and creates a model for private-sector EIP development.

19.3.4 Stakeholder involvement

Stakeholder involvement began with the careful construction of an advisory board for the park. Nancy Hirshberg, chief environmental officer for Stonyfield, sat down with town staff and created a 'dream team' of advisors. Representatives from Businesses for Social Responsibility, the Environmental Business Council, Londonderry citizens and elected officials, representatives from the New Hampshire Department of Environmental Services and the US Environmental Protection Agency (EPA) Region I along with academics and non-profit organisations were purposefully chosen to be on the board. This broad circle of representation and involvement helped to make this project one of state-wide and regional importance. When the park's first tenants sought approval for their projects they found a receptive audience in state and federal regulators who knew of the park and who were aware of the community's efforts. A small town staff (a one-person Planning and Economic Development Department) limited proper implementation of the stakeholder involvement plans. A series of neighbourhood teas to introduce the project to neighbourhood groups never got off the ground. In hindsight, this proved to be a problem, and thorough integration of the park's residential neighbours into park planning and development, no matter how distant those neighbours (in this case $\frac{1}{4}$–$\frac{1}{2}$ mile away), should not be overlooked by others interested in creating a park.

19.3.5 Community Stewardship Board

The purpose of the Community Stewardship Board is to ensure public accountability through the review and release of an annual environmental and ecological report. The Board is charged with ensuring adherence to the Londonderry Ecological Industrial Park's vision statement, to facilitate communication between the community and the park and to operate as a Section 501C3 non-profit organisation to secure grants to further these goals.

19.3.6 Londonderry Ecological Industrial Park by-laws

The by-laws establish the relationship between the developer (Sustainable Design and Development) and the park tenants. They establish the performance requirements for locating a firm within the park, spell out the details of the annual report, establish the review of the project by the town's Ecological Development Team of Consultants and ensure town representation on the park management board.

19.3.7 Performance requirements

Performance requirements are: specify adherence to the vision statement; participation in public accountability procedures, such as filing the annual ecological report signed by the firm's chief executive officer; underscore the firm's commitment to seek synergies; ensure compliance with applicable environmental, health and safety laws; and require adoption of an environmental management system equivalent to the ISO 14000 series requirements.

19.3.8 The AES plant

A press release in 2000 described the plant as follows (AES 2000):

> Primarily fueled by clean natural gas, the AES facility will be one of the cleanest, most energy efficient facilities in the world and will provide many benefits to the region, including clean, low-cost power, approximately $5 million in annual property taxes to Londonderry and improved regional air quality.

19.3.9 Other tenants

AES is not the only tenant in the Londonderry Ecological Industrial Park. Sustainable Design and Development has attracted Gulf Southern, a medical supply distribution firm to the site; Rideaway Vans of Londonderry has expanded its existing adjacent facility into the park; and a computer software firm is building in the park as well. In fact, Sustainable Design and Development is holding the remaining few lots off the market until AES gets under way in order to effectively market the synergies AES brings to the table, such as district heating and cooling. Sustainable Design and Development will be pricing these lots at premium prices (i.e. above the market rate).

19.3.10 Funding

The park concept was developed with very minor direct costs on the part of the town. Stonyfield dedicated a tremendous amount of time to the project, through its chief environmental officer, Nancy Hirshberg. Town personnel also dedicated a good portion of their time toward the project. Two internships were utilised by the town to forward the park's development, and Stonyfield and the town spent US$5000 to hire the Conway School of Landscape Design to develop visuals to compare conventional development of the site to its development as an EIP. The town and Sustainable Design and Development also funded an environmental mediator to assist with the final phases of developing park governance documents. In addition, town legal staff spent a great deal of effort on the project. On a personal note, if we had known how much work was involved and how long it would take us, we might not have ventured down the primrose path. Fortunately for us, 'fools rush in where angels fear to tread', and the reader may benefit from our experience.

19.3.11 Privatisation

The New Hampshire State motto is 'live free or die', and 'don't spend taxpayer dollars foolishly' is the unpublished caveat. The town was unlikely to fund the infrastructure development necessary to successfully market and develop the park. Therefore town staff recommended that it be sold to a developer with governance structures and provisions in place that ensured the area would be developed as an EIP. Developing these provisions was the park's darkest hour. The dynamic tension between regulators and businesses came to the forefront as both sides engaged in hard-nosed negotiations to balance the needs for environmental protection and marketability. This resulted in both sides being somewhat dissatisfied with the final results, and yet it was probably one of the advisory board's greatest accomplishments. The developer bought the park, at a subsidised price, the back taxes and interest. It took out a loan to fund the infrastructure needed to develop the rest of the park and repaid the loan in full within six months. Sustainable Design and Development is now one of the biggest advocates for EIPs in the USA.

19.4 Londonderry Ecological Industrial Park: the future

With AES coming online in 2002, the park's success will be assured. Its true potential lies in its ability to influence development in the 1,000 acres of industrially zoned land it abuts. The advisory board created an associate membership position open to all companies willing to adhere to the park's vision and willing voluntarily to adhere to its performance requirements. AES has indicated it will sell hot water and steam to companies within its 'throw' range at prices below those of its competition. AES also offers price breaks on electricity for companies locating in Londonderry. Further price breaks may be offered to park associate members. This provides the opportunity for creating a green industrial base for Londonderry and southern New Hampshire. Welcome to the intersection of economic development and environmental protection. Its address is Londonderry Ecological Industrial Park, Londonderry, New Hampshire.

20
THE RED HILLS INDUSTRIAL ECOPLEX
A case study

Ron Forsythe
Pickering Inc., USA

It all began with a phone call. In April 1993, former governor of Mississippi, Kirk Fordice, received a phone call from now deceased Jake Mills. Mills and his close friend, Jack Sistrunk, were playing golf in Florida. Sistrunk asked Mills what it would take to get an audience with the governor to discuss a potential project in Mississippi. That prompted the phone call. The project was to be the Red Hills Power Project. In the ensuing years, the project has come to life. This is the story of what transpired to make the project a reality and to bring life to the Red Hills EcoPlex, the industrial park of the future.

In late April 1993, the governor hosted a meeting in the Mansion in Jackson, Mississippi. At the meeting were officials of the Phillips Coal Company and the former CRSS Capital (now Tractebel, North America), both Texas-based energy companies. As a representative of the Energy Division of the Department of Economic and Community Development, I also attended the meeting to advise the governor on the technical merits of the project. The project was proposed to be a mine-mouth lignite mine and a state-of-the-art electrical generating station. It was proposed to be a partnership arrangement between the mining company, the electrical generating company, the State of Mississippi, Choctaw County and the Tennessee Valley Authority (TVA). The US Department of Energy and the US Economic Development Authority subsequently became partners in the effort as funding agencies for parts of the project involving the Energy Division and Choctaw County. At the conclusion of the presentation by Phillips Coal Company and CRSS capital, the governor asked my opinion of the proposal. I responded, 'Don't stop with just one, governor.' The lifeblood of the Red Hills Power Project started to flow.

In August 1993, Fordice, Mills, Sistrunk, the Energy Division Director, Chester Smith, and I were taken on a tour of the TNP-1 lignite-fuelled electrical generating station near Calvert, TX. This facility is a relatively small circulating fluidised-bed electrical generating station. Phillips Coal Company was the owner of the lignite mine. After touring the station, which exhibited absolutely no visible emissions, the governor was fully convinced that his initial decision to support the project was correct. He became one of the project's most vocal supporters and advocates.

Meetings were conducted with representatives at all levels of the TVA. The attitude of TVA was initially very cool toward the project. TVA was not at all excited about having an independent power producer own and operate a generating station in its service territory, and even less eager to purchase the power from that facility.

Meetings were also initiated with the citizens of Choctaw County in an attempt to gain their support. The initial attitude of the citizens of Choctaw County and the surrounding counties was one of scepticism and reluctance. Phillips Coal Company was seen as a company that would come in, develop a typical strip mine, take the coal and walk away from the project. It would take years and some really creative public relations to gain the support of the people of north-east Mississippi that the project would bring tremendous economic opportunities. In December 1994, Phillips and CRSS Capital held an old-fashioned county-fair-like event at the armoury in Ackerman, county seat of Choctaw County. Presentations were made on all aspects of the project. Small working-group discussions were conducted and the public was given the opportunity to directly interact with the project developers. Support for the project began to grow. Several of the key selling points regarding the project were the fact that Mississippi, despite being the first state to become a member of the TVA, had no generating capacity and was subject to power losses; the number of jobs to be created was projected to be approximately 180, and those jobs would be given mainly to local people; the jobs would pay very well; the tax base of the county would be increased substantially; and many people would be handsomely rewarded with the royalties they would receive from the coal. Additionally, the people were told that the project would create stable electrical cost, as the cost of the fuel would not be subject to large fluctuations as was the case with other fuel sources.

In the months that followed, at the recommendation of the Energy Division, citizens and officials of Choctaw County were taken to Calvert, TX, and taken on a tour of the facilities there. Dan Bollner of Phillips Coal, one of the most instrumental people in making the project become a reality, orchestrated the trip. It was Dan that really made the project come to life. Elected officials were met by their Texas counterparts and told of the impacts of the project on the community and the surrounding area. Employees at the Texas facility told their story and of the positive impact it had made on their lives and on their families. The project gained considerable support as a direct result of those initial events, but it did not end there. Citizens and elected officials from the surrounding counties were given the opportunity to visit the Texas facility, and further support and enthusiasm for the project developed.

Despite all the support, TVA officials were still very reluctant to discuss the project, much less support it. During one of the initial discussions with TVA officials, the issue of cost of power was raised. The wholesale cost to TVA was cited at about five cents per kWh. The TVA officials thought that the price quote was a joke. TVA is one of the largest public power companies in the country, and at the time was subsidised by the federal government. Their average cost of producing power for its customers was about two cents. Much of the power TVA generates is from hydroelectric facilities, nuclear plants and coal-fired generating stations, all of which produce reasonable cost energy. But TVA was caught in somewhat of a dilemma. It was having to make some decisions about providing enough power for the rapidly growing customer base. TVA created a committee to conduct a study of the need for power and how that power was to be provided over the next 25 years. Chester Smith was named to sit on that committee, and the Red Hills Power Project gained an inside track on the outcome of the study.

The never-completed Bellefonte Nuclear Station in northern Alabama came into the picture as competition for Red Hills. TVA had spent an enormous amount of taxpayer dollars on the project, a project that was only partially completed. The board of directors of TVA and many of its highest-ranking officials wanted to see the project converted to an advanced natural gas-fuelled electrical generating unit. The politics regarding Red Hills jumped to a higher level. Members of the Mississippi congressional delegation interceded on behalf of the Red Hills Power Project. Mayors, supervisors and other elected officials of north-east Mississippi spoke out on behalf of the project. Negotiations continued on the cost of power, and another phone call helped to keep the Red Hills Power Project alive. That phone call was to advise the Energy Division that the project developers were going to have to get their costs in the 'low three cent' range to have a chance. That same phone call resulted in the Mississippi Energy officials being told that it might be advantageous to have a thermal host to purchase excess steam. That phone call initiated a conversation between the project developers and the Energy Division that resulted in the concept of the Red Hills EcoPlex being born. A tip about a project that was under consideration for a community in south Mississippi led to a discovery that a large pulp mill was being considered. That tip led to conversations between Energy Division staff, the project developers and Southern Paper Company.

Meetings were arranged to discuss changing the location to Choctaw County. The biggest challenge to be overcome was the water requirements for the pulp mill. Southern Paper Company had a potential solution for that situation. It could use process water from the power plant and manufacture pulp by using a thermo-mechanical process. The mill would be a zero-discharge facility. Several of the elements of eco-industrial technology had fallen into place. The Energy Division was advised that there was a facility in Kalundborg, Denmark, that had mastered the concept of waste and materials exchange and that this possibility needed investigation. What was determined was that we had the potential of creating the first 'greenfield' eco-industrial park in the world. But what would we call it? At an informal staff meeting of the Policy and Planning Bureau with the Energy Division director, several names were tossed about. The name Red Hills Power Project came from the then chancery clerk of Choctaw County, Donald Nunn. One of the staff members of the Policy and Planning Bureau, Carl Burnham, had driven by a local car dealership on Interstate Highway I-55 earlier that day, the Herrin Gear AutoPlex. The name Red Hills EcoPlex came out of that meeting. This was in early 1996.

The political battles heated up as officials of TVA continued to oppose the project. They demanded the endorsement of all the electrical co-operatives in their service area and imposed a requirement that they would be guaranteed not to withdraw from their rolling ten-year contracts. That resulted in a lawsuit being filed by the 4-County Electric Power Association. That was a near show-stopper. During the 1996 general session of the Mississippi Legislature, a meeting of all the electrical co-operatives was convened by the board of directors of TVA in Tupelo, TVA's first member city. Through the assistance of Representative Charlie Smith and the Mississippi Highway Patrol, Energy Division staff were allowed to attend that meeting and engage in dialogue with the general managers of the co-operatives. Support was expressed by the overwhelming majority of the general managers. There was not total support, however, and political intervention was again required to make sure the playing field was kept level. It was at a meeting between the co-operatives and the Energy Division staff that the first rendition of the EcoPlex was revealed. It depicted a hydroponic farming operation, a cement manufacturer, the pulp mill, a pharmacy, an oil filter recycling plant and the power plant. By that time, the idea of using waste biomass as a fuel supplement to the lignite was introduced. Jobs numbering in the hundreds were suggested, and public support for the project became stronger.

The Energy Division was approached by a local boiler manufacturer, requesting assistance in convincing the project developers to use locally manufactured boilers and other integral plant components. The company was Babcock & Wilcox (B&W), located in West Point, MS, less than 50 miles from the project site. As it turned out, B&W could not compete, price-wise, with the eventual boiler manufacturer, Alstom, a French company. By early 1996, the acquisition of CRSS Capital by Tractebel Power, based in Brussels, Belgium, had been completed, and Tractebel created its US subsidiary, Tractebel, North America. Several of the original members of the project development team departed. Frank Giacalone was replaced as project manager by Paul Margaritis, a highly skilled negotiator. The cost of wholesale power was still a bit of a problem, but Margaritis and Bill Utt, president of Tractebel, North America, through those negotiation skills were able to overcome debate after debate with TVA officials and others. Ultimately, Phillips Coal Company was disbanded by its parent company, and Dan Bollner was reassigned to the Phillips Gas Company. He is now in South-East Asia working on projects for Phillips Petroleum.

The ongoing feud between 4-County Electric Power Association and TVA continued to fester, and it was announced at the grand opening of the new TVA Customer Service Centre in Starkville, MS, that TVA would not agree to a facility being built in the service territory of a non-loyal customer. That move was publicly announced the following day at the court house in Webster County. Webster County is located due north of Choctaw County, but is in the service area of Natchez Trace Electric Power Association, a loyal TVA customer and strong supporter of the project. That announcement caused a near riot by the customers of 4-County Electric Power Association, particularly those in Choctaw County. They demanded to be heard by the board of directors of 4-County Electric Power Association. A 'listening meeting' was called by the board of directors of 4-County Electric Power Association, and the people of Choctaw County gave them an earfull. That meeting caused the board of directors to withdraw from its opposition to the new contract and withdraw the lawsuit. The suit was actually heard in court and was ruled in favour of the TVA, so the withdrawal was a moot point. TVA's demands had been met.

On 11 July 1996, on the Courthouse lawn in Ackerman, TVA board chairman Craven Crowell and Governor Kirk Fordice signed a proclamation of agreement to the Red Hills Power Project, conditioned on the successful completion of the terms of a power purchase agreement and a environmental impact statement (EIS), triggered by the major federal action. Conducting the work required for the EIS took over a year to complete and was one of the most intensive environmental investigations ever conducted in the State of Mississippi. US Army (retired) general Harry Walters, president of Southern Paper Company, was also on hand to 'place his flag in the ground' as the initial tenant of the Red Hills EcoPlex. Unfortunately, one of the three operational thermo-mechanical pulping plants of the same design as that planned for the EcoPlex had a serious system failure, resulting in a major environmental discharge of untreated water. Southern Paper Company was forced to withdraw from the partnership.

The Mississippi Department of Environmental Quality and the Mississippi Legislature were charged with the enormous task of rewriting the laws regarding the surface mining of coal and with preparing and getting the approval of the Mine Safety and Health Administration for the new coal-mining regulations. To their great credit, both tasks were completed prior to the completion of the EIS. The permit preparation by the project developers was also a huge task and involved thousands of hours gathering the requisite information for the application and for the review and comment on the application.

The Energy Division began the task of recruiting tenants for the Red Hills EcoPlex as a partner with Choctaw County. One of its first efforts was to engage in a contract with a consortium of firms to prepare a master plan for the Red Hills Power Project and the EcoPlex. The consortium included several Mississippi engineering firms, the University Research Consortium and the renowned planning group, Arthur Andersen. A principal task assigned by the Energy Division was to conduct a formal industrial cluster analysis and to identify the proper mix of tenants for the EcoPlex given all the known physical amenities to be associated with the Red Hills Power Project. The development of an Internet website was an integral part of the contract. The site was to be used by the various parties involved in the preparation of the EIS to exchange information and ideas. A formal marketing brochure was developed for use by the Energy Division, Choctaw County and project managers with the other divisions of the Department of Economic and Community Development in promoting the Red Hills EcoPlex and in recruiting tenants.

The bulk of the master plan for the Red Hills Power Project was developed by Arthur Andersen. Its evaluation of the project resulted in a finding that an initial cluster of four tenant types would be enough to open the doors to satisfying the necessary symbiosis to make the EcoPlex a real example of what industrial ecology means. The recommendation coming from Arthur Andersen suggested that the first-round tenant mix should include:

- Intensive aquaculture
- Hydroponic greenhousing
- Poultry processing
- Manufacturing of compressed kenaf fibre panels

There was some resistance to the poultry-processing recommendation by local elected officials. As a result, the Energy Division set its sights on the other three categories of tenants. It was at about that time that the Cornell Work and Environment Initiative learned of the project and took a serious interest in it. I was persuaded that Cornell could help in the recruitment effort of tenants and that they would not insist on extreme notions of what is 'green'. In time, a contract was entered into by the Department of Economic and Community Development and Cornell. The contract called for Cornell to provide details on a strategy to attract the firms that matched those recommended by Arthur Andersen. Cornell looked at the project and developed various product and tenant 'trees' that branched off the major categories.

Most interesting is the fact that Arthur Andersen did not recommend the recruitment of firms that could take advantage of the ash generated by the power plant. The officials of Choctaw County saw that by-product as a potential 'cash cow'. They also saw the sale of ash as a means through which to maintain sustainability for the EcoPlex. This has become a major sticking point for Choctaw County, a subject to developed further in this chapter.

One of the factors that made the Red Hills Power Project a reality was a creative financing tool made available by the Mississippi Legislature, the 1996 Major Energy Projects Development Act. That act authorised up to US$50 million dollars toward the purchase of infrastructure that would be publicly owned and directly related to the project. The interpretation of the act created some tense moments between the project developers, the county officials in Choctaw County and representatives of the Mississippi Department of Economic and Community Development. It was the position of the county officials that the infrastructure include the upgrading of roads, new law enforcement capabilities and facilities, new fire-fighting equipment, new medical facilities, new schools and the provision of other traditional local government needs. Project developers felt that some portions of the electrical inter-ties should qualify for funding. In the end, it was the Finance Division of the Department of Economic and Community Development that made the determination on each component that was requested by the project developers for public financing.

Among the features of the Red Hills Power Project that were financed with use of the legislatively appropriated US$30 million dollars were the natural gas line to start the boilers, the land on which the Red Hills Power Project, including the EcoPlex, was located, a portion of the access road to the project, the limestone crushing unit, water treatment facilities and the water supply systems for potable and process water. The town of Weir, located in the western part of Choctaw County, asked for and received assistance from the Energy Division in getting authorisation to serve the natural gas needs for the Red Hills Power Project; therefore, Weir received the financial assistance to build, own and operate the 10 inch natural gas pipeline. The rest of the infrastructure is owned by Choctaw County.

Tax incentives were another contentious area that had to be overcome by the parties. The supervisors of Choctaw County had never offered a tax abatement for any industry locating in Choctaw County. When a request for a ten-year abatement was presented by the project developers, the board of supervisors balked. Again, creative negotiations and a lot of give and take by the project developers and the supervisors resulted in all parties being able to sign up to the project. Many of the citizens of Choctaw County felt that the board of supervisors had given in to the project developers, and several members of the board were replaced during the ensuing election.

A number of prospective tenants were referred to the Energy Division, and contacts were made. Most of the initial contacts were with hydroponic greenhouse operators, although there were several prospective tenants interested in using the fly ash for the manufacture of plasterboard. It was not possible at the time to guarantee the quality of the fly ash, including the amount of calcium sulphate (gypsum) that would be in the finished fly ash. That would have to wait until the plant was fully operational and the ash quality stabilised. Only then could the ash quality be known for certain.

In spring 1998, representatives of the Energy Division, a member of the Mississippi legislature, and the director of the Atlanta Regional Support Office of the US Department of Energy made a trip to Kalundborg, Denmark, to witness first-hand the industrial symbiosis that had evolved at that location. A comparative analysis of the eco-industrial complex and the proposed Red Hills EcoPlex was prepared by the Energy Division. There were a number of similarities identified, but there are many differences. Box 20.1 provides a portion of the report on the comparative analysis of the Red Hills EcoPlex and the facilities at Kalundborg to represent both the remarkable similarities and differences of the two facilities. For further details of the Kalundborg symbiosis, see Chapter 16.

THE FIRST GLIMPSE THAT WE HAD OF THE FACILITIES AT KALUNDBORG GAVE immediate rise to the thought that this is no way what is envisioned for the Red Hills EcoPlex. It was reminiscent of the industrial complexes in Baton Rouge and along the Louisiana and Texas coast. The facilities were massive and almost foreboding. The industries were located adjacent to a large harbour, capable of handling large, ocean-going vessels. It was not until we began to hear from Dr Jørgen Christensen and Mr Robert Rasmussen that the favourable comparisons between Kalundborg and the Red Hills EcoPlex became clear. Most of the comparisons are based not on the industries themselves but rather on the exchanges that occur between the industries and the frugal use of the available resources.

In this comparative analysis, I will endeavour to detail the areas in which the two complexes have likenesses and those areas where there are differences. For ease of presentation, we will attempt to present the differences first, then concentrate on the similarities and how best to capture the most favourable of the features of the Kalundborg facilities at the Red Hills EcoPlex.

The most stunning difference between the Kalundborg facility and the Red Hills EcoPlex is the mix of industries. The heart of both facilities is a power plant, and both plants use a form of coal as the fuel source. There the similarities end. The Asnæs plant is a massive five-unit base-load station rated at approximately 1,500 MW, capable of providing nearly 60% of the electrical needs of eastern Denmark. The largest of the power plants is the newest. It is rated at 695 MW. The power station has numerous environmentally friendly attributes, including low-nitrogen-oxide-emission burners, flue gas ash and dust removal systems and, in unit 5, a lime-water flue-gas desulphurisation plant. The fly ash is sold to Alborg Portland A/S for use in the manufacture of cement and concrete products. Some is used as a road-building material and for land reclamation. The lime-water flue-gas desulphurisation plant effectively removes about 95% of the sulphur from the flue gas. The precipitate is a very high grade of gypsum that is sold to Gyproc A/S for use in the manufacture of plasterboard. Acoustical damping has also been incorporated into the plant to reduce the noise levels.

In contrast, the Red Hills generating facility, a 440 MW machine, will employ a different combustion technology than that used at Asnæs. That at Red Hills will be the world's largest circulating fluidised bed machine. It will operate at temperatures that will keep airborne ozone precursors extremely low. By injecting dried limestone into the combustion bed with the lignite fuel, sulphur will be sequestered and become entrained in the bottom ash and the fly ash.

The fjord at Kalundborg was identified as an ideal location for the Asnæs power plant. The harbour was key to the siting decision. Coal shipments are handled at the facility and, by locating the facility on the fjord, the operators are able to get extremely low-cost fuel for the plant. The fuel that is currently being used at unit 5 is called Orimulsion, which is little more than dirt with a high concentration of crude oil. The fuel comes from the Orinoco River delta in Venezuela, although fuel from all over the world can be used in the plants.

Cooling water for the plants at Asnæs originally came from the fjord. Now, water for the plant and for the other industries at the site comes from a rather large, 750 acre, reservoir, Lake Tisso, located approximately ten miles east of the facilities. This is in contrast to the cooling water for the Red Hills generating facility. Water for Red Hills will be groundwater withdrawn from the Massive Sand Unit of the Tuscaloosa Aquifer. The three wells that will provide make-up water at Red Hills will be the deepest water wells in Mississippi. However, the water is slightly saline and has rather high concentrations of

---

Box 20.1 *A comparative evaluation and analysis of the industrial complex at Kalundborg, Denmark, with the proposed Red Hills EcoPlex in Choctaw County, Mississippi* (continued over)

> total dissolved solids, which will limit the use of the water by other tenants of the EcoPlex without extensive treatment. At the Kalundborg facilities, five water exchanges are incorporated into the symbiotic relationships between the various industries. The 'wastewater' is then collected into a central re-use basin capable of holding 200,000 m^3 of water. This augments the withdrawal from Lake Tisso.
>
> Steam and gas form other feedstock materials involved in the symbiosis at Kalundborg. The steam comes from the power plant and the gas is from the Statoil Refinery. The steam is provided to Novo Nordisk, a pharmaceutical manufacturer, and to the Statoil Refinery. After removal of the condensate from some of the steam, the heat is transferred to a large fish-farming operation and to the municipality of Kalundborg for district heating.
>
> Liquid sulphur is sold by the Statoil Refinery to a local fertiliser industry, and fly ash and clinker from the power plant are sold to a local cement industry. Those two industries are not presently participants in the Symbiosis Institute, nor are the numerous nearby farms that receive biomass and yeast slurry as fertiliser from the Novo Nordisk plant.
>
> It is noteworthy to again mention that the key to the success of the entire symbiotic relationship is water, which poses a potential problem for the Red Hills EcoPlex. At Red Hills the water, because of its salinity, limits the types of industries that can co-locate with the Red Hills generating facility. Efforts are ongoing to evaluate the construction of a fairly large surface-water impoundment near the EcoPlex to attract a more diverse tenant make-up for the EcoPlex. At present, efforts are directed toward recruiting tenants that will have relatively low fresh-water needs. Included among the most promising of the prospects is a very large, state-of-the-art greenhouse for growing ornamental and limited vegetable species for wholesale distribution. The symbiotic relationship between the power plant and the tenants of the EcoPlex is primarily due to the combustion by-products and treated water, steam, heated water and compressed air that may be needed by EcoPlex tenants. These items continue to be evaluated by Pickering, Inc.

Box 20.1 (continued)

Pickering Inc., the company with which I am currently affiliated, became a partner to the Red Hills EcoPlex by entering into a Memorandum of Understanding with the Choctaw County board of supervisors and the Choctaw County Economic Development District. Pickering has already made a sizeable contribution to the project and has been directly involved with tenant recruitment and the development of the business plan and marketing strategy for the EcoPlex

Additional financial assistance for the project came about when a grant in the region of US$1.5 million was requested from and awarded by the US Economic Development Authority. The grant was used for the some of the process-water system. A portion of the funds are to be used for a heat exchanger that will remove heat from the process water and supply EcoPlex tenants with heated water. Until such time as a tenant with hot-water needs commits to locating in the EcoPlex, there is no need for the heat exchanger. The Choctaw County officials were notified by the Mississippi Development Authority in January 2003 on clarification of what items can be financed with use of the match money from the grant. Pickering was requested to assist Choctaw County in discussions with the Mississippi Development Authority. Pickering and Choctaw County will be awarded US$1.443 million to develop the initial infrastructure needs for the EcoPlex.

By August 1998 all the necessary permits were ready and the final EIS had been issued. The Record of Decision on the final EIS was favourable, and ground clearance and site preparation activities began in earnest. The official ground-breaking ceremonies took place on 27 October 1998. A members of the boards of directors of Tractebel SA (the parent company of Tractebel, North America), Choctaw Generation, Inc. (the owner–operator of the power plant), Phillips Coal Company and its partner, North American Coal Corporation, were on hand. They were greeted by nearly 1,000 greatly appreciative residents and officials of Choctaw County. The ceremonies were also attended by Governor Fordice, US Senator Thad Cochran, Congressman Roger Wicker and Jimmy Heidel, Executive Director of the Mississippi Department of Economic and Community Development. The Red Hills Power Project and the Red Hills EcoPlex were alive and under way.

That was a banner day for north Mississippi. Over the course of the ensuing 27 months, Choctaw County has been the benefactor of a workforce of over 1,200 skilled and semi-skilled workers. However, 1 December 2000 was to have been the date for commercial operation, but, given that the Red Hills generating facility is the largest circulating fluidised bed electrical generating station in the world, there have been several 'bumps in the road' that have prevented the project developers from starting full-scale commercial operation. The plant became fully operational in April 2003.

On 19 July 2001, Choctaw County was notified by Tractebel, North America, of the successful completion of negotiations for the sale of power from a gas-fired unit to be located adjacent to the existing lignite-fuelled facility. It will be located on EcoPlex property. Proceeds from the sale of the land will be used to purchase additional land with better access roads. Tractebel, North America, has agreed to assist the county in the development of the EcoPlex properties. The monies from grant from the US Economic Development Authority will be used for the creation of infrastructure. By spring 2003, the fly ash will be fully characterised, and tenant recruitment will begin in earnest. The Choctaw County Economic Development District hired Alan Bates, as permanent Executive Director, in early 2002. Bates and Don Threadgill, Choctaw County Chancery Clerk, have been very active in helping get the money released so that the infrastructure for the EcoPlex could be put in place and the land available cleared for tenant occupancy. That activity has been ongoing since 2000 with little success. Finally, the Mississippi Development Authority prescribed a series of funding eligible items for which the money could be used. Choctaw County was able to sell a portion of the land to Tractebel for a second electrical generating station. It will use natural gas as its fuel source. The proceeds from the sale of the property to Tractebel allowed Choctaw County to purchased approximately 105 acres of contiguous property next to the 40 acres from the original property. The board of supervisors of Choctaw County has made a commitment to support the development of the EcoPlex. There are several EcoPlex prospects awaiting word that the EcoPlex is suitable for receiving tenants. This process is ongoing. Mr Bates prepared a Request for Proposals for management of the ash from the coal-fuelled power plant. The proposals are currently being evaluated and a contract will soon be awarded. That will provide the sorely need cash flow for the EcoPlex managers and allow for more aggressive tenant recruitment.

The story of the Red Hills EcoPlex development is one of politics, perseverance, persuasion and patience. It is a story of how difficult it can be to introduce a totally different type of approach to industrial development. It is a story of how critical it is to have dedicated people and organisations step up and take the needed leadership role in implementing the concepts and ideas of industrial ecology. The story of Red Hills is also a story of passion and compassion, a story of ups and downs, a story with joys and frustrations, but most of all it is a story of co-operation. The story of a partnership that was created in 1996 is a truly amazing story. In the years to come, the effectiveness of the co-operation and partnership will become evident.

Mississippi State University has become a player in the development of the EcoPlex through the Mississippi Co-operative Extension Service. In spring 2000, just before my retirement, Mississippi Co-operative Extension Service hosted a round-table of people involved in various eco-industrial development projects. It was a total success, and any readers of this chapter who attended that meeting will recall how magnificent the Red Hills Power Project was seen to be during a tour of the site and the Red Hills mine. At the time of this writing (February 2003), the issue of 'sustainability' is paramount to the continued development of the EcoPlex. Tractebel, North America, has made a commitment to Choctaw County to assist in the preparation of properties for tenants of the EcoPlex. Pickering, likewise, has made, as previously noted, a commitment to help in the development of the EcoPlex. Time, and time only, will tell if the county officials are willing to make the needed commitments to bring this eco-industrial project to fruition.

21
REGION-WIDE ECO-INDUSTRIAL NETWORKING IN NORTH CAROLINA

Judy Kincaid
Triangle J Council of Governments, USA

The Triangle J Council of Governments (TJCOG), a non-profit planning organisation serving 30 local governments in North Carolina, decided in the mid-1990s to promote eco-industrial development in its six-county region. Instead of focusing on a particular industrial park, which might have limited interest in the concept to those interested in one piece of real estate, TJCOG began promoting eco-industrial development region-wide. The goal was to transform the way the entire region thinks about the flows of materials, water and energy.

The TJCOG project actively promotes three levels of regional eco-industrial development. The first level is a region-wide network of by-product exchanges between facilities, facilitated by a database and geographical information system (GIS) maps. The second level concerns multifaceted partnerships between industries that are neighbouring tenants in existing industrial parks, facilitated by meetings convened to explore joint projects. The third level relates to new infrastructure for planned industrial parks, facilitated by promotional materials and model by-laws, covenants and restrictions.

21.1 Region-wide by-product exchanges

The project formally began when the US Environmental Protection Agency (EPA) agreed to fund a two-year TJCOG project to identify and match re-usable local industrial by-products with local industrial purchases for which these by-products might be substitutes.

In 1997, TJCOG formed a partnership with several other organisations in order to undertake this EPA-funded project. These included three local universities (Duke University, North Carolina State University and the University of North Carolina at Chapel Hill), six local economic development organisations (one from each county) and the state pollution prevention agency. TJCOG also convened a group of representatives from ten local industries to advise it on the best way to approach industries and gather information.

After developing a survey form for recording data, the project developed a list of 343 local businesses and institutions to contact regarding providing information. These businesses had one or more of the following characteristics:

- Discharged waste-water directly into surface waters or discharged significant amounts of waste-water into the sewer system
- Filed federally required reports on hazardous or toxic materials
- Employed a large number of people
- Had some past involvement with pollution prevention programmes
- Represented a variety of facility types and locations throughout the region
- Were located very close to several other industries

Each of the 343 facilities was telephoned and asked to participate in the project. Participation involved filling out a survey form and then having an hour-long interview at the facility.

During the initial phone call, representatives of 75% of the facilities either agreed to participate in the project or said they might be interested but either wanted to see the survey or speak with their bosses first. All these people received the survey form and were contacted later to schedule an in-plant interview. Contacting people and arranging interviews was very time-consuming, and some of the initially interested people were not able to schedule an interview during the limited initial project period. In the end, project representatives visited and collected survey responses from 182 facilities over a one-year period. This represented 53% of the 343 facilities originally contacted.

A database was created and facility locations were mapped by using a linked GIS. It was then possible to easily identify potential partners for resource sharing. From the 182 facilities in the database, potential partnerships were identified for 48% of the respondents (88 facilities) (Kincaid and Overcash 2001; Kincaid 1999). It should be noted that this figure does not include any partnerships regarding commonly recyclable materials such as cardboard and scrap metal.

Some of the potential partnerships revealed by the database were partnerships that would require a fair amount of time to implement. Examples include distribution of excess steam from a university co-generation plant to neighbouring commercial enterprises, the manufacture of absorbents from a variety of by-products and the use of coal ash as an ingredient in several local products.

Other potential partnerships seemed promising in the short term, and the project focused on bringing these potential partners together. The most successful by-product exchange that developed from this initial work involved a resin manufacturer that reported having a methanol by-product that it was paying to have treated and disposed of. In a neighbouring county, a public waste-water treatment plant reported buying new methanol to remove nitrogen from waste-water. The project facilitated discussions between these two facilities, resulting in a successful trial partnership. The resin manufacturer began delivering the methanol to the waste-water treatment plant, with an estimated annual cost saving of US$100,000 for the resin manufacturer and US$70,000 for the waste-water treatment plant.

21.2 The creation of partnerships in existing industrial parks

During the two years of EPA-funded work, the project worked particularly intensely with industries located in two industrial parks located in different counties. This work with close neighbours was of interest to the State of North Carolina's Energy Division, which agreed to provide funding for this type of work after the conclusion of the two-year EPA-funded portion of the project. This allowed the project to focus intensely on several other industrial parks in addition to the original two.

With each industrial park chosen for attention, the project convened several meetings of representatives from tenant industries to discuss potential opportunities for collaboration regarding resource use. For example, at one industrial park, a brainstorming session resulted in identifying four possible partnerships. Two company representatives thought their companies might be able to pool their cafeteria food waste and set up a joint composting site. Another company's representative asked whether other tenants had shipping pallets his company could use, and he got three offers. A third potential partnership involved co-ordinating long-distance freight deliveries, which could result in cutting delivery costs and reducing truck traffic in and out of the park. A final possible partnership identified by the group was use of excess thermal capacity from one plant by its neighbours. These partnership discussions also resulted in sharing information that could result in resource conservation: one company learned about a cardboard shredder that it could use for turning its waste cardboard into a substitute for packaging padding; another company learned about options for reducing and recycling its polystyrene waste; a third company was about to have an energy audit and offered to share the lessons learned with other tenants.

The region-wide database was a valuable resource that complemented the project's work with individual industrial parks. For example, at a meeting of representatives from one of the industrial parks, a tenant representative reported that her company had a very fluffy wood by-product containing urea formaldehyde. In the database was a company that was able to use this by-product in its composting operation. The composting company reported that the urea formaldehyde, which might have been a problematic contaminant for some re-uses, was a beneficial source of nitrogen for the compost.

21.3 Encouraging eco-industrial features for new industrial parks

As the TJCOG eco-industrial development project entered its fourth year, economic development commission staff from three of the region's counties asked the project for help. Each of the three counties was planning to develop a new industrial park in its county and was interested in sample eco-industrial park provisions for possible inclusion in the new park by-laws, covenants and restrictions.

Fortunately, the project had been networking for several years with other eco-industrial development projects throughout the USA and even beyond. This networking primarily occurred because of regular meetings hosted by Cornell University's Work and Environment Initiative. Owing to the familiarity of TJCOG project representatives with personnel from other projects that had developed similar legal documents, it was quick and easy for the TJCOG project to obtain ideas from others regarding by-laws, covenants and restrictions for the proposed North Carolina parks. Moreover, Cornell's *Handbook of Codes, Covenants, Conditions and Restrictions for Eco-industrial Development* was an invaluable resource for drafting these documents (WEI 2000).

The TJCOG project developed presentations and promotional material to assist the local developers of these three new industrial parks in understanding and explaining to others the value of incorporating principles of eco-industrial development into their parks. All three parks are being developed on county-owned land, so there is public interest and involvement in exploring park options. As an organisation funded and run by local governments, TJCOG is in a particularly good position to work on this with local governments.

21.4 The value of diversity

The TJCOG project has served as a facilitator for eco-industrial development throughout its home region. The project has not played the role of investor, developer or sponsor for any one particular cluster of industrial connections. By focusing on a variety of avenues to promote eco-industrial development, the TJCOG project has been able to survive and achieve successes that will help it continue to attract funding and support in the future.

The value of this multifaceted approach is analogous to the value of biological diversity. Some of the proposed partnerships may fail without jeopardising the larger project. In fact, if an entire constellation of partnerships at one industrial park fails, there are many other potential partnerships throughout the region from which success can spring.

Diverse aspects of the project also mean that there is a greater probability that any particular audience will find some aspect appealing. Significant audiences include industries, economic developers, elected officials and funding sources.

21.5 Funding the project in the future

In addition to the project's diversity, another feature that has added to its survival is the fact that it is managed by a 30-year-old non-profit organisation. TJCOG is a well-respected, well-known organisation in North Carolina and it has received millions of dollars in grant funding for regional programmes over its lifetime. TJCOG's board of directors consists solely of elected officials from the 30 local governments that are TJCOG members. This factor makes it attractive to state, local and private sources of funding.

In the project's initial stages, the formal partnerships with universities, the state pollution prevention office and local economic development offices helped to introduce the project to a wide audience, to give it additional credibility and to facilitate the process of working with graduate student interns. These formal partnerships lasted through the two years of EPA funding but lapsed into informal relationships thereafter. After two years, the project was stable and well known as a TJCOG project, and it was administratively less cumbersome to manage the project as a TJCOG project rather than as a project involving ten partner organisations.

Federal and state grant funding for the project to date has paid for a project manager, graduate student interns, some consultant time and a few miscellaneous other expenses, including travel. The only other financial support for the project has been a very small portion of local government TJCOG membership dues, which provided a tiny portion of the project manager's salary during the period between grant funding. The proximity of universities with graduate programmes in public policy, environmental studies, and city and regional planning has meant that the project can rely heavily on graduate student interns rather than on permanent staff to carry out some of the work. An annual budget for the project could be between US$100,000 and US$150,000.

The key financial issue for the project is how to make it self-sustaining over the long-term. The project has valuable services to sell to its industrial customers and to local and state government programmes promoting waste reduction and energy conservation. For example, in one week in 2000, the project received a call from a local business wondering whether anybody could use its unwanted, unopened containers of isopropanol, and another call from somebody who had liquid mercury to give away. This illustrates the value of the database for by-product exchange. This database is not currently web-based, maintained or easily accessible, but stable funding would make this possible.

The project fills a niche not taken by national waste exchanges or other state and local programmes. The project is a proactive facilitator of eco-industrial development in its various aspects. It performs a valuable service in bringing together diverse facilities and encouraging them to pursue joint projects that would otherwise not be priorities for those individual facilities. With a nudge from an advocate, partnerships occur that otherwise would remain undiscovered and unrealised.

The next phase of the project will involve developing a plan for supplementing federal and state grant funds with funds from local industries and local government solid-waste programmes. In spite of lean budget times, the project's various successes to date should justify support from these local stakeholders.

22
A CASE STUDY IN ECO-INDUSTRIAL DEVELOPMENT
The transformation of Burnside Industrial Park

Raymond P. Côté
Dalhousie University, Canada

Peggy Crawford
Eco-efficiency Center,
Burnside Industrial Park, Canada

In 1990, as many researchers, policy-makers and industrialists were grappling with the term 'sustainable development' and its implications for environment, economy and society, a number of concepts were converging in an office at Dalhousie University. That convergence led to the idea that industrial parks might be viewed as the human, technological equivalent of an ecosystem. On this basis they should be designed and operated with some of the essential characteristics of ecosystems in mind.

One of the basic tenets of mature ecosystems is that outputs are cycled in such a manner that they are re-used by the same or other species. Another feature of mature ecosystems is that the stability of the system is maintained by the diversity and interconnectedness of the species. The current industrial system is best described as a linear flow of materials and energy in which large amounts of resources are transformed into large amounts of waste. More efficient industries producing 'more with less', making better use of the inputs into the various processes, as well as products that are more readily re-used, remanufactured and recycled are needed. The hypothesis is that groups of industries that co-operate more closely reduce financial costs and environmental impacts. This co-operation would occur in an industrial ecosystem.

A multidisciplinary team of researchers was established and began looking for a laboratory. Burnside Industrial Park, Dartmouth, Nova Scotia, the largest industrial park in Atlantic Canada, and one of the largest in Canada, was selected as a viable site for such an investigation. The park is approximately 1,200 hectares in area, with 1,300 businesses in operation and 17,000 people employed. It is a very diverse park, with several dozen sectors represented and in which many of the sectors are represented by businesses of the same or similar 'species'. The flows of materials, energy and information are complex. The majority of the businesses in the park are small.

Small businesses are a significant feature of the Canadian economy, but, generally, they have not received much attention from the environmental agencies of governments. The idea of using an integrated approach to environmental and energy management has been neither presented nor considered. Generally, laws and regulations appear to be written with large companies in mind. Most managers and owners are preoccupied with survival and the day-to-day operations of running a small business. They have neither the money nor expertise to address the environmental issues facing them. In general, small businesses are difficult to reach and engage in any improvements to do with the environment. In many cities and towns, small businesses are grouped into industrial parks where their environmental impacts are cumulatively important. There are more than 1,000 industrial parks across Canada.

22.1 Background

As industrialised societies become more aware of their wastefulness and the environmental problems associated with it, some jurisdictions are taking a much more forward-looking approach. In 1995, after much public consultation, the province of Nova Scotia in Canada adopted a Solid Waste-resource Management Strategy. This strategy, which was enshrined in the Province's consolidated Environment Act, is designed in large part to:

- Achieve 50% diversion of its solid waste from disposal sites by the year 2000
- Improve environmental standards across the province
- Establish and support regional co-operative programmes to reduce implementation costs
- Encourage development of economic opportunities around solid waste resources

The catalyst to achieve the goals of the strategy has been stricter disposal standards. There are province-wide bans on the disposal of many materials, including: corrugated cardboard and newsprint; beverage, steel, tin and glass food containers; automotive batteries, antifreeze and used tyres; waste paint; selected plastics; and compostable organic materials.

At the same time, the Halifax Regional Municipality (HRM), Nova Scotia's capital and its most populous municipality, with 384,000 residents, was also under pressure by its citizens and regulators to manage its solid waste in a more environmentally acceptable manner. In the two decades preceding 1995, the experience with landfills had been very negative, and the siting of new landfills was becoming extremely difficult, time-consuming and costly. Again, through a process of community consultation, the municipality adopted a strategy that has since become the envy of many communities across Canada. The Halifax Integrated Waste Management Strategy emphasises separation and diversion. Separation occurs at source and begins at the homeowner level. The homeowner separates plastics, metals and glass into one container, and newspapers and cardboard into another. Compostable organic materials are separated and placed in a special container. The remainder can be placed in the 'old trash can'. Kerbside collection of all recyclable and organic materials and residual waste occurs on a weekly or bi-weekly basis. Hazardous waste such as paint and solvents can be taken to a special facility for transport to a treatment-and-destruction facility situated out of the province. Across the province, 50% diversion has now been recorded. Halifax's strategy has resulted in the diversion of more than 60% of waste from landfill.

The materials diverted from landfill are managed by a provincially established, non-government agency, called the Resource Recovery Fund Board Inc. (RRFB). The RRFB fosters Nova Scotia's value-added manufacturing sector and supports municipal waste management and public education efforts. Since the RRFB was established in 1996, more than 450 jobs have been created as a result of these programmes to reduce, re-use, recycle and compost. In some cases, diverted materials are sent to be recycled by companies outside the province but, increasingly, companies are being set up in Nova Scotia to turn these resources into new products. One of the most obvious products is compost from organic materials, but others include: paperboard and kitty litter from old newspapers and cardboard; cow mattresses and blasting mats from used tyres; and furniture from recycled plastics.

In addition: businesses are under increasing pressure to apply source control of materials discharged into sewers; the municipality has joined a national initiative to reduce greenhouse gas emissions; energy costs and water costs are rising, and the federal government is increasingly concerned about toxic substances.

While these provincial and municipal strategies were being developed, a group at Dalhousie University had also been investigating the concept of industrial ecosystems, groups of industries that would function in manner similar to species and communities in nature.

22.2 Burnside Industrial Park

Burnside Industrial Park was established in the early 1970s and continues to grow. Approximately two-thirds of the 1,200 hectares of the park are currently developed. The municipal government is responsible for the park, setting standards, selling land to developers, providing roads, water and sewers, and enforcing by-laws. The park itself is serviced by road, rail and sea. The private sector provides electricity, natural gas and advanced telecommunications services.

Approximately 90% of the businesses are small and medium-sized enterprises (SMEs), employing between 2 and 50 people. By sector, 10% of the businesses are in the manufacturing sector, 48% in sales and services, 11% in the construction industry, 9% in distribution and warehousing, 8% in retail and 14% in professional, financial and other business services. There are a large number of companies in the same categories, ensuring redundancy in terms of services and materials flow. For example, there are 18 businesses in the printing sector, 25 vehicle maintenance facilities, 20 companies involved in computer assembly and distribution, 17 companies in electronics sales and services, 17 businesses involved in chemical processing and distribution, 21 companies dealing in paints and coatings, 17 companies in the metal plating and finishing category and 36 trucking companies.

The city has strengthened its development standards over the 25-year period since the park was first established. The objectives of the standards and covenants, which apply to the park's development, are:

- To protect property values and enhance the investment of businesses located in the park by providing a well-planned and well-maintained development
- To create an attractive and efficient business environment through sound land-use, planning and environmental management standards
- To ensure harmonious relationships among uses

The covenants are intended 'to ensure that the park continues to be developed in a manner consistent with superior aesthetic and environmental protection standards and with the declared intention of creating a pleasant and harmonious environment for the Park's residents' (Halifax Regional Municipality 2000). These covenants apply to architecture, landscaping, signage, protection of natural areas (particularly streams, lakes and wetlands) and require buffer zones of undisturbed habitat or suitable areas of green space around all watercourses.

The municipality has expressed and demonstrates continuing interest in improving the environmental management of the park through enhanced co-operation and networking of various stakeholders. The monthly park newspaper, the *Burnside News*, has carried a column for the past seven years titled 'The Burnside Ecosystem' to inform and educate people working in the park. However, the nature and scope of Burnside is such that it will probably continue to serve primarily as a laboratory for testing strategies that will be applicable primarily to the design and operation of new and developing industrial parks.

22.3 The Eco-Efficiency Center as a catalyst

It appeared that the interests of provincial and municipal managers and researchers in enhancing Burnside's environment and investigating eco-industrial development strategies could converge. What was clearly needed was a catalyst. The Eco-Efficiency Center is a model for an integrated set of approaches that assists small business to achieve better environmental and economic performance through resource conservation, re-use, recycling, pollution prevention and general good environmental practices through individual and collective action. The Center is led by a team of public and private partners, with Dalhousie University, Nova Scotia's largest university, and Nova Scotia Power Inc., the province's electric utility, being the major partners. The environmental management and economic development agencies of the three levels of government have committed financial support to the Center for the first three years.

From Dalhousie University's point of view, the Center is an example of how research can be used to create a practical programme for the community. The university also uses the Center to support its educational role in providing students with a 'learning-by-doing' opportunity in eco-efficiency and industrial ecology. Nova Scotia Power Inc. has looked at the project as an opportunity to benefit its customers and to further demonstrate its strong commitment to the environment. Other environmental initiatives include adopting ISO 14001-equivalent environmental management systems across the company and producing an annual environmental report.

The Eco-Efficiency Center has employed a variety of strategies and has developed various tools to encourage and engage the 'greening' of businesses in this industrial park. Since June 1999 the principal vehicle the Center has used to encourage environmental improvement is Eco-Business Program Burnside. This is a voluntary programme in which companies are encouraged to be environmentally responsible and to commit to implementing waste minimisation and efficient resource use as part of their daily operations. 'Eco-businesses' adopt an environmental code and set reduction and conservation goals for their companies. When businesses join the programme, an environmental review of their facilities is usually undertaken and a written report highlighting opportunity areas is prepared. The Centre provides businesses with information, including: efficiency and conservation fact sheets and educational signage; access to resources and expertise and to library and Internet searches; quarterly newsletters; a website; a 'Greening Your Business' starter kit; and various educational and promotional activities.

An integral component is an award programme for outstanding businesses—the Eco-Efficiency Center Awards for Environmental Excellence. The awards profile leading-edge businesses—companies in Burnside that have improved their environmental performance and adopted innovative improvements in environmental management. Awards are offered in five categories: solid waste, liquid waste, water conservation, energy conservation and outstanding eco-business. To encourage companies that are not ready to compete for the awards, the Center acknowledges a number of companies for making substantial progress. The award-winners and the highlighted evolving companies are publicly celebrated during an annual breakfast held during Environment Week. The Center provides award-winners and all member companies with as much recognition and positive publicity as it can. The Center follows up with companies on a regular basis to identify: types and quantities of solid waste diverted from landfill; volume of water conserved; energy savings; types and quantities of materials diverted from sewers; cost savings or revenues; and so on. The Eco-Business Program is one of the first of its kind in Canada.

22.4 Accomplishments of the Eco-Efficiency Center

The Center has been in operation since September 1998 and has accomplished a number of things to date. They include the following:

- More than 100 companies have been enrolled in the Eco-Business Program. These companies have committed themselves to achieving specific reduction and conservation targets. 'Eco-businesses' include printers, automotive shops, trucking companies, retail paint stores, furniture manufacturers, machine and metal-working shops, chemical manufacturers, property management companies, companies providing environmental services and many others.

- More than 130 reviews of Burnside companies have been conducted. These 'walkabouts' result in reports being generated that provide companies with specific recommendations to improve operating efficiency and to assist them in setting goals for the Eco-Business Program. Companies have benefited financially and environmentally from the advice provided by the Center in areas of solid waste diversion, energy efficiency, water conservation and liquid and hazardous waste reduction.

- The first Eco-Efficiency Center Awards for Environmental Excellence were presented in May 2000. Since then, 14 companies have been recognised for their accomplishments.

- Useful materials and tools for businesses joining the Eco-Business Program have been developed. These include 34 fact sheets on issues from managing waste oil to working on win–win solutions for tenants and landlords, newsletters, a website, case studies and success stories and, most recently, a green 'toolkit'—a hands-on kit of materials, including an environmental green products or services coupon book, water and energy conservation and eco-friendly cleaning product samplers, posters, business checklists and resource listings.

- Partnerships among businesses in Burnside have been fostered. The Center has been successful in encouraging the co-operation of businesses in exchanging 'waste'. The most successful items for trading within the park have been wooden pallets, metals and packaging materials. Businesses are also accessing several local waste-exchange programmes, including an online, province-wide waste exchange, a technology recycling programme for computer hardware, a chemical exchange offered through Dalhousie University and a food rescue programme through a local food bank.

- Local companies that provide environmental and waste disposal products and services have been promoted. Burnside is home to many companies that fall into these sectors, including waste hauliers and recyclers, hazardous waste handlers, environmental consultants, paint remanufacturers, manufacturers of eco-friendly cleaners, and lighting and heating specialists. The Center makes businesses aware of the availability of these services and products.

A total of 100 companies with over 2,000 employees have now committed themselves to waste reduction and resource conservation by registering in the Eco-Business Program, and 130 reviews have been completed for Burnside companies. Although tracking has been difficult, the Centre is beginning to see companies undertaking better documentation of their environmental and economic results. A survey in 2001 has provided insight into some direct results. A total of 34 businesses have reported the following annual economic and environmental improvements:

- Revenue generated and cost savings: Can$90,602

- Solid waste diversion: 1,569 tons

- Liquid waste diversion: 99,413 litres
- Water reduction: over 11 million litres

Although many of these initiatives have occurred within companies, a few have been the result of symbioses, focused primarily on packaging.

22.5 Eco-efficiency success stories

Building on the successful implementation of other water-reducing initiatives, Farnell Packaging Ltd installed a closed-loop glycol-based cooling system. This replaced an older system that relied on city water to directly cool equipment. The older system used the water once then discharged it to the sewer systems. The new cooling system uses glycol instead of water and recirculates it. Implementation of the closed-loop glycol-based system resulted in substantial benefits including:

- Estimated savings of more than Can$5,000 per year
- An 85% reduced usage of the Halifax Regional Municipality water supply

Further, Farnell staff anticipate cost savings will pay for the capital investment in new equipment in just over one year.

Swedwood Canada Ltd has undertaken several projects that have resulted in environmental and economical improvement showing its dedication to the programme's philosophy. For example:

- It has installed a boiler to burn ground waste and dust for heat recovery. These waste-burning techniques have the amiable result of lowering the company's heating costs and reducing the level of waste sent to landfill.
- A lighting retrofit project has resulted in the replacement of 750–1,000 fluorescent bulbs. This not only saves money but also, more importantly, conserves energy. Further, a local company selling used building materials has expressed an interest in these bulbs, which would allow those bulbs to be re-used, thus keeping them out of landfill at least temporarily.
- The company is in the process of replacing its painting equipment with new technology based on the use of ultraviolet (UV) light. This new system reduces emissions of volatile organic compounds and uses the less-energy-intensive UV light to cure the paint. Further, the quality of the finished product has improved, dropping the rejection rate to less than 1% (i.e. the total number of products that do not meet the quality specifications is less than 1% of the entire production).

The use of wood chips and dust has reduced the use of heating oil by approximately 70%. The 1998 usage level of 110,000 litres had dropped dramatically to 35,000 litres by 1999. At a conservatively low cost of Can$0.40 per litre, this equates to fuel savings of Can$30,000, without considering the cost savings for reductions in fees to haul the solid waste away. Not to be ignored are the implications surrounding the lighting retrofit. The new lighting strategy will provide better-quality light in the plant, save energy and electricity and, ultimately, reduce greenhouse gas emissions and result in cost savings over the long term.

Miller Composting Corporation, which operates one of Canada's largest composting plants, received the Liquid Waste Reduction Award. Miller incorporated a special 'closed-cycle' leachate processing system in the building of its facility. No leachate water is discharged into the sewer. Some of the compost from the operation will be used for landscaping in the park.

22.6 Conclusions

Although the project and the Center are not the only impetus for change in Burnside, there is a general sense that they have contributed to a transformation in the park. Mature ecosystems are sustained by a large number of scavenger and decomposer species that facilitate the recycling processes. The conventional industrial system has tended to encourage a production–consumption nexus, ignoring the other two very important functions. As a consequence, air, water and soil have been polluted, habitat is being destroyed and species are disappearing at an alarming rate. In a 1998 study of businesses in Burnside, it was estimated that as many as 15% of the companies undertake one or more scavenging and decomposing functions (Geng and Côté 2002). Rental, re-use, repair, refurbishing, remanufacturing and recycling were included in these categories. In some cases, the cycling of materials occurs entirely within the boundaries of the park. This is believed to be occurring at a higher rate than in the municipality and the province generally. The diversification, the cycling of materials and the increased co-operation among businesses would suggest that a more mature industrial system is evolving. Thus we are beginning to see the emerging features of an industrial ecosystem.

23
ECO-INDUSTRIAL DEVELOPMENT AS A DEFENCE CONVERSION STRATEGY
A case study of the Louisiana Army Ammunition Plant Re-use

Joshua L. Tosteson
Columbia University, USA

Victor A. Guadagno
Northwest Louisiana Commerce Center, USA

The Northwest Louisiana Commerce Center (NWLCC) is an eco-industrial park located on the grounds of the Louisiana Army Ammunition Plant (LAAAP), a deactivated 15,000 acre ammunition facility approximately 20 miles to the east of Shreveport, LA. The NWLCC is the research, education, marketing and community development programme for the LAAAP re-use mission. Valentec Systems Inc. a private facility use contractor, has developed the re-use programme under the federal Armament Retooling and Manufacturing Support (ARMS) initiative, which provides funding and programmatic infrastructure for the re-use of deactivated government-owned, contractor-operated ammunition facilities.

The NWLCC programme is rooted in two related sciences of sustainable design: industrial ecology (Frosch and Gallopoulos 1989) and permaculture (Mollison and Holmgren 1978). Permaculture is 'the design and maintenance of agriculturally productive ecosystems that have the diversity, stability and resilience of natural ecosystems' (Mollison 1988: ix). Education, demonstration and application of permaculture design principles form the core of the NWLCC community and small-business development programme. Industrial ecology views industrial production systems as 'ecosystems', treating the by-products of industrial activity as potentially productive resources rather than as cost-generating waste. Industrial ecology seeks value-added uses for industrial by-products, either as feedstock for existing operations or as inputs for new production facilities. The NWLCC eco-industrial development programme, the re-use strategy for unneeded manufacturing and warehouse facilities at the LAAAP, represents a broad application of this emerging science.

In this chapter we present the history of the LAAAP re-use programme, an overview of permaculture and its relationship to industrial ecology, and the conceptualisation and development of the NWLCC eco-industrial park.

23.1 The Louisiana Army Ammunition Plant: background and history

Prior to construction of the LAAAP in 1941, the site consisted of numerous homesteads and small farms (see Fig 23.1), with cotton as the major crop. During the Second World War, the facility produced intermediate-calibre ammunition, mines, bombs, shells and grenades. After that war, the LAAAP experienced cycles of 'boom and bust' as the nation came in and out of times of armed conflict. Peak employment during the Second World War was 10,000 people, and by the end of the war LAAAP had developed the capability to produce and load 155 mm projectiles. The plant was deactivated in 1945 and then reactivated in 1951 to support the Korean War. During the Korean conflict LAAAP focused on construction of 155 mm projectiles, and peak employment during this period was about 5,100 people. The site was again deactivated from 1958–61, then reactivated to support South-East Asia munitions requirements. Peak employment during this time was 7,800 people. Employment rose and fell through the 1970s, 1980s and early 1990s, until LAAAP was finally decommissioned and, in 1997, placed under the ARMS programme.

As the Cold War drew to a close, the costs of maintaining ageing, inactive and under-utilised ammunition plants were severely draining the Army's declining ammunition budget. The ARMS programme was developed to enable private facility use contractors to market government-owned ammunitions manufacturing plants for commercial business. The idea is to preserve the equipment and plant infrastructure of these facilities at no cost to the taxpayer while at the same time reducing the impact of plant closure on local communities by supporting private economic development of the idle plant facilities.

The goals of the ARMS programme are to:

- Provide an incentive for re-use by small and disadvantaged businesses
- Encourage commercial re-use of government-owned contractor-operated facilities and equipment
- Reduce adverse effects of defence conversion
- Create jobs and re-employment opportunities
- Foster economic stability
- Maintain readiness and a skilled workforce
- Promote free-market competition
- Relocate offshore production to the USA
- Be a model defence conversion programme

In December 1997 the US Army Operations Support Command (OSC; formerly known as the Industrial Operations Command) awarded the LAAAP facility use contract to Valentec Systems Inc., a privately held company specialising in metal parts manufacturing for the defence industry. The five-year contract calls on Valentec to manage the facilities, buildings and grounds of LAAAP and to attract rent-paying commercial tenants to meet the costs of these operations. The Army's primary interest, in addition to

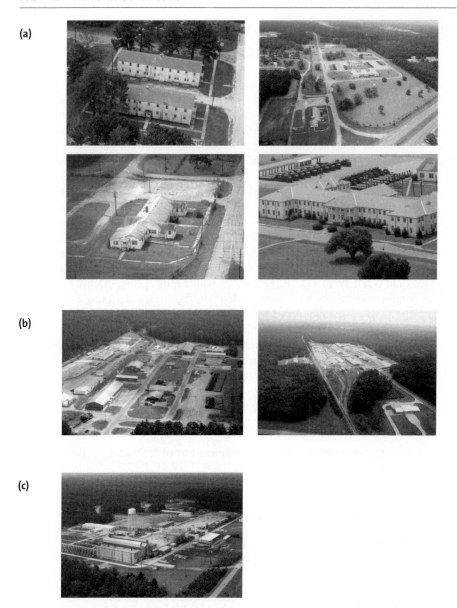

Figure 23.1 *The site of the Louisiana Army Ammunition Plant: (a) area A; (b) area B; (c) the 'Y-Line' area*

releasing itself from the cost and management burden, is to maintain the 'Y-Line' area (see Fig. 23.1c) of the site in layaway status. Y-Line is the metal parts manufacturing facility for 155 mm projectile metal parts, and Valentec's contract with the Army stipulates that Y-Line be maintained so that, within 18 months, the facility could again be placed into full production.

23.2 Re-use of the Louisiana Army Ammunition Plant: programme conceptualisation and development

Beginning in 1998, Valentec Systems assembled a team to develop and implement a model commercial re-use strategy for the LAAAP. Valentec called on Victor A. Guadagno (co-author of this chapter), founder of Full Circle Production, a small marketing company, to develop the re-use strategy for the LAAAP. The previous year Guadagno had been working in Guatemala, where he attended a course in permaculture design taught by Australian designer, educator and activist Geoff Lawton. Through this experience, Guadagno became an advocate for permaculture as a vehicle for sustainable system design. Originally developed in the 1970s by Australian scientists Bill Mollison and David Holmgren (1978), permaculture is a design science rooted in the observation that, in natural ecosystems, each living and non-living component of the system performs multiple functions that serve to benefit the health, function and resilience of the system as a whole. In natural systems there are no waste products, only resources. In Mollison's words (1988: ix), permaculture:

> is the conscious design and maintenance of agriculturally productive ecosystems which have the diversity, stability and resilience of natural ecosystems. It is the harmonious integration of landscape and people, providing their food, energy shelter and other material and non-material needs in a sustainable way. Permaculture design is a system of assembling conceptual, material and strategic components in a pattern which functions to benefit life in all its forms. The philosophy behind permaculture is one of working with, rather than against, nature, of looking at systems in all their functions, rather than asking only one yield of them and of allowing systems to demonstrate their own evolutions.

Inspired by Lawton's work and the design science of permaculture, Guadagno contacted Lawton when he became involved in the LAAAP re-use effort. The two worked closely with Valentec and the OSC to explore how the principles of sustainable design could provide the best blueprint for guiding the re-use of the LAAAP. By seeing waste and cost sinks as resources and opportunities, Lawton and Guadagno emphasised, true efficiency and greater profits could be realised—all while making positive contributions to local and regional environmental sustainability. They used this basic argument to build support for using permaculture design principles as an appropriate framework for the facility redevelopment strategy.

However, permaculture design focuses primarily on the use and re-use of land assets, whereas the ARMS programme objectives emphasise the re-use of industrial assets. So, while the re-use team cultivated programme support for permaculture as an overarching framework for the LAAAP re-use programme, they conducted additional research over the course of 1998–99 to refine the strategy. During this time the team became familiar with the science of industrial ecology and one of its primary fields of application, eco-industrial development, through the work of the Cornell University Work and Environment Initiative. It was evident to the team that eco-industrial development represents the industrial 'twin' of permaculture design (see Table 23.1) and would be the most appropriate strategy for redeveloping the substantial idle industrial assets at the LAAAP site (Box 23.1).

Permaculture	*Eco-industrial development*
BLUEPRINT FOR DESIGN SCIENCE	
Ecosystem	Ecosystem
FIELDS UNDERLYING DESIGN SCIENCE	
Systems ecology; hydrology; soil science and agronomy; economics and finance; traditional knowledge systems	Industrial ecology; engineering; economics and finance; organisational dynamics and change
PRIMARY DOMAINS OF CURRENT APPLICATION	
Land use and design; water management; energy systems; housing; economics, finance and human social organisation; education	Industrial design; regional economic and community development; regional and national material flow accounting and management; education
PRIMARY SCALES OF CURRENT APPLICATION	
Household, settlement, village	Industrial park, city, region

Table 23.1 *Examples of objectives, indicators and targets for an industrial park*

- Buildings and structures
  - Over 400 buildings, covering a combined area of 2.1 million ft^2
  - The largest building is 80,000 ft^2
  - 171 on-site, climate-controlled bunkers (336,000 ft^2)
- Transportation
  - 65 miles of railway on-site
  - 56 miles of paved road
  - 46 miles of unpaved road
  - Shreveport Regional Airport, 25 miles west
  - Interstate 20, 1 mile from site
  - Interstate 49, 20 miles from site
  - Navigable water access to the Gulf of Mexico

Box 23.1 *Industrial assets of the Louisiana Army Ammunition Plant*

Permaculture and eco-industrial development are design sciences that use basic principles of how ecosystems work as the blueprint for creating sustainable land-based and industrial production systems, respectively. A sustainable system may be defined as 'one that, over its lifetime, produces the resources required to build and maintain itself'.[1] (In current practice, permaculture design and eco-industrial development differ mainly in their primary domains of application [landscapes compared with industrial systems] and scales [household or settlement compared with community and regional industrial systems]).[2] Based on these observations, the re-use team successfully proposed to utilise permaculture design as the centrepiece of LAAAP's small-business, community development, outreach and education programme, and eco-industrial development as the primary strategy for redeveloping LAAAP's industrial assets. Although it is difficult to separate these two elements of the re-use programme in day-to-day practice, we focus below on development and implementation of the eco-industrial development programme.

23.3 Northwest Louisiana Commerce Center eco-industrial development programme: strategy and implementation

In autumn 2000 the NWLCC created a position for an eco-industrial development manager, with a goal of crafting and implementing an eco-industrial development strategy for the industrial assets of the LAAAP. This led to the establishment of a formal eco-industrial development programme, the characteristics of which we describe below.

The NWLCC eco-industrial development programme has four goals:

- To identify and recruit new companies to establish manufacturing operations on the NWLCC site, using regionally abundant by-products and/or by-products produced on-site

- To promote education, efficiency, cost-savings and waste reduction in Northwest Louisiana

- To promote eco-industrial development as a viable economic development strategy for Louisiana

- To develop information products and tools to support these eco-industrial development activities

In 2001, four objectives were established to advance these goals:

---

1 From a meeting between Bill Mollison and the authors at the LAAAP Permaculture Workshop, 10 October 2000.
2 It should be noted, however, that neither design science is theoretically constrained in its potential scale of application. Both aim to transform modes of agricultural and industrial production at the planetary scale.

- To identify and categorise all waste (by-products) produced by the major manufacturing firms in Northwest Louisiana and to estimate the associated disposal costs
- To identify value-added opportunities for new manufacturing facilities that use regionally abundant industrial by-products as feedstock
- To recruit companies to develop projects at LAAAP based on the above analyses and to facilitate them through the ARMS process
- To conduct education and outreach activities and to forge partnerships in order to develop the programme

Because existing tenants on the LAAAP site did not initially appear eager to forge eco-industrial linkages, the team decided to focus on recruiting new tenants who could profitably intervene in the existing network of regional materials flows, thus reducing the landfill pressure and disposal costs faced by the regional generators of by-products. In parallel, an effort was initiated to develop operating policies for the eco-industrial park to promote on-site linkages as the number of tenants attracted by the eco-industrial concept increase.

Having established programme goals and objectives, the re-use team worked quickly to move forward in 2001. A significant first milestone occurred in February, when the LAAAP hosted and co-sponsored an eco-industrial development round-table meeting with the Cornell Work and Environment Initiative. This event both promoted the developing NWLCC eco-industrial park and provided the team with useful assistance in structuring the by-product research and tenant recruitment activities.

Based in part on suggestions received at this meeting, the team decided to begin the regional by-product analysis by focusing on a promising target sector—the wood products industry. In March the team initiated a pilot study of this sector, focusing on characterising the stocks and flows of post-consumer, post-industrial waste wood in Northwest Louisiana (Figs. 23.2 and 23.3). The team developed a lengthy list of potential

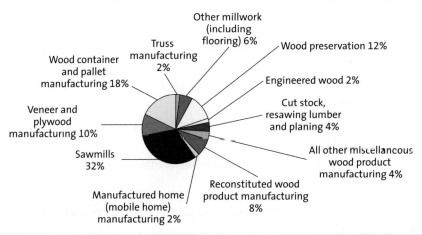

Figure 23.2 **Analysis of the wood product industry in Northwest Louisiana**

Source: County Business Patterns database, US Bureau of the Census

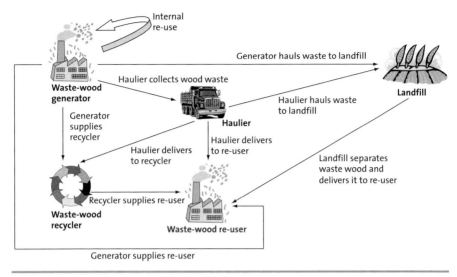

Figure 23.3 **Sources, flows and sinks of waste wood in a regional economy**

generators, hauliers, re-users, recyclers and disposal sites for waste wood, and conducted telephone interviews to develop a quantitative estimate of the annual production of waste wood in the region (Table 23.2). Based on a conservative estimate of 135,000 tons of waste wood generated annually, the team then sought out projects that had potential to use this resource profitably as a feedstock.

Source	Amount produced (tons per year)
Pallets	32,000
Mill waste	17,500
Wood waste from recyclers	15,000
C&D waste	70,000
LAAAP site	300
Total	134,800

C&D = construction and demolition   LAAAP = Louisiana Army Ammunition Plant

Table 23.2 **Post-consumer, post-industrial waste-wood production in Northwest Louisiana**

By mid-April negotiations were under way with a start-up company, Greentech Panels LLC. Greentech uses 100% post-consumer, post-industrial waste wood as the feedstock for manufacturing thin, fire-retardant particle board. Facilitated by the re-use team, Greentech leveraged over US$20 million in private investment under a loan guarantee programme jointly administered by the US Department of Agriculture (USDA) and the ARMS programme, and approximately US$3 million in federal ARMS incentive dollars (no-interest loans). In June the company signed a ten-year lease for 71,000 ft² of manufacturing space, and will employ 35 people, re-route 40,000 tons of waste wood each year from local landfills to value-added production and save waste-wood generators in the region approximately US$600,000 each year in avoided disposal costs. The deal represents US$1,400,000 in total rent revenue to the Department of Defense, for a 2001 (financial year) investment of approximately US$90,000 in the eco-industrial development programme.

Based on these results, the team developed plans to expand the by-product research programme. After reviewing company survey programmes of other eco-industrial development efforts, the team developed a by-product survey instrument adapted from the successful Triangle J Council of Governments programme (see TJCOG 2000; see also Chapter 21). The Minden–South Webster Parish Chamber of Commerce and the Northwest Louisiana Partnership for Economic Development agreed to distribute and collect the survey over the summer of 2001 and to assist in reviewing the results.

Meanwhile, the team had opened collaborative discussions with the chairperson of the Department of Renewable Resources at the University of Louisiana Lafayette (ULL), Charles Reith. These discussions resulted in the establishment of a collaborative internship programme in which four students from ULL spent the summer on the LAAAP site—two focused on the permaculture programme, two on the eco-industrial development programme. The eco-industrial development interns pursued two projects: (1) collecting survey data and analysing potential business opportunities and (2) developing a geographical information system (GIS) database for visualisation and analysis of survey results. One student developed a business plan for a compost and bio-gas facility that would use as feedstock horse manure from a local racetrack; this plan suggests that such a business could yield up to US$250,000 in annual profit. The re-use team has already attracted interest from local investors in this project.

At the time of writing this chapter (August 2001), discussions are under way to develop an 'anchor' eco-industrial development project that would integrate a materials processing centre, compost facility, plasma arc furnace for syngas production from waste tyres and an electricity generating centre (Fig. 23.4). In addition to a composting operation, described above, the proposed facility would take in 10 tons of used tyres an hour, strip and sell the steel and run the shredded tyres through a plasma arc furnace. The furnace produces hydrogen gas (H_2) and carbon monoxide (CO) in amounts sufficient to power, respectively, a 22 MW fuel cell and a 70 MW combined-cycle power plant. Water (H_2O), the only material input to the plasma furnace besides the feedstock, would be fully recycled via the fuel cells. The furnace produces an inert, non-leachable slag for use in road repair, roofing and insulation. The project would create at least 30 new jobs, attract US$45 million in private investment to the region and demonstrate a next-generation technology for clean energy that simultaneously solves a nagging disposal problem. This project represents the primary focus for the eco-industrial development programme during autumn 2001.

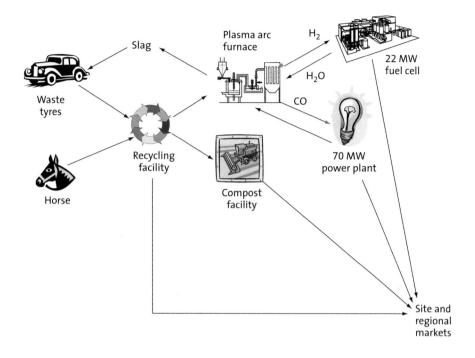

Figure 23.4 **The proposed industrial ecosystem facility in Northwest Louisiana**

23.4 Toward the future

The NWLCC re-use project has taken roughly four years to evolve from concept development to full implementation. This process has required a considerable upfront investment of time and energy simply in introducing the concepts of sustainable systems and their applicability to the ARMS defence conversion programme. 'Stakeholder education' has been a dominant feature of the effort, with groups ranging from Valentec Systems and the OSC to the regional political and financial communities, state government agencies and prospective tenants. In the future it is hoped that this upfront time and resource investment can be decreased in other areas, as documented economic and environmental benefits resulting from eco-industrial development begin to add up. It should be noted that success at the NWLCC can be largely attributed to the fact that it has introduced very little in the way of new innovation. Instead, it has focused its energies on identifying the best existing approaches and models and applying them to the local regional and organisational landscape. As a result, NWLCC has developed an eco-industrial park that shares many of the features of other efforts around the country but that is uniquely suited to its particular region of the USA.

Like other eco-industrial development efforts, the project itself is conceived as a self-organising natural system and must begin developing rich internal feedback processes as the number of components in the system increases. Anticipating this, the re-use team is in the process of crafting a site-wide environmental policy that will include eco-industrial park codes and covenants, waste-stream monitoring, incentive programmes and other features designed to promote internal synergies between site tenants. Also in process is work on a US$220,000 grant for the financial year 2001 from the US Army to establish a Permaculture Demonstration Centre.[3] This is a critical component of the overall re-use programme, as it will provide a living laboratory to illustrate the dynamics of agriculturally productive, natural systems. This Permaculture Demonstration Center will serve as the long-term hub for the re-use programme's education, outreach and community development efforts.

The NWLCC project demonstrates that the industrial needs of people can be met while enhancing the environment in which they reside. By applying an understanding of how natural ecosystems function to the design of landscapes and industrial systems in Northwest Louisiana, the NWLCC has proven a unique approach to defence conversion and has confirmed that sustainable development is an obtainable goal.

---

3   The grant was awarded in 2001 and is being managed by Minden, LA, non-profit, Cultural Crossroads.

24
ECO-INDUSTRIAL DEVELOPMENT IN ASIAN DEVELOPING COUNTRIES

Ernest Lowe
RPP International, USA

Eco-industrial development projects in developing countries are seeking to leapfrog the earlier errors of developed countries through robust application of industrial ecology. A number of initiatives in Asia surpass any comparable efforts in North America or Europe. Since 1997 the economic crisis in South-East Asia has led many industrial estate developers and managers to look for new strategies for increasing efficiency and competitiveness.

Eco-industrial development is finding acceptance because it offers concrete means of achieving these goals as well as realising the balance crucial to sustainable development: the balance between environmental protection and social and economic development. Industrial ecology may ultimately find its most comprehensive application in the developing world.

With strong collaboration between government and developers, innovators are working together to form eco-industrial networks in the Philippines and Thailand. Managers at a very large industrial park in Gujerat, India, are searching for opportunities to utilise a variety of by-products. A sugar company in southern China has put together 'eco-chains' of new plants that utilise virtually all the facility's by-products.

These projects involving eco-industrial parks, estates and networks and by-product exchange projects are reflecting a solid application of the principles and strategies for eco-industrial development that are recommended in other chapters of this book. Some Asian initiatives are the most ambitious ones to be found anywhere in developed or developing regions.

24.1 An eco-industrial network in the Philippines[1]

One of the first eco-industrial initiatives in Asia was an eco-industrial network seeking to create a by-product exchange and a resource recovery system to serve five industrial estates south of Manila in the Philippines. A petrochemical estate in Bataan is also involved, with the objective of becoming an eco-industrial estate. Project development started in 1998, with funding from the United Nations Development Programme (UNDP) to the Philippine Board of Investments. The PRIME (Private Sector Participation in Managing the Environment) Project (www.psdn.org.ph/prime) encompasses the work of four modules: (1) Agenda 21 for Business, (2) Industrial Ecology, (3) Environmental Management Systems and (4) Environmental Entrepreneurship. Each module has its own mission, but staff members seek opportunities for synergy among their teams. The Industrial Ecology Module planned its work through consultation and workshops with industrial estate managers and staff, industrial estate associations and other governmental agencies. Initially, the Board of Investments team focused on the formation of a by-product exchange at one estate. This was seen as the leading edge for introducing a broader range of industrial ecology initiatives: 'It is a relatively easy strategy to communicate, it requires active business leadership to achieve, and it offers relatively quick returns in cost savings and revenues' (www.iephil.com).

A crucial decision by the project staff was to follow a self-selection process with the six shortlisted estates and to include more than one site in the pilot project. This choice in the summer of 1999 led to a much more viable pilot project than would be achieved simply by selecting one of the sites. The variety of by-products generated at the different estates is broad enough to enable more matches between generators and users. The higher level of commitment on the part of some estates provides leadership to the others. All six estates on the shortlist decided to participate. Five industrial estates in Laguna and Batangas Provinces are participating in the by-product exchange: Laguna International Industrial Park (Binan, Laguna), Light Industry Science Park (Cabuyao, Laguna), Laguna Technopark (Santa Rosa, Laguna), Carmelray Industrial Park (Canlubang, Laguna) and Lima Technology Centre (Malvar, Batangas). The sixth site, in Bataan Province, is owned by Philippine National Oil Company.

In a planning workshop during June 1999 each team formed its action plan, defined its project investment and indicated its willingness to be responsible for data collection on by-products available from tenants. This commitment was based on industry's growing concern about the costs of solid waste disposal and pressure from the Laguna government. Through 2000 the PRIME staff has supported the estate teams in the analysis of the data and in facilitating information flows regarding potential exchanges. This analysis has included identification of potential uses and users of specific by-products, reprocessing and conveyance technologies and opportunities opened for businesses by the by-product exchange. The recommendations include strategies for waste reduction in the factories. By autumn 2000 a workshop for estate tenants drew 70 participants, who began working in interest groups around specific groups of by-products such as used oils, water and packaging materials.

The PRIME team discovered that the four estates in Laguna Province had started working together to meet demands the provincial government made for management of

1 See also Lowe 1999; PRIME Industrial Ecology Module 1999–2000.

their tenants' solid waste. The PRIME staff has co-operated with them in conducting a feasibility study for a resource recovery system, including a facility that could incorporate businesses accumulating and processing by-products from their tenants. This system would improve the performance of the by-product exchange and could be the site for an environmental business incubator.

The Philippine National Oil Company operates the industrial estate in Bataan, which is too far from the other sites to participate in their by-product exchange. This estate management team has created a strategy for becoming an eco-industrial estate, including a regional by-product exchange, a green chemistry business incubator and a programmatic environmental impact assessment (PEIA). This form of impact assessment is an innovation in Philippine regulatory structures, enabling a site to receive an environmental permit that covers the property as a whole, with a list of potential tenant companies that fall under that umbrella permit. The estate manages community enhancement programmes, including housing, training and micro-enterprise development components. It has also applied for ISO 14000 status.

The PRIME Industrial Ecology Module has also analysed policy barriers to the work of this eco-industrial network. These include:

> the arbitrary assignment of values for 'scrap' materials, imposing of taxes for 'scrap' materials being hauled out of economic zones if the customs officer knows that it will be used as a production substitute and taxing them again when they come back in to the factories that will be using them as feedstocks.[2]

The Industrial Ecology Module has also maintained an active outreach and communications programme. In June 1999 staff organised a one-day briefing in Manila for 100 private-sector and public-sector stakeholders. In November 1999 the Board of Investments hosted a working conference on eco-industrial estates for real estate developers and managers and other stakeholders in industrial estate development. The PRIME team provided both conceptual and logistical support to the organisation of a second conference and workshop in April 2001 for eco-industrial project teams from 11 Asian countries.[3]

24.2 An initiative for green industrial estates in Thailand[4]

The Industrial Estate Authority of Thailand (IEAT) has launched an initiative to make the 28 estates it manages eco-industrial estates. IEAT governor Avanchalee Chavanich invited the German Technical Co-operation Organisation (GTZ) to support development of a programme that will begin with five estates as pilot sites. They include: Map Ta Phut

---

2 Personal communication with G. Pascual-Sison is project manager for the Industrial Ecology Module, Board of Investments, 2001.
3 For updates on the progress of this project, see www.iephil.com
4 See also Industrial Estate Authority of Thailand (IEAT), initiative updates at www.geocities.com/iethai2001; Koenig 2000; Lowe 2000.

Industrial Estate, a petrochemical park; Eastern Seaboard Industrial Estate, with automotive and electronics factories; Amata Nakorn; and two estates built in the 1980s with very diverse sets of factories—Bangpoo Industrial Estate and Northern Industrial Estate. The pilot sites reflect both newer and older estates and a good cross-section of industries in Thailand. Andreas Koenig and the author conducted site visits and interviews with managers of four of the IEAT estates during November 2000.

The governor of IEAT envisions an initiative incorporating by-product exchange, resource recovery, cleaner production, community programmes and the development of eco-industrial networks linking estate factories with industry outside the estates. GTZ will assist through capacity development for IEAT headquarter staff and for personnel at the estates and through technical transfer and policy development. GTZ will co-ordinate with the IEAT initiative its other programmes in Thailand, such as its energy conservation project with the Bureau of Energy Regulation and Conservation.

This eco-industrial initiative appears to be the most far-reaching eco-industrial effort in developed or developing countries. It promises ultimately to impact the environmental, social and economic performance of all industrial estates managed by IEAT as well as the operations of stand-alone plants surrounding the estates. IEAT is in a unique position to demonstrate the principles and strategies of eco-industrial development. The proposed vision for the initiative is as follows: 'Through the technical co-operation project it is intended to improve the environmental performance of selected IEAT industrial estates and develop a policy for eco-industrial development for Thailand' (Chavanich 2001).

The first step will be for the management of each of the pilot estates to form its individual vision and business plan and budgeting investments required for specific projects. Estate managers have identified utilisation of by-products as an early concern, but they are aware that opportunities for exchanges among the factories at any one estate are limited. As they develop their estate plans they will start to explore opportunities for building an eco-industrial network between their companies and suppliers outside the estates. Four of the pilot sites are in the Eastern Seaboard area south-east of Bangkok, which includes 11 estates within a 50 km^2 area. This provides a significant opportunity for creating an eco-industrial network across this area. The inter-estate networks would complement the links from each estate to its surrounding stand-alone factories.

IEAT and the individual estate teams will consider creating potential supporting structures for their initiative, including:

- An integrated resource recovery system or, possibly, a resource recovery park
- A system for encouraging and managing the exchange of by-products
- Training and services in all aspects of eco-industrial development
- A co-ordinating unit to manage eco-industrial network relationships
- A community enhancement office to manage projects with neighbouring communities
- One or more business incubators
- Public-sector support in research and development (R&D), policy development, access to investment and information management

Managers of the individual pilot industrial estates highlighted a number of serious barriers that they will encounter in their effort to become eco-industrial estates and to form eco-industrial networks. Policy and regulations for the operation of estates and individual factories are defined by several agencies, including IEAT, the Department of Industrial Works and the Ministry of Science, Technology and Environment. Eco-industrial development requires harmonisation of these policies and adaptation to enable by-product utilisation and exchange. For instance, a re-refining company near Rayong is limited to receiving only 5 of the possible 20 solvents that it is designed to recycle. This limits the plant's return on investment, the cost saving for factories seeking to use recycled products and the diversion of hazardous materials from the limited landfill devoted to them. Such regulatory barriers are both economically and environmentally damaging.

Estate and factory managers and recyclers all reported a need for R&D to identify new technologies to manage the many by-products not currently usable in a resource recovery system. They also need guidance on increasing the energy efficiency of their operations and possible incorporation of renewable energy technologies, such as bio-gas from food, agriculture and sewage by-products.

External project funding for eco-industrial estates and networks will be required to augment the investments IEAT, the pilot estates and factories will be able to make. Fortunately, the Thai eco-industrial development concept allows the integration of now independent projects, such as the Japanese-based energy efficiency project at Map Ta Phut and GTZ's own energy conservation project that supports implementation of Thailand's Energy Conservation Act.

The long-range plan for this initiative in Thailand includes work at the policy level within IEAT and among the different ministries to address the needs that estate managers have raised. In addition, it will support development of more effective emergency management systems in the estates, capacity development to help IEAT staff to improve organisational performance within an eco-industrial concept and transfer of technologies required for greater efficiency and cleaner production.[5]

24.3 Industrial metabolism studies in India

Ramesh Ramaswamy and Suren Erkman have played a central role in introducing industrial ecology into India through field research, conferences and workshops. Their organisation, the Institute for Communication and Analysis of Science and Technology (ICAST), organised a major workshop on industry and environment in Ahmedabad in 1999, working in collaboration with the Confederation of Indian Industries and the Indo-Dutch Project on Alternatives in Development (Pangotra et al. 1999). The groundwork for this was laid when ICAST organised an earlier meeting at Kalundborg, Denmark, that attracted industrial estate managers and other leaders in environmental management from India.

5   For updates, see the IEAT website at www.geocities.com/iethai2001.

ICAST has conducted four industrial metabolism studies on different industrial systems in India. These include a cotton-clothing production centre at Tirapur, foundries in Haora, the leather industry in Tamil Nadu and a complex integrating a paper mill and a sugar mill. Below, in Sections 24.3.1–4, we extract information on four case studies from the paper that these two industrial ecologists presented at a Cleaner Production conference in 2000 organised by the United Nations Environment Programme (Erkman and Ramaswamy 2000). More detail on the studies described below may be found in Erkman and Ramaswamy 2001 and in Ramaswamy 2000. The first-mentioned paper, available on the Web, also provides an excellent discussion of the methods for conducting regional resource flow studies and strategies for improving their efficiency while reducing pollution.

24.3.1 Case study of Tirupur Town

Tirupur is a major centre for the production of knitted cotton garments in southern India. Some 4,000 small production units in the town specialise in different aspects of the manufacturing process, with a collective output of around US$700 million. Much of the produce is exported, bringing in very valuable foreign exchange. Water is scarce and the heavily saline effluents from processing textiles have rendered the groundwater unusable (90 million litres per day is also contaminated with a variety of chemicals). Water is brought in by trucks from remote groundwater sources as far as 50 km away, at an enormous cost. A massive US$30 million project is under way to treat the Tirupur wastewater at central effluent treatment facilities, which will release still unusable water.

Ramaswamy (2000) carried out a resource flow analysis (RFA) for the town of Tirupur as an example of how a regional RFA could be effectively used. Only when the figures from this detailed RFA were aggregated did the industrialists realise that they were collectively spending over US$7 million annually on buying water. They also could see that the annual maintenance cost of the effluent treatment plant would be an enormous burden. The aggregate figures immediately showed that water could be recycled profitably. On the basis of the study, a private entrepreneur developed a relatively low-cost water-recycling system that could be installed in each dyeing unit. The system used the waste heat from the boilers already working in the dyeing units for the recycling process.

The RFA study also quantified the high calorific value of the solid waste because of the large quantities of textile and paper waste. This could be used effectively to partially replace the 500,000 tons of scarce firewood being used in the town (there is grave concern over rapid deforestation in India [Ramaswamy 2000]). Since the use of the firewood is distributed over nearly 1,200 points, it was not obvious that such large quantities of firewood were being used. The possibility of setting up a central steam source (needed by some of the industries) is also under serious consideration in order to reduce the consumption of firewood.

24.3.2 Case Study of Seshasayee Paper and Board Ltd

The case study of Seshasayee Paper and Board Ltd illustrates a corporate growth strategy by which the viability of a paper mill was guaranteed by establishing a local sugar industry whose by-product of bagasse would become the primary input to paper production.

Another by-product from the sugar mill, molasses, was used in a distillery for the production of ethyl alcohol. In order to ensure a regular supply of sugar cane, the company took interest in the cultivation of sugar cane by organising the farmers in the region. The company struck long-term agreements with the farmers to buy back their produce and, in turn, took the responsibility of supplying them with water. Part of the water supplied for cultivation was the treated waste-water from the manufacturing operations. The company also used bagasse pith (a waste from paper-making) and other combustible agricultural waste in the region as energy sources. This example could be viewed as an agro-industrial eco-complex (see Section 24.5, in which the Chinese Guitang Group case is described; here, a sugar-refining company existed first and then established the other parts of the complex).

24.3.3 The leather industry in Tamil Nadu

A case study by Ramaswamy (2000) of a leather-producing region in Tamil Nadu state recommended strategic relocation of this water-intensive and polluting industry to ensure its long-term survival.

24.3.4 Foundries in Calcutta

A metabolism study of a foundry district in the Haora suburb of Calcutta identified the technology needed to meet pollution control requirements within the economic limits of 500 plants (Ramaswamy 2000).

24.3.5 Naroda Industrial Estate, Gujarat, India[6]

Another research team worked with Naroda Industrial Estate, which is one of the largest sites for eco-industrial development in the world. Some 700 companies with 35,000 employees operate on 3.5 km² of land. Naroda was founded in 1966 by the Gujarat Industrial Development Corporation, which also oversees 256 other estates. Industries at this estate include chemical, pharmaceutical, dye and dye intermediates, engineering, textile and food production. Naroda Industrial Association (NIA), including 80% of the companies, has founded a charitable hospital and a bank and has constructed a common effluent treatment plant. It has also planted 30,000 trees.

The initiative at Naroda Industrial Estate is an industrial ecology networking project seeking a co-operative approach to achieve pollution prevention. Local leadership comes from the NIA and the local bureau of the Confederation of Indian Industry (CII). Researchers from the University of Kaiserslautern are providing technical assistance and guidance on eco-industrial principles and methods. The project is funded by the German Ministry for Education and Research.

The effort began with a baseline survey of NIA members, focusing on material, water and energy usage. Local university graduates conducted surveys of 477 respondents, and data was analysed in a geographical information system (GIS). This identified common

---

6 See also Naroda Industrial Association at www.niaindia.com; *Ahmedabad News* 2000.

environmental problems as a basis for designing individual projects for the participating companies. The NIA convened open meetings in which the companies explored their needs, using a broad eco-industrial network framework proposed by Ed Cohen-Rosenthal (as described in WEI 2000: ch. 11). In these discussions they identified focus points for projects and created project teams with managers. Subjects of the first four projects were recycling of spent acid, recycling of chemical gypsum, recycling of chemical iron sludge and re-use or recycling of biodegradable waste.

In the spent acid project, four chemical companies planned to collect their spent acid (H_2SO_4) to produce ferrous sulphate ($FeSO_4$). Their combined by-product outputs would yield enough material to attain the concentration necessary for the generation of ferrous sulphate. A fifth firm with the necessary technology and energy supply is doing the processing. The companies will pay an amount that is half the usual waste disposal fee for the recycling. The recycling firm will create ten new jobs.

The chemical gypsum project started with a company that discovered that this by-product could be used in concrete production instead of being transported to landfill, with the associated costs of transportation and landfill fees. Through project information channels, three other companies with the same by-product joined the initial project. This group set up the logistics and a drying area for handling their common output. They are now recycling 300 tons per month instead of adding that mass to landfill.

The iron sludge project involves producers of dyes and dye intermediates that generate large quantities of iron oxide in a form that is quite hazardous. The project team identified production process changes to reduce the volume of iron oxide and to reduce the amount of hazardous impurities. Through the network, this cleaner production solution has been shared with all the firms in this industry.

The food companies at Naroda, mostly small operations, collectively generate large volumes of food waste (about 100 tons per month). They have done a feasibility study to identify ways of utilising this output, possibly through fermentation processes. As a group they could act to handle a problem that no one firm could deal with.

After these first four projects started 15 firms in the ceramic industry formed a fifth project with a cleaner production approach to assuring the purity of their input materials. They are jointly investing in a testing laboratory.

The Naroda stakeholders are interested in establishing an eco-industrial networking centre to disseminate and share their experiences and to help individual companies handle some of their internal clean production issues. This would build on the improved access to information, easier project management and consideration of new recycling technologies that the eco-industrial network has already achieved. This initiative at Naroda Industrial Estate has demonstrated that once-isolated companies can work effectively in a collaborative approach and improve their environmental and financial performance.

24.4 Eco-industrial development in Indonesia

Michael von Hauff and Martin Wilderer have been working with Zona Industri Manis in Indonesia at the town of Tangerang on the outskirts of Jakarta (von Hauff and Wilderer

2000). Reports on projects in existing industrial estates in India and Indonesia can be found at www.cc.jyu.fi/helsie/proceed.html.

24.5 Eco-chains in China: the Guitang Group[7]

China produces 10.5 million tons of sugar annually from 539 sugar industries, the majority from sugar cane. Over the past few years, the sugar industry in China has experienced a significant economic decline. This industry needs to increase its productivity to remain competitive with Brazil, Thailand and Australia, three major sugar-producing countries. Low prices for sugar on world markets in recent decades have eliminated the industry in former leading countries, including Hawaii and Puerto Rico. Sugar production is becoming much less competitive in the Philippines.

The Guangxi Zhuang Autonomous Region, in the far south of China, is the country's largest source of sugar, producing more than 40% of the national output. The cost of producing sugar is high in Guangxi. Most farmers have small landholdings, productivity is low and the sugar content of the canes is low. Most refineries are smaller-scale and fail to utilise their by-products. This gap causes them to lose secondary revenues and generate high levels of emissions to air, water and land. The farmers burn the cane leaves every harvest season, generating emissions to air. Ning Duan (2001) estimates that there are 70,000 families growing sugar in the region and that there are 100 sugar mills. The economy of the town of Guigang is 50% dependent on sugar-related industries (Duan 2001).

The Guitang Group is a state-owned enterprise formed in 1954 that operates China's largest sugar refinery, with over 3,800 workers. (The author visited this complex in April 2001.) The Group owns 14,700 ha land for growing cane. Though the sugar industry in China is generally responsible for high levels of emissions, this company has created a cluster of companies in Guigang to re-use its by-products and thereby reduce its pollution. The complex includes an alcohol plant, a pulp and paper plant, a toilet-paper plant, a calcium carbonate plant, a cement plant, a power plant and other affiliated units. The goal of the initiative is 'to reduce pollution and disposal costs and to seek more revenues by utilising by-products' (Duan 2001). Figure 24.1 (Duan 2001: fig. 2) shows the present flows of materials and water. Duan identifies in these flows two primary eco-chains that Guitang has established, each of which has additional members and some internal feedback loops.

The annual output of the Guitang complex of companies is as follows: 120,000 tons of sugar; 85,000 tons of paper; 10,000 tons of alcohol; 330,000 tons of cement; 25,000 tons of calcium carbonate; 30,000 tons of fertiliser; and 8,000 tons of alkali. In the late 1990s the secondary products accounted for 40% of company revenue and nearly as large a portion of profits and taxes paid.

---

7 E-mail on EIP Projects in China, 23 February 2001, from X. Wei, who works on the Guigang initiative of the State Environmental Protection Agency. The author visited the Guitang Group complex in April 2001 and interviewed Group managers and city officials.

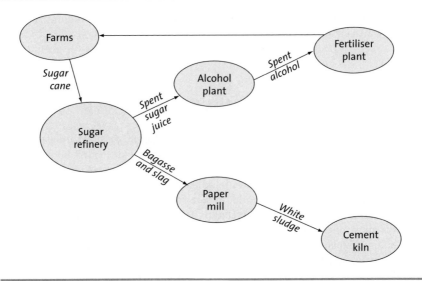

Figure 24.1 **Guitang Group eco-chains, Guangxi, Autonomous Region**
Source: Duan 2001

The Guitang Group's plans for the future include expansions of the industrial ecosystem and changes in processes at various stages. This innovative plan includes intentions to:

- Construct a new beef and dairy farm, with dried sugarcane leaves as feed for the livestock
- Construct a milk-processing factory to make fresh milk, milk powder and yoghurt for the local market
- Construct a beef-packing house to process beef, ox hide and bone glue
- Build a biochemical plant to make nutritional products based on amino acids as well as other bio-products with use of by-products from the beef-packing house
- Develop a mushroom-growing company with use of manure from the new dairy and beef farm
- Process residue from the mushroom base to use on sugarcane fields as natural fertiliser

China's expected entry into the World Trade Organisation poses a major threat to the economy of Guangxi. With barriers to lower-priced imports lowered, the economy of this region could be injured profoundly. Therefore Guitang Group's eco-industrial initiative has strategic importance for this and other sugar-producing regions in China.

Howard Nemerow describes an alternative pattern for a sugar refinery complex, with part of the bagasse and all of the sludge going to an anaerobic digester to generate methane, which is used as fuel for boilers. The remainder of the bagasse is burned in another boiler. The material output from the digester is filter cake that goes to farms for fertiliser (Nemerow 1995).

24.5.1 City of Guigang: plans to become an eco-industrial city

The Group's example has inspired the town of Guigang to adopt a five-year plan to become an eco-industrial city. The heavy dependence of its economy on the sugar industry makes it important to improve the efficiency of its many processing plants. The plan calls for smaller sugar producers to send their by-products to Guitang's eco-industrial complex and sets targets for high by-product utilisation. Targets for the city are for utilisation of sugarcane slag to reach more than 80%, the use of spent sugar juice reach 100% and the use of spent alcohol to reach 100%.[8] The plan also calls for consolidation of cane-growing land into larger holdings. This will require a transition for small farmers into other crops or into industrial employment. The plan also includes training of industry and government managers in eco-industrial principles and methods and broader dissemination of cleaner production strategies. Some of the long-term goals of this plan are to:

- Develop an eco-sugarcane park to: enable planting of organic cane, increase the sugar content of canes, increase the production per hectare of land and to extend the harvest period

- Enlarge the paper mill, with a goal of increasing production to 300,000 tons per year, in three phases

- Switch some production from sugar to fructose, which has a strong market

- Build a facility to produce fuel alcohol from spent sugar juice and sugar (with a capacity of 200,000 tons per year); this will help reduce air pollution from vehicle exhausts

- Adopt a low-chlorine technology to bleach pulp; paper made by this technology will be much whiter than the paper made by traditional technologies

Guitang and the leadership of the town are supported by China's State Environmental Protection Bureau (SEPA) and the China National Cleaner Production Centre (CNCPC). Ning Duan, deputy president of the Chinese Research Academy of Environmental Sciences, has been a key advisor to the Guitang Group. Financing is from the financial bureau of Guigang City. The local tax administration will return 50% of the agricultural tax to construction of irrigation systems for sugarcane farms.

8 Site visit by the author, 2001.

24.6 Eco-industrial development in other Asian developing countries

Stakeholders in eco-industrial projects gathered in Manila in early April 2001 for a conference and workshop: Strategies for Industrial Development: Learning from Pioneer Experiences in Eco-industrial Networking in Asia. Participants came from China, the Philippines, Taiwan, Malaysia, Thailand, Singapore, Indonesia, Sri Lanka, India, Pakistan and Nepal. Their discussions were supported by international experts from Asia, North America and Europe.

In the workshops, participants reported on their far-reaching and varied projects and sought ideas for improving their performance. In addition to the Philippine, Thai, Indian and Chinese initiatives and other projects summarised in this chapter, participants reported on further projects in Japan, Indonesia, Nepal, Sri Lanka, Pakistan, Malaysia and Singapore. They agreed to continue collaboration through an Asian Network for Eco-industrial Development, which will be co-ordinated by the Philippine PRIME Project.

Meeting sponsors and organisers included the UN Environment Programme, the UNDP, the Institute for Communication and Analysis of Science and Technology, Research EIT, the Asian Development Bank, the Philippine Board of Investments, the University of Kaiserslautern and RPP International-Indigo Development.[9] This meeting of Asian leaders in eco-industrial development was a strong indication of the appeal of eco-industrial development strategies in developing countries.

---

9   For reports on this eco-industrial gathering and network, see www.iephil.com.

BIBLIOGRAPHY

Adriaanse, A., S. Bringezu, A. Hammond, Y. Moriguchi, E. Rodenburg, D. Rogich and H. Schutz (1997) *Resource Flows: The Material Basis of Industrial Economies* (Washington, DC: World Resources Institute).
AES (2000) 'AES Announces New Construction Team' (press release; Londonderry, NH: AES, 3 August 2000).
AFL–CIO (American Federation of Labor–Congress of Industrial Organizations) (1998) *Economic Development: A Union Guide to the High Road* (Human Resources Development Institute, AFL–CIO).
Ahmedabad News (2000) 'Your Waste, My Gain: CII, IIM Suggest New Recycling Process', http://ahmedabad.com/incity/aug/7rec.htm.
Allen, D.T., and N. Behamanesh (1994) 'Wastes as Raw Materials', in B.R. Allenby and D.J. Richards (eds.), *The Greening of Industrial Ecosystems* (Washington, DC: National Academy Press): 68-96.
Allenby, B., and D.J. Richards (eds.) (1994) *The Greening of Industrial Ecosystems* (Washington, DC: National Academy Press).
Amato, I. (1997) *Stuff: The Materials the World is Made Of* (New York: Basic Books).
Anderson, G. (1994) 'Industry Clustering for Economic Development', *Economic Development Review*, Spring 1994: 26-32.
Anderson, R.C. (1998) *Mid-course Correction: Toward a Sustainable Enterprise* (White River Junction, VT: Chelsea Green).
Andrews, C.J. (1999) 'Putting Industrial Ecology into Place: Roles for Environmental Planners', *Journal of the American Planning Association* 65.4: 364-75.
Andrews, C., D. Rejeski, R. Socolow and V. Thomas (1998) 'Linking Industrial Ecology to Public Policy', workshop report, http://radburn.rutgers.edu/andrews/projects/iepp/iepprpt.pdf.
Angel, D.P., and M.T. Rock (eds.) (2000) *Asia's Clean Revolution: Industry, Growth and the Environment* (Sheffield, UK: Greenleaf Publishing).
APO (Asian Productivity Organisation) (1997) *Green Productivity: In Pursuit of Better Quality of Life* (Tokyo: APO).
Argyris, C. (1992) *On Organisational Learning* (Cambridge, MA: Blackwell Business).
Ausubel, J.H. (1998) 'Industrial Ecology: A Coming of Age Story', *Resources* 130 (Winter 1998): 14 (http://phe.rockefeller.edu/RFF_IE).
Ayres, R.U. (1996) 'Creating Industrial Ecosystems: A Viable Management Strategy?', *International Journal of Technology Management*: 12: 5-6.
—— (1998) 'Eco-thermodynamics: Economics and the Second Law', *Ecological Economics* 26: 191.
—— and L.W. Ayres (1996) *Industrial Ecology: Toward Closing the Materials Cycle* (Brookfield, MA: Edward Elgar).
—— and U.E. Simonis (1994) *Industrial Metabolism: Restructuring for Sustainable Development* (New York/Tokyo: United Nations University Press).

Azar, J. (2001) 'Xerox: Environmental Leadership Programme', in P. Allen (ed.), *Metaphors for Change: Partnerships, Tools and Civic Action for Sustainability* (Sheffield, UK: Greenleaf Publishing): 48-54.

BCSD (Business Council for Sustainable Development) (1997) *By-product Synergy: A Strategy for Sustainable Development. A Primer* (Gulf of Mexico: BCSD).

Bechtel Corporation (1997) *Development of Technically Feasible Materials Exchange Scenarios: Brownsville/ Matamoros Regional Industrial Symbiosis Project* (final report to the Brownsville Economic Council; San Francisco: Bechtel Corporation).

Benyus, J.M. (1997) *Biomimicry: Innovation Inspired by Nature* (New York: William Morrow).

Bevill Amendment (1980) 42 USC Section 6921(b)(3)(A)(ii) (Washington, DC: US Government Printing Office).

Boorstyn, N.B. (1998) *Boorstyn on Copyright* (New York: Clark, Boardman, Callaghan).

Bosworth, B.R., and J. Rogers (1997) *Using Regional Economic Analysis in Urban Jobs Strategies* (Regional Technology Strategies Inc. and Centre on Wisconsin Strategy, February 1997).

Broadway, B. (2001) 'Good for the Soul: And for the Bottom Line', *The Washington Post*, 19 August 2001: A01.

Brown, L.R., M. Renner and B. Halwell (2000) *Vital Signs: The Environmental Trends that are Shaping Our Future* (New York: W.W. Norton).

Bunge, J., E. Cohen-Rosenthal and A. Ruiz-Quintanilla (1996) 'Employee Participation in Pollution Reduction: Preliminary Analysis of the Toxics Release Inventory', *Journal of Cleaner Production* 4.1: 9-16.

CAA (Clean Air Act) (1990) 42 USC s/s 7401 *et seq.* (Washington, DC: US Government Printing Office).

Cappello, S. (1974) 'The Industrial Recycling Park: Cash from Trash', *Industrial Development* 143.

CERCLA (Comprehensive Environmental Response, Compensation, and Liability Act, or 'Superfund') (1980) 42 USC s/s 9601 *et seq.* (Washington, DC: US Government Printing Office).

Cernea, M.M. (1994) 'The Sociologist's Approach to Sustainable Development', in I. Serageldin and A. Steer (eds.), *Making Development Sustainable: From Concepts to Action* (Washington, DC: The World Bank).

CFED (Corporation for Enterprise Development) (1999) *Strategic Planning for Economic Development: Moving beyond the Overall Economic Development Program* (Washington, DC: CFED).

Chavanich, A. (2001) 'Thailand's Eco-Industrial Development', conference proceedings from New Strategies for Industrial Development: An International Conference and Workshop, Manila, Philippines, 3-6 April 2001 (CD-ROM; Bangkok: Industrial Estate Authority of Thailand; see also www.ecobizthai.net).

Chertow, M.R. (1988) 'Waste, Industrial Ecology and Sustainability', *Social Research* 65.1 (Spring 1998): 33.

—— (2000) 'Industrial Symbiosis: Literature and Taxonomy', *Annual Review of Energy and the Environment* 25: 313-37.

Chiaro, P.S., and G.F. Joklik (1998) *The Extractive Industries in the Ecology: Sectors and Linkages* (Washington, DC: National Academy Press).

Chisholm, R.F. (1998) *Developing Network Organisations: Learning from Practice and Theory* (New York: Addison Wesley Longman).

Chisum, D.S. (1978) *Chisum on Patents* (Newark, NJ: Mathew Bender).

Clavelle, P. (1997) 'Burlington's Riverside Eco-industrial Park: Developing a Real-World Model of Sustainability', prepared for Brownfields '97: 'Back to the Future with Eco-industrial Parks', Kansas City, MO, 7 August 1997.

Cohen-Rosenthal, E. (1998a) 'Eco-industrial Development: New Frontiers for Organisational Success' (Seminar Series on Industrial Ecology; Ithaca, NY: Johnson Graduate School of Business, May 1998).

—— (1998b) 'Eco-industrial Development: New Frontiers for Organisational Success', in *Proceedings of the Fifth International Conference on Environmentally Conscious Design and Manufacturing*, Rochester, NY, June 1998 (New York: Work and Environment Initiative, www.cfe.cornell.edu/wei/eipnewfrontiers.htm).

—— (1998c) 'Working Ecologically', paper presented at 'Connecting Business and the Environment: Seminars in Industrial Ecology', Johnson Graduate School of Management, Cornell University, Ithaca, NY; www.cfe.cornell.edu/WEI/papers/work_ecol_files/working_eco.htm.

—— and T. McGalliard (1996) 'Designing Eco-industrial Parks: The US Experience', *Industry and Environment* 19.4 (December 1996): 14-18.
Commonwealth of Virginia (2002) 9 Va. Administrative Code (Eagan, MN: Thompson West).
Considine, D.H., and G.D. Considine (1984) 'Compounds', in *Van Nostrand Scientific Encyclopedia* (New York: Van Nostrand, 5th edn).
Côté, R.P. (1991) 'The Nature and Scope of the Toxic Chemicals Issue', in R.P. Côté and P.G. Wells (eds.), *Controlling Chemical Hazards: Fundamentals of the Management of Toxic Chemicals* (London: Unwin Hyman).
—— (1995) 'Supporting Pillars for Industrial Ecosystems', *Journal of Cleaner Production*: 5.1–2: 67-74.
—— (1998) 'Thinking Like an Ecosystem', *Journal of Industrial Ecology* 2.2: 9-11.
—— (2000) 'Exploring the Analogy Further', *Journal of Industrial Ecology*: 3.23: 11-12.
—— and F. Balkau (1999) *Environmental Management Systems for Industrial Parks* (discussion paper; Halifax, Nova Scotia, Canada: School for Resource and Environmental Studies, Dalhousie University; Paris: United Nations Environment Programme).
—— and E. Cohen-Rosenthal (1998) 'Designing Eco-industrial Parks: A Synthesis of Some Experiences', *Journal of Cleaner Production*: 6.3–4: 181-88.
—— and J. Hall (eds.) (1995) *The Industrial Ecology Reader* (Halifax, Nova Scotia, Canada: Dalhousie University, School for Resource and Environmental Studies).
——, R. Ellison, J. Grant, J. Hall, P. Klynstra, M. Martin and P. Wade (1994) *Designing and Operating Industrial Parks as Ecosystems* (Halifax, Nova Scotia, Canada: School for Resource and Environmental Studies, Dalhousie University).
CRA (Community Reinvestment Act) (1977) US Code, Title 12, Chapter 30 (Washington, DC: US Government Printing Office).
Culshaw, B. (1997) 'Smart Structures and Materials', *Journal of Electronic Defense* 1 (January 1997): 76.
CZMA (Coastal Zone Management Act) (1972) 16 USC 1451 *et seq.* (Washington, DC: US Government Printing Office).
De Geus, A. (1997) *The Living Company* (Boston, MA: Harvard Business School Press).
DeSimone, L.D., and F. Popoff (1997) *Eco-efficiency: The Business Link to Sustainable Development* (Cambridge, MA: The MIT Press).
Desrochers, P. (2000) *Eco-industrial Parks: The Case for Private Planning* (Bozeman, MT: Political Economy Research Center).
Dobers, P., and R. Wolff (1996) 'Managing the Learning of Ecological Competence', in W. Wehrmeyer (ed.), *Greening People: Human Resources and Environmental Management* (Sheffield, UK: Greenleaf Publishing): 271-88.
Drexler, E., and C. Peterson, with G. Pergamit (1991) *Unbounding the Future: The Nanotechnology Revolution* (New York: William Morrow).
Duan, N. (2001) 'Make Sunset Sunrise: Efforts for Construction of the Guigang Eco-industrial City', paper presented at the April 2001 Workshop on Strategies for Industry, Manila; Chinese Research Academy of Environmental Sciences, Beijing, www.iephil.com.
Duany, A., E. Plater-Zyberk and J. Speck (2000) *Suburban Nation* (New York: Farrar, Straus & Giroux).
Dunphy, D., and A. Griffiths (1998) *The Sustainable Corporation: Organisational Renewal in Australia* (St Leonards, Australia: Allen & Unwin).
Eddington, A. (1958) *The Nature of the Physical World* (Ann Arbor, MI: University of Michigan Press).
Ehrenfeld, J., and N. Gertler (1997) 'Industrial Ecology in Practice: The Evolution of Interdependence at Kalundborg', *Journal of Industrial Ecology* 1.1: 67-79.
Emergency Planning and Community Right-to-Know Act (1986) 42 USC 11011 *et seq.* (Washington, DC: US Government Printing Office).
Emery, F.E., and E.L. Trist (1975) *Towards a Social Ecology: Contextual Appreciations of the Future in the Present* (London: Plenum).
Energy Conservation Act (1992) (Bangkok, Thailand: Ministry of Energy).
Engberg, H. (1993) *Industrial Symbiosis in Denmark* (New York: New York University, Stern Business School).

ENN (Environmental News Network) (2001) 'New York State Creates the First Clean-Energy Technology Park in the US', ENN, 23 August 2001 (www.enn.com/news/enn-stories/2001/08/08232001/cleanenergy_44710.asp).

Environment Act (1994–95) c1, s1 (Halifax, Nova Scotia, Canada).

Erkman, S. (1998) *Vers une écologie industrielle* (Paris: Charles Leopold Mayer).

—— and C. Francis (2001) *Application of Industrial Ecology to the Management of Industrial Estates* (Paris: UNEP).

—— and R. Ramaswamy (2000) 'Cleaner Production at the System Level: Industrial Ecology as a Tool for Development Planning (Case Studies in India)', *Cleaner Production 2000 Conference*, Montreal, Canada, 16–17 October 2000, www.agrifood-forum.net/db/cp6/ArticleErkman.doc.

—— and —— (2001) *Industrial Ecology as a Tool for Development Planning: Case Studies in India* (New Delhi/Paris: Sterling Publishers).

Esty, D., and M.E. Porter (1998) 'Industrial Ecology and Competitiveness', *Journal of Industrial Ecology* 2.1: 35-43.

Ferri, M.R., and J. Cefola (undated) 'A Case for Eco-industrial Development', *The Cap Gemini Ernst & Young Centre for Business Innovation Journal* 4, www.cbi.cgey.com/journal/issue4/features/ecoindustrial/index.html.

Field, E. (1996) 'The Future of Garbage', *Illinois Issues*, February 1996: 34-35 (www.lib.niu.edu/ipo/ii960234.html).

Fisher, H., and D. Findlay (1995) 'Exploring the Economics of Mining Landfills', *World Wastes* 38.7 (July 1995): 52.

Fleig, A.-K. (2000) *Eco-industrial Parks as a Strategy towards Industrial Ecology in Developing and Newly Industrialised Countries* (report prepared for the German Technical Co-operation Organisation [GTZ], Berlin).

Foecke, T. (2000) 'Identifying Opportunities for Investors in Companies that Use Natural Capitalism Principles to Gain Competitive Advantage', in *Conference Proceedings. The Greening of Economic Opportunity: Natural Capitalism and Eco-Industrial Development in Firms, Industries and Communities*, St Paul, Minnesota, 4–5 October 2000 (Minneapolis: Green Institute).

Forward, G., and A. Mangan (1999) 'By-product Synergy', *The Bridge* 29.1 (Spring 1999): 12-15 (www.hatch.ca/Sustainable_Development/Articles/Forward_mangan-bps.htm).

Frej, A., J. Goose, M. Christensen, W. D'Ellia, M. Eppli, L. Howlad, J. Musbach, F. Spirk and D. Verdor (2001) *Business Park and Industrial Development Handbook* (Washington, DC: Urban Land Institute).

Frosch, R.A. (1996) 'Toward the End of Waste: Reflections on a New Ecology of Industry', *Daedalus* 125.3: 199-212.

—— and N.E. Gallopoulos (1989) 'Strategies for Manufacturing', *Scientific American*, September 1989: 144-52.

Fujita, T. (1999) 'Planning Environmentally Sound Urban Renovation Strategies and Estimation of their Performance', in *Proceedings from the CCP International Workshop on Construction of ReCycle-oriented Industrial Complex Systems with Environmental Sound Technology at Societal Experimental Sites Project*, Osaka, Japan: 188-97.

Gardiner, R. (2002) 'Sustainable Finance: Seeking Global Financial Security', *Towards Earth Summit 2002, Sustainable Finance Briefing Paper*, Economic Briefing 2, UNED, www.earthsummit2002.org.

Gardner, G., and P. Sampat (1998) 'Mind over Matter: Recasting the Role of Materials in our Lives', *Worldwatch Paper 144* (December 1998): 18.

Geng, Y., and R.P. Côté (2002) 'Scavengers and Decomposers in an Eco-industrial Park', *International Journal of Sustainable World Ecology* 9: 330-40.

Gertler, N. (1995) 'Industrial Ecosystems: Developing Sustainable Industrial Structures', www.sustainable.doe.gov/business/gertler2.shtml.

Giannini-Spohn, S. (1997a) 'Eco-industrial Parks Offer Sustainable Base Redevelopment', *ICMA Base Reuse Consortium Bulletin* (The International City/County Management Association [ICMA], May 1997): 2, 3, 8.

—— (1997b) 'Eco-industrial Parks: One Strategy for Sustainable Growth', *Developments*, 15 January 1997 (www.smartgrowth.org/library/ecoind_strategy.html).
Gilson, J. (1974) *Trademark Protection and Practice* (New York: Mathew Bender).
Goldman, B.A. (1995) *Sustainable America: New Public Policy for the 21st Century* (prepared for the US Department of Commerce, Economic Development Administration; Cambridge, MA: Jobs and Environment Campaign).
Graedel, T., and B.R. Allenby (1995) *Industrial Ecology* (Englewood Cliffs, NJ: Prentice Hall).
Grimble, R., and M.K. Chan (1995) 'Stakeholder Analysis for Natural Resource Management in Developing Countries', *Natural Resources Forum* 19.2: 113-24.
Grogan, P.L. (1996) 'To Burn or Not to Burn', *Biocycle* 1: 86.
Halifax Regional Municipality (2000) *Burnside Business Park Site Development and Building Standards* (Halifax, Nova Scotia, Canada: Halifax Regional Municipality).
Hamner, B. (1998) 'Industrial Ecology in East Asia', *Journal of Industrial Ecology* 1.4: 6-8.
Han, S. (1998) 'ECO Investment Promotion for Environmentally Sound Technology for the Taihu Basin', *China Newsletter* 2:3.
Hart, M. (1999) *Guide to Sustainable Community Indicators* (North Andover, MA: Hart Environmental Data).
Hauff, M., and M. Wilderer (1999) 'Industrial Estates: The Relevance of Industrial Ecology for Industrial Estates', *Green Business Opportunities* 5: 2.
Hawken, P. (1993) *The Ecology of Commerce: A Declaration of Sustainability* (New York: HarperBusiness).
——, A. Lovins and L.H. Lovins (1999) *Natural Capitalism: Creating the Next Industrial Revolution* (Boston, MA: Little, Brown).
Henton, D., and K. Walesh (1998) *Linking the New Economy to the Livable Community* (Palo Alto, CA: Collaborative Economics, April 1998).
Hollander, J.B. (2000) *Analysing the Effectiveness of the Eco-industrial Park to Promote Sustainability* (MRP thesis, University of Massachusetts Amherst, Amherst, MA).
Hopfenbeck, W. (1993) *The Green Management Revolution: Lessons in Environmental Excellence* (Hemel Hempstead, UK: Prentice Hall).
HUD (US Department of Housing and Urban Development) (1999) *Empowerment Zone and Enterprise Community Tax Incentives: A Guide for Business* (Washington, DC: HUD, www.ezec.gov/Pubs/taxincentives.pdf).
—— (2001) 'Notice Inviting Applications: Designation of Forty Renewal Communities', *Federal Register* 66.152 (7 August 2001): 41,432-38.
ICMA (International City/County Management Association) with G. Anderson (1998) 'Executive Summary', in *Why Smart Growth? A Primer* (Washington, DC: ICMA, July 1998).
IDeA (Information Design Associates) (1997) *Cluster-Based Economic Development: A Key to Regional Competitiveness* (report prepared for the US Economic Development Administration; Washington, DC, October 1997).
IHDP (International Human Dimensions Programme on Global Environmental Change) (1999) *IHDP Update, February 1999* (Bonn, Germany: IHDP).
ISO (International Organisation for Standardisation) (1996) *ISO 14001: Environmental Management Systems: Specification with Guidance for Use* (Geneva: ISO).
IWGIEMEF (Interagency Workgroup on Industrial Ecology, Material and Energy Flows) (1998) *Materials* (Washington, DC: IWGIEMEF)
Jacobs, J. (1984) *Cities and the Wealth of Nations: Principles of Economic Life* (New York: Random House).
Jelinksi, L.W., T.E. Graedel, R.A. Laudise, D.W. McCall and C.K.N. Patel (1992) 'Industrial Ecology: Concepts and Approaches (Proceedings of the National Academy of Science conference, February 1992)', *Industrial Ecology* 89: 793-97.
Katz, B., D. Nguyen and R. Lang (1999) *A Rise in Downtown Living* (Washington, DC: Brookings Centre on Urban and Metropolitan Policy).
Katz, K. (1994) *The New Urbanism: Toward an Architecture of Community* (New York: McGraw–Hill).
Kendig, L. (1999) *Traffic Sheds, Rural Highway Capacity and Growth Management* (Report 485; Chicago: Planning Advisory Service, American Planning Association).

Kimura, M., and M. Taniguchi (1999) 'Zero Emission Clustering; Achieving Zero Emission', paper presented at the Eco 1999 International Congress, Paris.

Kincaid, J. (1999) *Industrial Ecosystem Development Project Report* (Research Triangle Park, NC: Triangle J Council of Governments, ftp://mail.tjcog.org/pub/solidwst/ieprept.pdf or www.tjcog.dst.nc.us/indeco.htm).

—— and M. Overcash (2001) 'Industrial Ecosystem Development at the Metropolitan Level', *Journal of Industrial Ecology* 5.1: 117-26.

Kiuchi, T. (1999) 'What I Learned in the Rainforest' (Sacramento CA: The Future 500).

Klee, R.J. (1999) 'Zero Waste in Paradise', *BioCycle* 40.2: 66-67.

—— and R. Williams (1999) *Emerging International Eco-industrial Projects. Casebook: Asia, The Pacific and Africa* (Report for the US Environmental Protection Agency, Office of Policy, Planning and Evaluation, Washington, DC).

Koenig, A. (2000) *Development of Eco-industrial Estates in Thailand: Project Development and Appraisal, June to December 2000* (German Technical Co-operation Organisation [GTZ], on behalf of the Industrial Estate Authority of Thailand).

Koestler, A. (1967) *The Ghost in the Machine* (London: Hutchinson).

Kollmer, T. (2000) 'Species-pool Affect Potentials (SPEP) as a Yardstick to Evaluate Land-use Impacts on Biodiversity', *Journal of Cleaner Production* 8.4: 293-312.

Krause, M. (2001) 'The Green Institute', presentation to the Eco-industrial Development Roundtable, Minden, LA, 26–27 February 2001.

LeRoy, G. (1994) *No More Candy Store: States and Cities Making Job Subsidies Accountable* (Washington, DC: Good Jobs First, Institute on Taxation and Economic Policy).

Lin, N. (2001) *Social Capital: A Theory of Social Structure and Action* (Cambridge, UK: Cambridge University Press).

Loftness, V., I. Oppenheim and J. Shankavaram (1998) *NSF Workshop Research Needs on Building Systems Integration for Performance and Environmental Quality* (Pittsburgh, PA: NSF Industry–University Co-operative Research Center [IUCRC], Center for Building Performance and Diagnostics, School of Architecture, Carnegie Mellon University, draft, January 1998).

Lowe, E.A. (1997) 'Creating By-product Resource Exchanges for Eco-Industrial Parks', *Journal of Cleaner Production* 4: 4.

—— (1998) *Fieldbook for the Development of Eco-industrial Parks* (Research Triangle, NC: Research Triangle Institute).

—— (1999) *Final Report on PRIME Project by International Consultant to Philippine Board of Investments* (Emeryville, CA: RPP International).

—— (2000) *Final Report to GTZ on Industrial Estate Authority of Thailand Eco-industrial Initiative* (Bangkok: Indigo Development).

—— (2001) *Eco-industrial Park Handbook for Asian Developing Countries* (report to Asian Development Bank Environment Department; Emeryville, CA: RPP International, http://indigodev.com/ADBHBdownloads.html).

—— and J. Warren (1996) *The Source of Value: An Executive Briefing and Sourcebook on Industrial Ecology* (Richland, WA; Battelle).

——, —— and S.R. Moran (1997) *Discovering Industrial Ecology: An Executive Briefing and Sourcebook* (Columbus, OH: Batelle Press).

——, S.R. Moran and D.B. Holmes (1998) *Eco-industrial Parks: A Handbook for Local Development Teams* (Oakland, CA: Indigo Development).

Lowitt, P.C. (1998) 'Sustainable Development with a Local Focus', in *Proceedings of the 1998 Conference of the American Planning Association* (Boston, MA: American Planning Association): 1.

—— (2001) 'Londonderry Eco-industrial Park', *Cornell Work and Environment Initiative: EIDP Update*, May 2001: 5.

Makower, J. (1993) *The E Factor: The Bottom-Line Approach to Environmentally Responsible Business* (New York: Times Books).

Manahan, S. (1999) *Industrial Ecology: Environmental Chemistry and Hazardous Waste* (Washington, DC: Lewis Publishers).

Manki, I.S. (1997) *Dictionary of Professional Management* (Los Angeles: Systems Research Institute).

Maskell, P. (2000) 'Social Capital, Innovation and Competitiveness', in S. Baron, J. Field and T. Schuller (eds.), *Social Capital: Critical Perspectives* (Oxford, UK: Oxford University Press): 111-23.

Mathews, J. (1996) *Catching the Wave: Workplace Reform in Australia* (Ithaca, NY: ILR Press).

Matthews, J. (1995) 'Organisational Foundations of Intelligent Manufacturing Systems: The Holonic Viewpoint', *Computer-integrated Manufacturing Systems* 8.4.

Mayor Energy Projects Development Act (1996) §57-1-251, MS Code of 1972, annotated (Washington, DC: US Government Printing Office).

McDonough, W., and M. Braungart (1998) 'The Next Industrial Revolution', *Atlantic Monthly*, 282.4 (October 1998): 82-92.

McGalliard, T., B. Clemens, M. Gresalfi, B. Fabens and E. Cohen-Rosenthal (1997) 'Eco-industrial Development and Re-industrialisation in Oak Ridge', in *Proceedings from the Fifth Annual Conference on the Recycling and Re-use of Radioactive Scrap Metal*, Knoxville, TN, www.cfe.cornell.edu/wei/BR97_fin.htm.

McVeigh, J., D. Burtraw, J. Darmstadter and K. Palmer (1999) 'Winner, Loser or Innocent Victim? Has Renewable Energy Performed As Expected?' (Discussion Paper 99-28; Washington, DC: Resources for the Future, www.rff.org/disc_papers/PDF_files/9928.pdf).

Meyer, P.B., and H.W. van Landingham (2000) *Reviews of Economic Development Literature and Practice. I. Reclamation and Economic Regeneration of Brownfields* (Washington, DC: US Economic Development Administration).

Mollison, B. (1988) *Permaculture: A Designer's Manual* (Tyalgum, Australia: Tagari Publications).

—— and D. Holmgren (1978) *Permaculture One: A Perennial Agriculture for Human Settlements* (Tyalgum, Australia: Tagari Publications).

Morikawa, M. (2000) *Eco-industrial Developments in Japan* (Working Paper 9; Emeryville, CA: Indigo Development).

Musnikow, J., and M. Schlarb (2001) *Eco-industrial Development Community Participation Manual* (Ithaca, NY: National Centre for Eco-industrial Development).

MWCG (Metropolitan Washington Council of Governments) (1997) *Stormwater Management Design Manual for Montgomery County, Maryland* (Washington, DC: MWCG)

NAHB (NAHB Research Centre Inc.) (2000) *A Guide to Deconstruction* (prepared for the US Department of Housing and Urban Development, Office of Policy Development and Research; Upper Marlboro, MD: NAHB, February 2000).

Nash, J., and J. Ehrenfeld (1997) 'Codes of Environmental Management Practice: Assessing their Potential as a Tool for Change', *Annual Review of Energy and the Environment* 22: 487-535.

Nattrass, B., and M. Altomare (1998) *The Natural Step For Business: Wealth, Ecology and the Evolutionary Corporation* (Gabriola Island, BC, Canada: New Society).

NCUED (National Council for Urban Economic Development) (1995) *The Planning and Development of an Urban Industrial Park* (Washington, DC: NCUED).

Nelson, A.C. (1993) 'Manufacturing Trends: Analysis of Urban Manufacturing Employment Trends', in *Urban Manufacturing: Dilemma or Opportunity?* (Washington, DC: National Council for Urban Economic Development).

Nemerow, N.L. (1995) *Zero Pollution for Industry: Waste Minimization through Industrial Complexes* (New York: John Wiley).

North, J.M., and S. Giannini-Spohn (1999) 'Strategies for Financing Eco-industrial Parks', *Economic Development Commentary* 23.3 (Autumn 1999): 5-13.

OECD (Organisation for Economic Cooperation and Development) (2000) *Strategic Waste Prevention: OECD Reference Manual* (Paris: OECD, Environment Directorate).

Orée and UNEP (United Nations Environment Programme) (1998) *Guide de management environnemental des zones d'activités* (Paris: Orée).

OSTP (Office of Science and Technology Policy) (1994) *Technology for a Sustainable Future* (Washington, DC: OSTP, The White House, July 1994).

Packard, V. (1960) *The Waste Makers* (New York: D. McKay).

Paehlke, R.C. (1991) 'Occupational and Environmental Health Linkages', in R.P. Côté and P.G. Wells (eds.), *Controlling Chemical Hazards: Fundamentals of the Management of Toxic Chemicals* (London: Unwin Hyman).

Pangotra, P., S. Erkman and H Singh (1999) 'Industry and Environment', in *Proceedings of a Workshop Held at the Indian Institute of Management, Ahmedabad, 5–6 February 1999* (Geneva: Institute for Communication and Analysis of Science and Technology [ICAST]).

Pauli, G. (1998) *Upsizing. The Road to Zero Emissions: More Jobs, More Income and No Pollution* (Sheffield, UK: Greenleaf Publishing).

PCSD (President's Council on Sustainable Development) (1996a) 'Eco-efficiency Task Force Report', http://clinton2.nara.gov/PCSD/Publications/TF_Reports/eco-top.html.

—— (1996b) *Eco-efficiency Task Force Report, Chapter 3: Summary of Demonstration Projects* (Washington, DC: PCSD).

—— (1996c) *Eco-industrial Park Workshop Proceedings, Cape Charles, VA, 17–18 October 1996* (http://clinton2.nara.gov/PCSD/Publications/Eco_ Workshop.html).

—— (1996d) *Sustainable America: A New Consensus* (Washington, DC: PCSD, February 1996).

—— (1997) *Town of Cape Charles, VA* (Washington, DC: PCSD, February 1997).

Pelletiere, D. (2000) 'Eco-restructuring and the "Friction of Distance" ', in J. Kohn, J.Gowdy and J. Van der Straaten (eds.), *Sustainability in Action: Sectoral and Regional Case Studies* (Cheltenham, UK: Edward Elgar).

Piasecki, B., and P. Asmus (1990) *In Search of Environmental Excellence* (New York: Simon & Schuster).

Pollard, P.S. (2001) 'Trade in the Neighbourhood', *International Economic Trends* (St Louis, MO: Federal Reserve Bank of St Louis, July 2001).

Porter, M. (1998) 'Clusters and the New Economics of Competition', *Harvard Business Review*, November/December 1998: 77-90.

PRIME Industrial Ecology Module (1999–2000) *Framework and Development Pla.* (Manila, Philippines: Philippine Board of Investments).

Putnam, R.D. (2000) *Bowling Alone: The Collapse of and Revival of American Community* (New York: Earth Pledge Foundation).

PWF (Pax World Funds) (2001) *Socially Responsible Mutual Funds in the United States: A Look Back . . . and Ahead* (Portsmouth, NH: PWF, August 2001).

Ramaswamy, R. (2000) 'Relevance of Industrial Ecology in Developing Countries' (draft paper; Bangalore, India: Institute for Communication and Analysis of Science and Technology [ICAST]).

Rejeski, D. (1999) 'Learning before Doing: Simulation and Modeling in Industrial Ecology', *Journal of Industrial Ecology* 2,4: 29-43.

Resource Conservation and Recovery Act (1976) 42 USC Section 6901 et seq. (Washington, DC: US Government Printing Office).

RMI (Rocky Mountain Institute) (1998) *Green Development: Integrating Ecology and Real Estate* (Billings, MT: RMI/New York: John Wiley).

Robins, R. (1991) *Waste Not: Garbage as an Economic Resource for the Northeast* (Boston, MA: Conservation Law Foundation, December 1991).

Rohatgi, P., K. Rohatgi and R.U. Ayres (1998), 'Materials Futures: Pollution Prevention, Recycling and Improved Functionality', in R.U. Ayres and P.M. Weaver (eds.), *Eco-Restructuring: Implications for Sustainable Development* (Tokyo/New York/Paris: United Nations University Press): 120.

Ross, B., and M. Sasser (1998) 'Reduce, Re-use, Recycle: Water Conservation and Industry', *Consulting-Specifying Engineer*, October 1998: 10.

Rossi, P.H., and H.E. Freeman (1993) *Evaluation: A Systematic Approach* (Newbury Park, CA: Sage, 5th edn).

RTI (Research Triangle Institute) (1996) *Eco-industrial Parks: A Case Study and Analysis of Economic, Environmental, Technical and Regulatory Issues* (executive summary; Research Triangle Park, NC: RTI, www.rti.org/units/ssid/cer/parks.cfm).

Scanlan, P.M. (1998) *The Dolphins are Back: A Successful Quality Model for Healing the Environment* (Portland, OR: Productivity Press).

Schlarb, M. (2001) *Reviews of Economic Development Literature and Practice. VIII. Eco-industrial Development: A Strategy for Building Sustainable Communities* (Washington, DC: US Economic Development Administration, Department of Commerce, www.cfe.cornell.edu/wei/papers/EID_litreview.pdf).

—— and E. Cohen-Rosenthal (2000) *Eco-industrial Developments* (Sustainable Architecture White Papers; New York: Earth Pledge Foundation).
—— and B. Keppard (2001) *Springfield Eco-industrial Baseline Study* (report submitted to the City of Springfield, MA, and the Chelsea Centre for Recycling and Economic Development; Ithaca, NY: Work and Environment Initiative, Cornell University).
—— and J. Musnikow (2001) 'Green Productivity and Environmental Management of Industrial Estates: Applications in Asia', in *Proceedings from the International Symposium on Green Productivity*, Tokyo, Japan (Asian Productivity Organisation, www.apo-tokyo.org/gp.new/Aogpdp.htm).
Schwarz, E.J., and K.W. Steininger (1997) 'Implementing Nature's Lesson: The Industrial Recycling Network Enhancing Regional Development', *Journal of Cleaner Production* 5: 1-2.
SIF (Social Investment Forum) (2001) *Trends Report: Report of Responsible Investing Trends in the US 2001* (Washington, DC: SIF).
SIR (Society of Industrial Realtors) (1984) *Industrial Real Estate* (Washington, DC: SIR, 4th edn).
Slone, D.K. (2001) 'Overcoming Impediments to Implementation of the New Urbanism', in *New Urbanism: Comprehensive Report and Best Practices Guide* (Ithaca, NY: New Urban Publications): 9-16.
Smith, L., J. Means and E. Barth (1995) *Recycling and Re-use of Industrial Wastes* (Columbus, OH: Battelle Press).
Smolenaars, T. (1996) 'Industrial Ecology and the Role of the Cleaner Production Centre', *Industry and Environment Review*, October/December 1996.
Snell, R., and A. Man-kuen Chak (1998) 'The Learning Organisation: Learning and Empowerment for Whom?', *Management Learning* 25.3: 337-64.
So, F., and J. Getzels (eds.) (1988) *The Practice of Local Government Planning* (Washington, DC: ICMA Training Institute, 2nd edn).
Socolow, R., C.C. Andrews, F. Berkhout and V. Thomas (1994) *Industrial Ecology and Global Change* (Cambridge, UK: Cambridge University Press).
Stacey, R.D. (1996) *Complexity and Creativity in Organizations* (San Francisco: Berrett-Koehler).
Stahel, W.R. (1998) 'The Product-life Factor', in R.U. Ayres, K.J. Button and P. Nijkamp (eds.), *Global Aspects of the Environment* (Cheltenham, UK: Edward Elgar).
Stamps, D. (1998) 'Learning Ecologies', *Training*, January 1998: 32-38.
State of Florida (1998) Fla. Admininistrative Code 62-660.803: General Permit for Car Wash Systems.
State of Mississippi (2002) Code of Miss. Rules 08-030-007: Wastewater Regulations for NPDES Permits, UIC Permits, State Permits, Water Authority Based Effluent Limitations and Water Quality Certification.
State of North Carolina (2002) 15A North Carolina Administrative Code 13A.0119: Standards for Universal Waste Management.
State of South Carolina (2002) 61 South Carolina Code of Laws, Regulations 58.1: Construction and Operating Permits.
Stead, E.W., and J.G. Stead (1996) *Management for a Small Planet: Strategic Decision Making and the Environment* (Thousand Oaks, CA: Sage).
Steger, U., with R. Meima (1998) *The Strategic Dimensions of Environmental Management* (London: Macmillan, English edn).
Steinhilper, R. (1998) *Remanufacturing: The Ultimate Form of Recycling* (Stuttgart: Fraunhofer IBC Verlag).
Sternlicht, B. (1979) 'Recapturing BTUs from Waste Heat', *IIE Solutions*: 11-12.
STPP (Surface Transportation Policy Project) (2001) *Easing the Burden: A Companion Analysis of the Texas Transportation Institute's Congestion Study* (Washington, DC: STPP, May 2001).
Szreter, S. (2000) 'Social Capital, the Economy, and Education in Historical Perspective', in S. Baron, J. Field and T. Schuller (eds.), *Social Capital: Critical Perspectives* (Oxford, UK: Oxford University Press): 56-77.
Takahashi, M. (2000) 'A Planner's Perspective on Eco-industrial Development', paper presented at the National American Planning Conference, New York, April 2000.
Tax Relief Act (1997) (Washington, DC: US Government Printing Office).

Thomas, J.C. (1995) *Public Participation in Public Decisions: New Skills and Strategies for Public Managers* (San Francisco: Jossey-Bass).
Tibbs, H. (1992) 'Industrial Ecology: An Environmental Agenda for Industry', *Whole Earth Review*, Winter 1992: 4-19.
—— (1998) 'Humane Ecostructure: Can Industry Become Gaia's Friend?', *Whole Earth*, Summer 1998: 63.
TJCOG (Triangle J Council of Governments) (2000) 'Partnership Opportunities for Re-using Waste Materials, Water and Energy: A Survey for Triangle Region Facilities', www.tjcog.dst.nc.us/indeco.htm.
UNEP (United Nations Environment Programme) (1996) 'Environmental Management of Industrial Estates', *Industry and Environment* 19.4 (October–December 1996).
—— (1997) *The Environmental Management of Industrial Parks* (Technical Report 39; Paris: UNEP, Industry and Environmental Parks).
UNESCO (United Nations Educational, Scientific and Cultural Organisation) (2003) 'Man and the Biosphere Programme', www.unesco.org/mab/brlist.htm.
United States of America (2002) 40 Code of Federal Regulations (Washington, DC: US Government Printing Office).
United States v. Monsanto Co. 858 F.2d 160 (4th Cir. [1988]).
US EPA (US Environmental Protection Agency) (1997a) 'Remanufactured Products: Good as New', *Waste Wise Update* 5: 3.
—— (1997b) *Municipal Solid Waste Factbook, 1997* (Washington, DC: US EPA Office of Solid Waste).
—— (1998a) *Funding Brownfield Remediation with the Clean Water State Revolving Fund* (EPA 832-F-98-006; Washington, DC: US EPA, Office of Water, October 1998).
—— (1998b) *Targeted Brownfields Assessments* (EPA 500-F-98-251; Washington, DC: US EPA, Office of Solid Waste and Emergency Response, November 1998).
—— (1999a) *Guidebook of Financial Tools: Paying for Sustainable Environmental Systems*, www.epa.gov/efinpage/guidbk98/index.htm, April 1999.
—— (1999b) *Private-Sector Pioneers: How Companies are Incorporating Environmentally Preferable Purchasing* (EPA742-R-99-001; Washington, DC: US EPA, Office of Pollution Prevention and Toxics, Environmentally Preferable Purchasing Program, June 1999).
—— (1999c) *Tools for Eco-industrial Development Planning* (Version 1.3; Cambridge, MA: Industrial Economics Inc.).
—— (2000) 'EPA Brownfields Economic Redevelopment Initiative', www.epa.gov/swerosps/bf/html-doc/econinit.htm.
—— (2001a) *Brownfields Tax Incentive* (EPA 500-F-01-339; Washington, DC: US EPA, Office of Solid Waste and Emergency Response, August 2001).
—— (2001b) *The United States Experience with Economic Incentives for Protecting the Environment* (EPA-240-R-01-001; Washington, DC: US EPA, www.epa.gov/economics).
—— (2002) 'Public Involvement at EPA', www.epa.gov/ stakeholders.
USDT (US Department of Treasury) (1999) *Tax Incentives for Empowerment Zones and Other Distressed Communities* (publication 954; Washington, DC: USDT, Internal Revenue Service, www.irs.ustreas.gov, rev. February 1999).
USGBC (US Green Building Council) (2002) LEED (*Leadership in Energy and Environmental Design*) (Washington, DC: USGBC).
Vale, A.V. (1996) *Environmental Awareness Training: A Strategy for Change* (South Melbourne, Australia: Macmillan Education Australia).
Van Der Ryn, S., and S. Cowan (1996) *Ecological Design* (Washington, DC: Island Press).
von Hauff, M., and M.Z. Wilderer (2000) 'Eco industrial Networking: A Practicable Approach for Sustainable Development in Developing Countries', paper presented at the *Helsinki Symposium on Industrial Ecology and Material Flows*, Helsinki, 31 August 2000–3 September 2000.
Von Weizsäcker, E., A.B. Lovins and L.H. Lovins (1997) *Factor Four: Doubling Wealth, Halving Resource Use. The New Report to the Club of Rome* (London: Earthscan Publications).
Wackernagel, M., and W. Rees (1996) *Our Ecological Footprint: Reducing Human Impact on the Earth* (Gabriola Island, BC, Canada: New Society Publishers).

Walker, J., and P. Desrochers (1999) 'Recycling is an Economic Activity, Not a Moral Imperative', *The American Enterprise* 1–2: 75.
Wallner, H.P. (1999) 'Towards Sustainable Development of Industry: Networking, Complexity and Eco-clusters', *Journal of Cleaner Production* 7.1: 49-58.
WCED (World Commission on Environment and Development) (1987) *Our Common Future* (Brundtland Report; Oxford, UK: Oxford University Press).
Wehrmeyer, W. (ed.) (1996) *Greening People: Human Resources and Environmental Management* (Sheffield, UK: Greenleaf Publishing).
WEI (Work and Environment Initiative) (2000) *Handbook of Codes, Covenants and Restrictions for Eco-Industrial Development* (ed. M. Deppe [Takahashi] and E. Cohen-Rosenthal; Ithaca, NY: WEI, Cornell University).
—— (2001) 'Minutes: EIDP Roundtable on Financing EID', US Department of Commerce, Herbert C. Hoover Building, Washington, DC, 12–13 October 2001, www.cfe.cornell.edu/wei/EIDP/EIDPMinutes.htm.
Weisbord, M.R. (1992) *Discovering Common Ground* (San Francisco: Berrett-Koehler).
Weiss, P.A. (1968) 'The Living System: Determinism Stratified', in A. Koestler and J.R. Smythies (eds.), *Beyond Reductionism: New Perspectives in the Life Sciences* (London: Hutchinson).
Weitz, K.A., and S.A. Martin (1995) 'Regulatory Issues and Approaches for Encouraging Eco-industrial Park Development', draft paper presented at *Designing, Financing and Building the Industrial Park of the Future: An International Workshop on Applications of Industrial Ecology to Economic Development*, Institute of the Americas; University of California, San Diego, CA, 4–5 May 1995.
Wellesley-Miller, S. (1977) 'Towards Symbiotic Architecture', in M. Katz, W.P. Marsh and G. Gordon Thompson (eds.), *Earth's Answer: Explorations of Planetary Culture at the Lindisfarne Conferences* (New York: Harper & Row): 9.
Wernick, I.K., R. Herman, S. Govind and J.H. Ausubel (1996) 'Materialisation and Dematerialisation: Measures and Trends', *Daedalus* 125.3: 171-98.
Wheatley, M.J. (1992) *Leadership and the New Science: Learning about Organization from an Orderly Universe* (San Francisco: Berrett-Koehler).
—— and M. Kellner-Rogers (1996) *A Simpler Way* (San Francisco: Berrett-Koehler).
World Bank (2000) *Greening Industry: New Roles for Communities, Markets and Governments* (Washington, DC: World Bank).
Zannes, M. (1997) 'The Driving Power Behind Waste to Energy', *World Wastes* 4: 40.
Zialcita, F.N., et al. (1995) *People's Participation in Local Governance: Four Case Studies* (Quezon City, Philippines: Ateneo de Manila University).

ABBREVIATIONS

AFL–CIO	American Federation of Labor–Congress of Industrial Organizations	CFC	chlorofluorocarbon
		CFED	Corporation for Enterprise Development
AFV	alternative-fuel vehicle	CHP	combined heat and power
APO	Asian Productivity Organisation	CII	Confederation of Indian Industry
ARMS	Armament Retooling and Manufacturing Support (USA)	CLF	Conservation Law Foundation
		CNCPC	China National Cleaner Production Centre
ATSDR	Agency for Toxic Substances and Disease Registry (HHS)	CO	carbon monoxide
B&W	Babcock & Wilcox	CO_2	carbon dioxide
B2B	business-to-business	CRA	Community Reinvestment Act (USA)
BCRLF	Brownfield Clean-up Revolving Loan Fund (US EPA)	CWSRF	Clean-Water State Revolving Fund (US EPA)
BEDI	Brownfield Economic Development Initiative (HUD)	CZMA	Coastal Zone Management Act (NOAA)
BIC	business information centre (SBA)		
BPX	by-product exchange	DCDP	Division of Community Demonstration Programs (OCS)
BTS	Building Technology, State and Community Program (US DOE)	DDT	dichlorodiphenyltrichloroethane
BTU	British thermal unit	DIET	Designing Industrial Ecosystems Tool (US EPA)
C&Rs	covenants and restrictions		
CAA	Clean Air Act (USA)	DOE	US Department of Energy
CAB	community advisory board	DOI	US Department of Interior
CBOT	Chicago Board of Trade	EC	enterprise community
CC&Rs	codes, covenants and restrictions	EDA	Economic Development Administration (US DOC)
CDBG	community development block grant (HUD)	EDI	Economic Development Initiative (HUD)
CDC	Certified Development Company (SBA)	EERE	Office of Energy Efficiency and Renewable Energy (US DOE)
CDFI	Community Development Finance Institution (USDT)	EFAB	Environmental Financial Advisory Board (USA)
CEDEIP	Computer and Electronics Disposition Eco-Industrial Park	EFC	environmental finance centre
		EFIN	Environmental Financing Information Network (USA)
CEDS	comprehensive economic development strategies		
CERCLA	Comprehensive Environmental Response, Compensation and Liability Act (USA)	EH&S	environment, health and safety
		EID	eco-industrial development

Abbreviation	Definition
EIDC	Eco-Industrial Development Council
EIN	eco-industrial network
EIP	eco-industrial park
EIS	environmental impact statement
EMAS	Eco-management and Audit Scheme
EMS	environmental management system
ENN	Environmental News Network
EPA	Environmental Protection Agency (USA)
EPP	Environmentally Preferable Purchasing (US EPA)
EREN	Energy Efficiency and Renewable Energy Network (US DOE)
EZ	empowerment zone
FaST	Facility Synergy Tool (US EPA)
FCC	Federal Communication Commission (USA)
FeSO$_4$	ferrous sulphate
FGD	flue gas desulphurisation
GIS	geographical information system
GNP	gross national product
GTZ	Gesellschaft für Technische Zusammenarbeit (German Technical Co-operation Organisation)
H$_2$SO$_4$	sulphuric acid
HHS	Department of Health and Human Services (USA)
HRM	Halifax Regional Municipality (Canada)
HUD	Department of Housing and Urban Development (USA)
HVAC	heating, ventilation and air conditioning
I&I	Inventions and Innovations (OIT)
ICAST	Institute for Communication and Analysis of Science and Technology
IDA	industrial development authority
IDeA	Information Design Associates
IEAT	Industrial Estate Authority of Thailand
IHDP	International Human Dimensions Programme on Global Environmental Change
IHS	Indian Health Service (HHS)
IRR	internal rate of return
ISO	International Organisation for Standardisation
ISTEA	Intermodal Surface Transportation Efficiency Act (USA)
IWGIEMEF	Interagency Workgroup on Industrial Ecology, Material and Energy Flows
JIT	just-in-time
JVSV	Joint Venture Silicon Valley
LAAAP	Louisiana Army Ammunition Plant
LADWP	Los Angeles Department of Water and Power
LAL	limulus amebocyte lysate
LEED	Leadership in Energy and Environmental Design (USGBC)
LEIP	Londonderry Ecological Industrial Park
LEM	location-efficient mortgage
LRT	light-rail transport
MBDA	Minority Business Development Agency (US DOC)
MEP	manufacturing extension project (NIST)
MOU	Memorandum of Understanding
MWCG	Metropolitan Washington Council of Governments
NAAQS	National Ambient Air Quality Standards
NAFTA	North American Free Trade Agreement
NASA	National Aeronautics and Space Administration (USA)
NCUED	National Council for Urban Economic Development (USA)
NEA	National Endowment for the Arts (USA)
NGO	non-governmental organisation
NIA	Naroda Industrial Association
NICE3	National Industrial Competitiveness through Energy, Environment and Economics (OIT)
NIMBY	'not in my back yard'
NIST	National Institute of Standards and Technology (US DOC)
NNSR	Non-attainment New Source Review
NOAA	National Oceanic and Atmospheric Administration (USA)
NSF	National Science Foundation (USA)
NTIA	National Telecommunications and Infrastructure Administration (USA)
NWLCC	Northwest Louisiana Commerce Center
NYSERDA	New York State Energy Research and Development Authority

O&M	operation and maintenance	SBDC	small business development centre (SBA)
OCRM	Office of Ocean and Coastal Resource Management (NOAA)	SEP	State Energy Program (US DOE)
OCS	Office of Community Services (HHS)	SEPA	State Environmental Protection Bureau (China)
OECD	Organisation for Economic Co-operation and Development	SIF	Social Investment Forum
		SIR	Society of Industrial Realtors
OEHE	Office of Environmental Health and Engineering (IHS)	SJF	Sustainable Jobs Fund
		SME	small or medium-sized enterprise
OIT	Office of Industrial Technologies (US DOE)	SO_2	sulphur dioxide
		SRI	socially responsible investment
OPPE	Office of Policy, Planning and Evaluation (US EPA)	STEP	Saratoga Technology Energy Park
		STPP	Surface Transportation Policy Project
OPT	Office of Power Technologies (US DOE)	TANF	Temporary Assistance to Needy Families (USA)
ORR	Office of Response and Restoration (NOAA)	TBIC	tribal business information centre (SBA)
OSC	Operations Support Command (US Army)	TEA-21	Transportation Equity Act for the Twenty-first Century (USA)
OSCS	one-stop capital shop (SBA)		
OSTP	Office of Science and Technology Policy (USA)	TEEX	Texas Engineering Extension Service
OTT	Office of Transportation Technologies (US DOE)	TFM	Total Facility Management
		TIF	tax increment financing
P2	pollution prevention	TJCOG	Triangle J Council of Governments
PCB	polychlorinated biphenyl		
PCSD	President's Council for Sustainable Development (USA)	TVA	Tennessee Valley Authority
		ULL	University of Louisiana Lafayette
PEEC	Phillips Eco-Enterprise Center	UN	United Nations
PEIA	programmatic environmental impact assessment	UNDP	United Nations Development Programme
POA	property-owners' association	UNEP	United Nations Environment Programme
PRIME	Private Sector Participation in Managing the Environment (Philippines)	UNESCO	United Nations Educational Scientific and Cultural Organisation
PROPER	Programme for Pollution Control, Evaluation and Rating (Indonesia)	USDA	US Department of Agriculture
		US DOC	US Department of Commerce
PSD	Prevention of Significant Deterioration (USA)	USDT	US Department of Treasury
PV	photovoltaic	USEAC	US export assistance centre (SBA)
PWF	Pax World Funds	USGBC	US Green Building Council
R&D	research and development	UV	ultraviolet
RCRA	Resource Conservation and Recovery Act (USA)	VCP	Virginia Coastal Programme
		W2W	Welfare to Work (USA)
RFA	resource flow analysis	WBC	women's business centre (SBA)
RLF	revolving loan fund	WCED	World Commission on Environment and Development
RMI	Rocky Mountain Institute		
ROA	return on assets	WEI	Work and Environment Initiative, Cornell University
RRFB	Resource Recovery Fund Board Inc.		
		WIA	Workforce Investment Act
RUS	Rural Utilities Service (USDA)	YWCA	Young Women's Christian Association
SBA	Small Business Administration (USA)		

BIOGRAPHIES

Dennis Alvord is an economic development specialist with the US Department of Commerce's Economic Development Administration (EDA). He co-ordinates EDA initiatives related to brownfield redevelopment, sits on the editorial board of the *Economic Development Digest* of the National Association of Development Organizations, serves on Federal Brownfields Interagency and Environmental Justice Working Groups and is an advisor to the Eco-Industrial Development Council (EIDC). In the past, he has represented the EDA on the President's Council on Sustainable Development, the White House Task Force on Liveable Communities and the American Heritage Rivers Initiative. Before joining the EDA, Dennis was project manager in the Infrastructure Finance Group of the environmental consulting firm Apogee Research Inc., based in Bethesda, MD.
Dennis_Alvord@msn.com

Corey Brinkema is a principal and founder of Trillium Planning & Development, a development services firm specialising in eco-industrial development and sustainable energy. Mr Brinkema formerly served as Project Manager of the Phillips Eco-Enterprise Center (PEEC), directing the facility's four-year development effort including the planning, design, marketing, financing and construction of the facility. As a tenant in the PEEC, Mr Brinkema and Trillium have aided in the evaluation of the facility's performance and have partnered with the Institute in the sustainable redevelopment of real estate surrounding the PEEC. In addition to his development work, Mr Brinkema provides technical and business management guidance for distributed energy, resource-efficient manufacturing and recycled product market development. He also has ten years of technical and managerial experience in the area of brownfields site investigation and remediation. Mr Brinkema holds a Master of Business Administration from the University of Michigan, Ann Arbor, and is also a graduate of Dartmouth College with a BA degree in earth sciences.
cbrinkema@trilliumplan.com

Marian Chertow, director of the Industrial Environmental Management Programme at the Yale School of Forestry and Environmental Studies since 1991, focuses her teaching and research on industrial ecology, environmental technology innovation and business–environment issues. From 1995–99 Marian led Environmental Reform: The Next Generation Project at the Yale Center for Environmental Law and Policy, which focused on the future direction of environmental policy. Marian is on the editorial board of *BioCycle* magazine and the *Journal of Industrial Ecology* as well as on the advisory board of the Alliance for Environmental Innovation of Pew Memorial Trusts and the Environmental Defense Fund. She is a gubernatorial appointee to the Connecticut Council for Environmental Quality and was recently appointed to the Connecticut Clean Energy Fund. Marian is a charter board member of the Eco-Industrial Development Council.
marian.chertow@yale.edu

Ed Cohen-Rosenthal, editor of this book, was Director of the Work and Environment Initiative at the Cornell Centre for the Environment in Ithaca, New York. He was on the faculty of the School of Industrial and Labor Relations as an Associate in Programs for Employment and Workplace Systems. He was appointed as a US Delegate to the Special UN General Assembly for EarthSummit +5 as the private-sector representative of workers and trade unions. Prior to joining Cornell, he was assistant to the President of the International Union of Bricklayers and Allied Craftworkers for education and labour–management relations. As director of the productivity/quality of work life programme of the International Masonry Institute, he worked with contractors, union officials and building professionals to design methods for increasing performance in the building industry. He was highly involved with finding practical ways for workers and their unions to contribute to environmental improvement as well as seeking innovative approaches to the greening of workplaces. These included research on employee participation in resource conservation and toxic use reduction in the United States and Japan. Ed was the author of several books and many articles on labour–management relations, worker education and environmental issues. He was a graduate of Rutgers College, where he later returned to join the faculty of the Rutgers Labour Education Center after receiving his master's degree from Harvard Graduate School of Education.

Raymond Côté is a professor of environmental studies, a former director of the School for Resource and Environmental Studies and the Marine Affairs Program at Dalhousie University and a former senior civil servant at Environment Canada. He has developed a local and international profile in the emerging field of industrial ecology as a consequence of the work of a multidisciplinary team he has been leading on industrial parks as ecosystems. This profile has resulted in his appointment as an advisor to the Technology, Industry and Economics Division of the United Nations Environment Programme on environmental management of industrial parks. Raymond Côté is the co-editor of four books: *Controlling Chemical Hazards: Fundamentals of the Management of Toxic Chemicals* (Unwin Hyman, 1991), *Business Meets the Environmental Challenge* (Lancelot Press, 1993), *Law and the Environment: Problems of Risk and Uncertainty* (Canadian Institute for the Administration of Justice, 1994) and *The Industrial Ecology Reader* (Dalhousie University, 1997). He has published extensively on various aspects of industrial ecology.
rcote@is.dal.ca

Peggy Crawford has been full-time coordinator of the Eco-Efficiency Center in Burnside since it opened in 1998. She brings a variety of perspectives to the Center, including her experience as an educational programmer, researcher, office manager and small-business owner. Peggy was employed at Acadia University for many years. She worked at the Division of Continuing and Distance Education, where she did educational and special event planning for adults and youth, with a special focus on science education. She played a key role in developing Acadia's Science Outreach Programmes, which were recognised nationally in 1994 with the Michael Smith Award for Science Promotion. Peggy has participated on a number of boards and committees, including Scientists and Innovators in the Schools, and the Atlantic Provinces Association of Continuing Education.
kpcrawfo@is.dal.ca

Ron Forsythe, a veteran of over 31 years in Mississippi State Government, serving his final years with the Energy Division of the Mississippi Department of Economic and Community Development, recently retired and joined Pickering Inc. as an energy and environmental specialist. Pickering works directly with Choctaw County in the development of the Red Hills EcoPlex industrial park. Forsythe was a leading proponent of the development of the Red Hills Power Project and of the Red Hills EcoPlex. He continues to serve in the development of the EcoPlex and is currently involved in the development of a large, surface-water impoundment to serve the water needs of the Red Hills Power Project and the EcoPlex tenants.
Rforsythe@pickeringinc.com

Suzanne Giannini-Spohn, PhD, is in the Office of International Activities (OIA), Office of Technology Cooperation and Assistance, of the US Environmental Protection Agency (EPA), where she is project leader for industrial eco-efficiency. She represents US EPA on the Federal Interagency Working Group on Materials and Energy, formed in response to recommendations of the President's Council on Sustainable Development. She contributed to and co-edited the report of this working group, which appears at www.sdi.gov. Suzanne is currently programme director for the Sino–US Partnership in Industrial Pollution Prevention and Energy Efficiency, a co-operative agreement between the US EPA and the China State Environmental Protection Administration. Prior to joining OIA, Suzanne was team leader for eco-industrial development at the US EPA's Office of Policy. She has co-directed several projects on eco-industrial networking, including eco-industrial network decision-support models, the *Handbook on Codes, Covenants and Restrictions that Promote Environmentally Responsible Development* and a directory to agencies

financing eco-industrial developments. She has published several articles about eco-industrial development. Many of these products, including the decision-support models Designing Industrial Ecosystems Tool and Facility Synergy Tool, can be accessed from the eco-industrial web pages on the Smart Growth Network website at www.smartgrowth.org.
giannini-spohn.suzanne@epa.gov

Vic Guadagno is currently the programme manager for the Northwest Louisiana Commerce Center (NWLCC), the eco-industrial park in development at the deactivated Louisiana Army Ammunition Plant. He has been employed in all aspects of industrial operations, from programme management and engineering, to systems analysis, manufacturing and inspection. He started Full Circle Production in 1994, a marketing and production company located in Boulder, Colorado. Full Circle Production has led the design and development of the NWLCC. With the goal of creating interactive media, a trip to Guatemala in 1996 introduced Vic to permaculture. He has been committed to education in permaculture as a sustainable design science. The NWLCC represents an opportunity to demonstrate permaculture as a design science for ecological repair and a key element of industrial ecology.
vguadagno@hotmail.com

Timothy E. Hayes serves as Director of Sustainable Development for Northampton County, VA, USA. He was the first executive director of the Sustainable Technology Park Authority, developer of the Cape Charles Sustainable Technology Park. He has a degree in Environmental Design from the University of Colorado and is currently pursuing a Master of Business Administration from Regent University in Virginia Beach, VA. Under his leadership, Northampton County has established itself as a model for sustainable development.
Thayes@esva.net

Bracken Hendricks is a senior associate with E^2 Inc. (Ecology & Economics), based in Charlottesville, VA, where he works on issues of Superfund site redevelopment, community involvement, clean production and labour and environmental issues. Hendricks served in the Clinton Administration as a special assistant to the Department of Commerce in the National Oceanic and Atmospheric Administration. He represented the Department of Commerce on the White House Liveable Communities Task Force, the President's Council on Sustainable Development, the Interagency Climate Change Working Group and Vice-President Gore's National Partnership for Reinventing Government. He has served as an economic analyst with the AFL–CIO Working for America Institute, where he focused on workforce and economic development and regional partnerships for social equity. Hendricks has worked extensively on issues of Smart Growth and sustainable development, trade and the environment and geographic information policy. He is a member of the Cornell Eco-industrial Development Project Roundtable. He holds a master's degree in public policy and urban planning from Harvard University's John F. Kennedy School of Government, and lives in Bethesda, MD, with his wife and two children.
bracken@thirdbridge.com

Judy Kincaid is the director of the solid waste and material resources programmes for the Triangle J Council of Governments, a regional planning organisation. She practised and taught law in North Carolina from 1975 to 1990, and she has been a solid waste planner at the Triangle J Council of Governments since 1990. She was a recipient of a North Carolina Recycling Association merit award in 1993 and was named the Carolina Recycling Association Green Builder of the Year for 1998. A manual that she co-authored with architects Greg Flynn and Cheryl Walker, *WasteSpec: Model Specifications for Construction Waste Reduction, Reuse, and Recycling* (TJCOG, 1995), won the 1998 National Recycling Coalition award for outstanding recycling innovation. She was the manager of a project funded by the US Environmental Protection Agency (EPA) to map local materials flows and initiate partnerships between local businesses to promote re-use of materials, water and energy. She also served on another project team funded by US EPA to produce reports on eco-industrial parks. She was the recipient of private foundation awards to participate in international workshops on industrial symbiosis in Kalundborg, Denmark, in 1996 and in Troyes, France, in 1999, and, funded by the European Union, she was a consultant to an industrial symbiosis project in the Tuscany region of Italy in 2000. She also presented her work on industrial ecosystem development at the 2002 International Conference on Territorial Development in Paris as the guest of the French government.
jkincaid@tjcog.org

Michael Krause has been executive director of the Green Institute since May 1996. He has five years of experience in policy positions with state and local governments. From 1982–90 he was editor and publisher of a chain of weekly newspapers in the western suburbs of Minneapolis and still writes editorial columns for several Twin Cities publications. Krause serves on the board of the Twin Cities Economic Development Group, Clean Water Action Alliance, Minnesota Environmental Initiative and is the immediate past chair of the Minneapolis Consortium of Community Developers. He is an appointed member of the Metropolitan Transportation Advisory Board and the Minneapolis Planning Commission. Krause was recently elected to co-chair the newly formed Eco-Industrial Development Council with members across the US and Canada. Krause is a graduate of Macalester College in St Paul, MN, and holds a law degree from the University of Minnesota.
michaelk@greeninstitute.org

As project manager of the symbiosis institute in Kalundborg, Denmark, **Noel Brings Jacobsen** is involved in the development of by-product synergies among large companies in Denmark and abroad. Prior to joining the institute, he was a research assistant at the University of Copenhagen and Roskilde University, where he worked with industrial networking between small and medium-sized enterprises in production networks. He is presently formulating his dissertation on eco-industrial networking from a managerial level, by comparing network characteristics between by-product networks and traditional industrial clusters (production networks).
Kalundborg@symbiosis.dk

Ernest Lowe is chief scientist and director of Sustainable Systems Inc., Indigo Development. He is the lead author of Indigo's two publications: *Discovering Industrial Ecology: An Executive Briefing and Sourcebook* (1997) and *Eco-Industrial Parks: A Handbook for Local Development Teams* (1995, 2001). In 1991 Ernest Lowe was co-director of the Change Management Network, an industry–academic research partnership at the Engineering School of Old Dominion University. He played a central role in the creation and development of this programme and in the recruitment of *Fortune* 100 companies as sponsors. From 1989 to 1991 he was president of Viable Systems International, a consulting firm applying the 'viable system model' to the design of large corporation management infrastructure. He was a member of the founding team of the International Centre for Organisational Design, where he explored improvements for total quality management practices. He has given several workshops and talks on industrial ecology throughout the past ten years and is presently the co-author of a strategic plan for environmental business development in Berkeley, California.
Ernielowe@indigodev.com

Peter C. Lowitt, AICP, is director and land-use administrator for the Devens Enterprise Commission, the agency charged with permitting the redevelopment of the former Fort Devens in Massachusetts. Prior to coming to Devens, Peter Lowitt served as director of planning and economic development for the Town of Londonderry, NH (1993–99), where he developed the Londonderry Ecological Industrial Park and the award winning Sustainable Londonderry programme. Peter Lowitt has built on his experience as town planner for Grafton, MA (1988–93) where he helped develop the Grafton Biotech Park in conjunction with the Worcester Business Development Corporation and the Tufts University School of Veterinary Medicine (now known as Cen Tech Park). He was also employed by the Town of Acton and the City of Waltham in junior planner positions. Peter Lowitt has been active in the planning profession, serving as chairperson of the Economic Development Division of the American Planning Association (APA), vice-president of the Massachusetts Chapter of APA and president of the New Hampshire Planners Association. He has published a number of articles, including 'Sustainable Londonderry' (1998, in the proceedings of the APA National Conference) and 'Bringing in Biotech', *Landletter* 2.3 (Winter 1992). He is a founding member of the Eco-Industrial Development Council.
peterlowitt@devensec.com

Judy Musnikow, assistant editor of this book, is currently attending Simmons College Graduate School of Library and Information Science, studying toward a master's in library science. Prior to entering Simmons, Judy was a labour organiser with the United Food and Commercial Workers (UFCW) International Union. She holds a BA from the State University of New York at Binghamton in environmental economics, and a master's in environmental management from Cornell University. While at Cornell, Judy

was on the staff of the Work and Environment Initiative, where she helped develop several projects, including a manual titled *Eco-industrial Development Community Participation Manual*.
jbm25@cornell.edu

Astrid Petersen (now Astrid October) was a Hubert Humphrey Fellow at Cornell University in Ithaca, New York, for the academic year 2000–2001. This is a professional enhancement programme for mid-to-high-level career professionals. She is employed as environmental management specialist for Eskom Distribution, which supplies electricity in South Africa. Prior to this, she was employed as senior environmental scientist at an environmental consulting firm in Cape Town, South Africa, focusing on environmental assessments and the design and implementation of environmental management systems (such as ISO 14001) for multi-land-use developments. She also worked as an environmentalist at an oil refinery for six years. She holds an honours science degree in environmental and geographical science from the University of Cape Town, South Africa.
astrid.october@eskom.co.za

Mary Schlarb is Associate Director of the Cornell Work and Environment Initiative (WEI). As project manager of the National Center for Eco-Industrial Development, a programme of WEI and the USC Center for Economic Development, Mary Schlarb manages research and outreach programmes for community-based eco-industrial development projects. She has worked on community development in Indonesia and the Philippines, and has been a consultant in the area of environmental management for the United Nations Food and Agriculture Programme, the German Technical Service (GTZ), and the Asian Productivity Organisation. She received a master's degree at the Department of International Agriculture and Rural Development, Cornell University, focusing on public participation in environmental policy and collaborative decision-making for watershed management in Cebu City, Philippines. She is also a graduate of Stanford University's International Relations programme.
mhs13@cornell.edu

Dan Slone is a partner in the 600-attorney law firm of McGuireWoods LLP, a full-service firm with offices throughout the USA and abroad. Dan is the co-team leader for the firm's 19-member Environmental Solutions Group. He co-ordinates land-use and environmental permitting and approval for large-scale projects. He also assists clients in the approval and application of new approaches, techniques or technologies in order to implement responsible development strategies. He is counsel for the Congress for New Urbanism, Greening America, the US Green Building Council and the Virginia Housing and Environment Network. He is a member of the Eco-Industrial Development Council's board as well as the board of the Seaside Institute. Dan and the McGuireWoods team have extensive experience with power-production facilities and eco-industrial development. He is involved in the development of more than six eco-industrial facilities in six states, and his team has drafted the first set of eco-industrial covenants and restrictions to measure environmental and social sustainability. Among his clients are the developers of the environmentally sensitive new urban towns of Haymount, VA, and Coffee Creek, IN, as well as the Cape Charles Sustainable Technology Park.
dslone@mcguirewoods.com

Mark Smith created Pario Research in 1995 after 13 years of experience with market and financial feasibility with the Sanford Goodkin Research Corporation and KPMG Peat Marwick's national real estate practice. Pario Research's mission is the integration of sustainable design and programming into mainstream real estate investment and development. Since forming Pario Research, Mark Smith has written articles and presented papers on sustainable building topics, including economics, design and implementation. He is quoted in and has provided counsel for the Rocky Mountain Institute book *Green Development: Integrating Ecology and Real Estate* (1998). He has also made presentations on the greening of real estate development at several conferences. In February 1999 Mark Smith was retained by the Canadian Consulate General and Industry Canada to develop the curriculum for a series of sustainable development education seminars for senior executives. He is presently writing a book on the business integration of sustainability into mainstream real estate development. Mark Smith is on the development team of a planned sustainable community in San Diego, CA. Additionally, in partnership with the late Ed Cohen-Rosenthal of Cornell, he formed Quantum Connections, a new system of design and programming for business environments, linking ecology, real estate, business operations and regional economic development.
marksmith@pario.org

Maile Takahashi is currently a project approvals manager at Harvard University's Planning and Real Estate Department. She previously worked as research director for the Work and Environment Initiative (WEI) at the Cornell Centre for the Environment, in Ithaca, NY. She worked with communities while she was with WEI to develop alternative economic development plans. She also co-wrote the *Handbook of Codes, Covenants, Conditions and Restrictions for Eco-industrial Development* (WEI, Cornell University, 2000). Maile Takahashi has a master's degree from Cornell University's Department of City and Regional Planning, and her undergraduate degree in environmental planning and biology at the University of California, Davis. Prior to going to Ithaca, she was an environmental consultant in Sacramento, CA. Her project experience includes conducting eco-industrial baseline studies, economic development plans and a variety of environmental assessments of development projects including environmental impact reports, environmental impact statements, initial studies and mitigation plans.
maile_takahashi@harvard.edu

Josh Tosteson is a PhD student at the Columbia University Earth Institute and is a consultant to the Northwest Louisiana Commerce Center (NWLCC). His programme of study integrates physical Earth science and public policy, and his research currently focuses on the NWLCC as a case study in the design of transitions toward sustainability. He comes to the NWLCC via the Permaculture Research Institute, where he obtained his Permaculture Design Certificate from Geoff Lawton. Having earned his BA in environmental science and public policy from Harvard in 1994, Tosteson has been a consultant to a variety of projects and institutions involved in interdisciplinary Earth system science and education, including Biosphere 2, the American Museum of Natural History, the Environmental Defense Fund, the Dalton School and the New Jersey Higher Education Partnership for Sustainability. He has developed electronic curricula and taught middle-school, high-school, undergraduate and master's degree students and has authored peer-reviewed papers on topics ranging from science education to the development and application of El Niño forecasts around the world.
Jlt34@columbia.edu

INDEX

4-County Electric Power Association 309-10

Accidents
   see Emergencies
Acid mine drainage 151
Adam Joseph Lewis Center, USA 262
Advisory committees, community 110, 183-84
AES Corporation 201-202, 265, 266, 301-302, 305
Agency for Toxic Substances and Disease Registry (ATSDR) 232, 242
Agriculture
   aquaculture, integration with 272, 290, 299
   fertilisers for 273, 314
   'seaside ecological farm' project 298-99
   and urbanisation 210-11, 300-301
   see also Composting; Fish farming; Greenhouses; Permaculture; Rural development
Agriculture, US Department of (USDA) 84, 242
   funding by 210, 216, 229-30, 338
Air pollution 142, 158
   emission credits 74-75
   and transportation 81, 95
Alaska 232
Aluminium 47
Amato, I. 31, 39, 44-45
Ammunition facilities, re-use of 330-40
Anderson, Ray 208
Aquaculture
   see Fish
Argyris, Chris 64
ARMS (Armament Retooling and Manufacturing Support) 330-31, 338, 339
Arthur Andersen Group 311
A/S Bioteknisk Jordrens/SOILREM 272, 273
Ash, coal 145-46, 272, 311-15
Asia, developing countries of
   eco-industrial development in 341-52
Asian Development Bank 352
Asian Network for Eco-industrial Development 352
Asnæs power plant, Denmark 272
Atlantis Energy Systems 291, 298
Atomic architecture 45
Audits 197

Audubon Society 261
Aurora Project, New York 200
Austria 62
Autogenesis 39
Avtex Superfund 210
Awards, environmental 326
Ayres, Leslie 42, 52
Ayres, Robert 42, 47, 49, 52

Babcock & Wilcox (B&W) 310
Balkau, F. 189, 190
Baltimore, USA 56, 79
Bayshore Concrete Products Corporation 298
Bechtel Corporation 52, 218
Benyus, Janine 53
Benzene 145
Bevill Amendment 146
Bielagus, Justin 201
Biocycle 46
Biodiversity 150
Biological technologies 211, 297
Biomimicry 51, 53
Bioproducts 25
BioSpace Development Company 255
Black liquor 130
Bonds 214
Braungart, Michael 207
Bremer Bank of St Paul 279
Brightfields programme 84, 229, 242
Brinkema, C. 203
British Plasterboard 273
Brown, John Seely 63
Brownfield Initiative programmes 221, 222-24, 242
Brownfield redevelopment
   assessment of 223
   community benefits of 104-105, 204
   financing 85, 102, 204, 221, 222-24, 299
   at Intervale Food Center 210-11
   job training for 102, 224, 282
   and remediation 94, 149, 204, 223-24
   'seaside ecological farm' project 298-99
   tax incentives 224, 239

Brownfields Economic Redevelopment Initiative, EPA 102
Brownsville, Texas EIP 52, 79, 218, 251
Brownsville Economic Development Council 218
Brundtland Report 122
Building Energy Efficiency Research Project 261
Buildings 161
  adaptive re-use of 203
  construction of 152-55, 161, 281
  deconstruction of 40, 156, 203, 278
  energy efficiency of 228
  green 27, 143, 302
    Cape Charles 295-96
    costs/returns of 248
    evaluation of 261-62
    LEED standards for 124, 143, 295
    see also Phillips Eco-Enterprise Center
  heating/cooling 34
  on landfills 47
  planned densification of 254-57
  refurbishment of 156-58, 203
  'smart' 39
  see also Housing; Materials; Waste
Building Technology, State and Community Programmes, Office of (BTS) 228
Burlington, City of, USA 237
  see also Intervale Food Center
Burlington Community Land Trust 237
Burlington Electric 211
Burnside Eco-Business Program 326-28
Burnside Industrial Park, Nova Scotia 150, 156, 159
  businesses in 151, 153-54, 155, 157, 324-25
  case study 322-29
  Eco-Efficiency Center 325-29
  management of 325
  Waste Management Strategy 323-24
Businesses, greening 326
Businesses for Social Responsibility 304
Business networks
  see Networks
Business opportunities, new 157
  locating/identifying 244-46
Business process redesign 124-26
Business systems, role of 57-60
By-product recovery
  see Waste

Calcutta, India 347
California, USA 80, 134, 145, 165-66
Calvert, Texas 308
Canada 323-24, 328
  see also Burnside Industrial Park
Cape Charles Sustainable Technology Park, USA
  and architectural charrettes 56, 293
  building, design of 295-96
  case study of 288-99
  and community development 79, 96, 105, 109, 207
  covenants and restrictions 143-44, 266
  development company for 293-94
  funding for 84, 214, 218, 228, 233-34, 241, 292-93, 296
    payback/return on 298
  future plans 298-99
  habitat preservation at 266, 288-89, 294, 296
  infrastructure of 294-95, 296
  and international trade 210
  land assembly 294
  management of 60, 139, 293-94
  operation 297-98
  planning 288-95
  success of, evaluating 263-64, 265-66
  sustainable technology features 295-96
  target market of 296-97
Capital investment
  see Finance; Investment
Cappello, S. 46
CEDEIP (Computer and Electronics Disposition Eco-Industrial Park), Texas 218
Center of Excellence for Sustainable Development 80, 228, 242
Center for Neighborhood Technology 238, 241
Cernea, Michael 51
Certified Development Company (CDC) 231
Chaparral Steel 262
Charrettes (design workshops) 56, 109, 266, 293
Chattanooga, USA 79, 139, 163, 237
Chemicals, health hazards of 148-49
Chesapeake Bay, USA 288-89, 293, 294, 297, 298-99
China 349-51, 352
Chisholm, R.F. 61, 63
Chlorine 48
Chlorofluorocarbons (CFCs) 148-49, 261
Choctaw County, USA
  see Redhills EcoPlex
Choctaw Generation Inc. 315
Cities 25
Clavelle, Peter 211
Clean Air Act (CAA 1990) 142
Clean Energy Incubator Program 233
Clean technologies 206, 233, 297-98, 338
Clean-Water State Revolving Fund (CWSRF) 224
Closed-loop systems 64
Clusters, business 61-62, 93-94, 165-66, 249, 250-52
  Guitang Group, China 349-51
  see also Networks; Red Hills EcoPlex
Coal ash 145-46
Coastal protection 219, 298
  see also Chesapeake Bay
Coastal Zone Management Act (CZMA) 84
Codes, Covenants, Conditions and Restrictions for Eco-Industrial Development, Handbook of 78, 124, 320
Codes, covenants and restrictions (CC&Rs) 138, 140, 143-44, 192, 320
Cohen-Rosenthal, Ed 2, 9, 100-101, 103, 291, 296, 348
Co-location
  see Clusters; Networks
Commerce, US Department of (US DOC) 71, 78, 85, 87, 215-20, 242
Communication systems 176-77, 194-95
Communities
  and brownfield redevelopment, benefits of 104-105, 204
  and business networks 180-81
  development/re-creation of 25-26, 29, 76-78
    and Cape Charles 79, 96, 105, 109, 207
    and Green Institute 76, 105, 110, 263, 277-78, 282-83
  and EID, potential benefits of 101-107, 206
  engaging, strategies for 108-10
  funding for 79, 85-87, 207, 209, 220-22
    see also Public financing
  group processes with 56-57
  involvement of 100-11, 267, 277-79, 304, 308

INDEX 375

open spaces in 96, 207, 283
and planners, role of 97-99
surveys of 109-10
visioning by 108-10, 290-91
workshops, grants for 220
see also Health; New Urbanism; Smart Growth; Welfare; Workers
Community advisory boards (CABs) 110, 183-84
Community Builders programme 86
Community development block grants (CDBGs) 87, 220
Community Development Finance Institution (CDFI) 86, 236
Community Reinvestment Act (CRA 1977) 79, 86, 204
Community Services, Office of 232, 242
Commuters 95-96, 172-73
Companies, as machines/living 55
Competitive advantage 207-208
Complexity, working with 31-32
Composting 75, 319, 324, 329, 338
Comprehensive economic development strategies (CEDS) 217
Comprehensive Environmental Response, Compensation and Liability Act (CERCLA) 139
Computer and Electronics Disposition Eco-Industrial Park (CEDEIP), Texas 218
Concrete manufacture 145-46, 262, 298, 313, 314, 348
Confederation of Indian Industries 345
Conservation Law Foundation (CLF) 301
Construct America 2003 253
Consumers, green 23
Consumption, of resources 28, 34, 36
Contaminated land 149
see also Brownfield redevelopment
Conway School of Landscape Design 305
Copper 47
Copying machines 41, 156
Copyrights 147
Cornell University 201, 311
 Work and Environment Initiative 9, 38, 124, 163, 244, 251
 and Cape Charles 296, 299
 and NWLCC eco-industrial park 334, 336
 and TJCOG project 320
Corporate sponsors 237
Corporation for Enterprise Development (CFED) 77
Costs
 of remediation, insurance for 147
 savings in 102-103, 122-23
Côté, Ray 100-101, 189, 190
Cotton 346
Creativity, role of 66
Credits, trading 74-75, 84, 130
CRSS Capital 307, 308, 310

Dalhousie University, Canada 159, 322, 324, 325-26, 327
Dallas City, USA 217
Dams 16
Databases 296, 317, 318, 321, 338
Data/information management
 and government, role of 79-80
Decision-making
 and local planners 97-98
 public involvement in 100-11
De Geus, A. 54-55, 63
Delisheries 298

Dell Computers 255
Dematerialisation 35-37
Denmark
 see Kalundborg
Department of Economic and Community Development 307
 see also Red Hills EcoPlex
Design
 competitions 233
 for disassembly 41
 for durability 39-40
 guidelines for 160-61
 permaculture 330, 333-35, 340
 of Phillips Eco-Enterprise Center, Minneapolis 56, 279-81, 284-85
 prototype eco-park 249-50, 253-57
 for repair/re-use/remanufacture 40-41
 software tools for 224-25
 sustainable 115-27, 207
  payback/return on investment 247-48, 285, 286
  and planned densification 254-57
 see also Buildings; Planners; Products; Remanufacturing
Design charrettes 56, 109, 293
Designer's bias 116-17
Designing Industrial Ecosystems Tool (DIET) 225
Detrivore technologies 43
Developers
 education of 116-17
 and EID, adoption of 112-27, 253-56
 private sector 202, 301, 304, 305-306
Developing countries, Asia
 eco-industrial development in 341-52
Development
 government influence on 69
 infill 204, 254
 rural 25-26, 229-30, 232, 288-90, 300-301
 urban 25, 112-27
 see also Buildings; Communities; Design; Developers; Eco-industrial development; New Urbanism; Sites; Smart Growth
Development, eco-industrial
 see Eco-industrial development
Devens Enterprise Commission 98, 99
Disassembly, of products 41
Dobers, Peter 65
Duke University 317
Dunphy, Dexter 57

Eco-Business Program Burnside 326-28
Eco-Efficiency Center, Nova Scotia 325-29
Eco-Industrial Development, National Center for 9
Eco-Industrial Development Council (EIDC) 291
Eco-industrial development (EID)
 in Asian developing countries 341-52
 benefits of 89-90, 101-107
 definition of 18-20, 22-25, 68-69, 100, 259
 density of 238, 253-57
 as framework 20-22
 and government, role in 68-88, 94-96
 guidelines for 160-61, 253-57
 infrastructure 85, 133-34, 141, 150-52, 160, 202
 life-cycle of 152
 multi-scale approach to 20-21
 and permaculture design 333-35
 policies to support 81-88

software tools 224-25
success of, evaluating 258-67
and value, enhancement of 243-57
see also Clusters; Eco-industrial parks; Finance; Industrial ecology; Networks
Eco-Industrial Development Roundtable 9
Eco-industrial integration teams 184
Eco-industrial networking
see Kalundborg; Networks; Triangle J
Eco-industrial Park Handbook 124
Eco-industrial parks
codes, covenants & restrictions (CC&Rs) 138, 140, 143-44, 192, 320
definition of 18-20, 259
establishment, guidelines for 160-61
and Kalundborg industrial symbiosis, differences 18-19
Red Hills EcoPlex, comparison with 313-14
and local government, role of 89-99
management structures 182-85
and networks 18-20, 166-81
structure of 185
prototype for 249-50, 253-57
total number of 151
zoning 140-41
see also Burnside; Cape Charles; Clusters; Eco-industrial development; Industrial ecology; Londonderry; Networks
Ecological model 24-25, 303, 322
scavenger/decomposer functions 157, 329
see also Eco-industrial development; Industrial ecology
Ecology of Commerce, The 18
Eco-management and Audit Scheme (EMAS) 22
Economic Adjustment programme (EDA) 218-19
Economic development
existing strategies for 91-94
'high road' strategies for 77-78
Economic Development Administration (EDA) 93, 242
funding programmes 85, 87, 210, 215-19
Economic Development Authority (US) 307
Economic Development Initiative, HUD 221
Economic efficiency 102-103
Ecosystems
and development sites 84, 149-50, 160
industrial 303, 322
see also Ecological model; Industrial ecology
Eco-towns 71
Eddington, Arthur 32
Education/training 76, 78, 85-86, 104, 176, 179, 181
for brownfield redevelopment 102, 224, 282
of developers 116-17
environmental 194, 224, 282
by Green Institute 203, 224, 282
Electronics industry 153, 155-56
Emergencies, preparing for 178, 195-96
Emission credits, trading 74-75, 84
Employment
see Jobs; Workers
Empowerment zones (EZ) 221-22
Energy 128-35
cascading 38, 83, 128-30, 171
issues in 133-34
efficiency 83-84, 170-71
geothermal 129, 130
harvesting 129, 131-32
renewable 80, 83-84, 129, 131-32, 171
renewable energy farm project 299

solar 80, 129, 131-32, 228, 229, 233
architecturally integrated systems 295, 298
subsidies/funding for 80, 226-29
from waste 45-46, 130-31, 328-29, 338
agricultural 346, 347
waste of 17, 34
wind 129, 297, 298, 299
see also Omnitility; Power generation
Energy, US Department of (US DOE) 70, 78, 200, 210, 242, 299
funding by 215, 226-29, 307, 312
Energy and Environmental Design, Leadership in (LEED) 124, 143, 295
Energy Conservancy 278
Energy Efficiency and Renewable Energy, Office of (EERE) 226-29, 239
Energy Efficiency and Renewable Energy Network (EREN) 228, 242
Energy Star programme 84, 239, 276
Enterprise communities (EC) 221-22
Entropy 32-33
decreasing, strategies for 38-48
Environment, health and safety (EH&S) 177-79
Environment, US Department of (DOE) 80, 84
Environmental Business Council 304
Environmental Education, Office of, EPA 225-26
Environmental elitism 26
Environmental Finance Programme, EPA 225
Environmental goals 94-96, 105-106
Environmental Health and Engineering, Office of (OEHE) 232
Environmental impacts
assessment of 343
of buildings construction 152-55, 161
cumulative, reduction of 159-62, 191
and environmental management systems 191
of infrastructure installation 150-52
of operations 155-58
regional 151, 158
of site clearance 149-50
upstream/downstream 151-52
see also Environmental management systems
Environmental issues 148-62, 204
Environmentally Preferable Purchasing (EPP) programme 239
Environmental management systems
ISO 14001 187-99
policies on 84
role of 60-61, 186-87
see also ISO 14000 series; ISO 14001
Environmental monitoring 196
Environmental News Network 233
Environmental permits
see Permits
Environmental Protection Agency, US (US EPA) 41, 52, 68, 71, 75, 242
Energy Star programme 84, 276
funding by 78, 210, 215, 222-26, 318
on procurement policies 239
public involvement policy 102, 111
Smart Growth initiative 78, 91
waste regulation by 146
Environmental reviews 189
E[4] Partners Inc. 203
Equity 27-28, 33, 90-91
Erkman, Suren 345, 346
Estuaries 84

INDEX

Esty, Dan   62
Evaluation   258-67
Exchange programmes, materials
    see Waste
EZ/EC tax credits   221-22, 239

Fabens, Bruce   163
Facility Synergy Tool (FaST)   78, 224-25
Fannie Mae   238, 241
Farnell Packaging Ltd   328
Federal Communication Commission (FCC)   146
Federal Housing Administration   222
Federal Reserve Bank of St Louis   209
Federal sources of funding   215-33
Finance
    Cape Charles Sustainable Technology Park   292-93
    creating viable projects for   201-205
    and environmental issues   204
    Londonderry Ecological Industrial Park,   305-306
    and market demand   202
    marketing projects for   206-9
    Northwest Louisiana Commerce Center   330, 331, 337-38
    Phillips Eco-Enterprise Center   234, 237, 279, 284
    Red Hills EcoPlex   217, 307, 312, 314, 315
    software tools   224-25
    sources of   210-11
        federal   215-33
        local   234
        private   202, 234-37
        public   211-15
        state governments   233-34
    Triangle J Council of Governments   320-21
    see also Developers; Investment; Private sector; Risk
'Financing Eco-industrial Parks, Strategies for'   235
Fines/penalties   213-14
Fish and Wildlife Service   210
Fish/shellfish
    cherrystone clams   299
    farming   211, 272, 290, 296, 299, 314
    and pharmaceutical industry   297
Flue gas desulphurisation (FGD)   46
Food production   210-11, 348
Ford Foundation   236, 241
Foreign-trade Zone Resource Centre   241
Foreign-trade zones   210, 265
Foundries   347
Front Royal, VA, USA   109, 210
Fuel cells   129, 132, 338
Full Circle Production   333
Funding
    see Finance
Future 500   64
Future search conferences   56

Gardiner, R.   208
Geothermal energy   129, 130
German Ministry for Education and Research   347
German Technical Co-operation Organisation (GTZ)   343
Giannini-Spohn, Suzanne   235
Glickman, Joan   200
Goals
    environmental   94-96, 105-106
    social, achieving   80-81
Goldman, Benjamin   206
Gold production   34

Government
    and data/information management   79-80
    development policies, influence of   69, 81-82
    financing by   211-15
    and history of EID   70-71
    local
        financing by   211-15
        planners, role of   97-99
        role of   89-99
    and market economy, role in   71-72
    and market failures, correcting   72-75
    policy descriptions, detailed   83-88
    procurement by   239-40
    and public goods, provision of   75-80
    regulation/enforcement by   72-73, 94-95
    and research, funding for   78
    as stakeholder   70
    subsidies/taxes   80-81
Grants/loans   71, 87, 214-15, 230-31, 233-34, 236
    loan guarantee programmes   87, 240-41
    see also Public financing
Green Building   114-15, 143
Green Building Council, US (USGBC)   124, 143, 295
Green Business Letter   23
Green Development: Integrating Ecology and Real Estate   261
Greenhouses   211, 218, 311, 312, 314
    see also Agriculture
Green Institute, Minneapolis
    community development by   76, 105, 110, 263, 282-83
    founding of   277-78
    fundraising by   234, 279, 284
    planning/marketing by   278-79
    ReUse Centre   40, 203, 277-78
    training by   203, 224, 282
    see also Phillips Eco-Enterprise Center
Green Lights programme   84
Green marketing   173-74
Greenpeace   48
Green Productivity   22
Greentech Panels LLC   337-38
Green twinning   262
Griffiths, Andrew   57
Grogan, Peter   46
Group processes   56
Guadagno, Vic   333
Guigang Group, China   349-51
Gujerat, India   347
Gulf Southern   305
Gyproc   272, 313
Gypsum   46, 273, 274, 312, 313, 348

Halifax Integrated Waste Management Strategy   323-24
Handbook of Codes, Covenants, Conditions and Restrictions for Eco-Industrial Development   78, 124, 320
Harris, Thomas   289
Harvard Business School   62
Hauge Technologies   297
Hawken, Paul   17, 18, 20, 34
Hayes, Tim   290
Hazardous waste   146-47, 156, 324, 345
    costs of   74, 139
Health
    of employees   175, 177-79
    hazards   148-49, 156
    programmes   231-32

Health and Human Services, US Department of (HHS) 216, 231-33, 242
Hewlett Packard 21
'High road' strategies 77-78
Hirshberg, Nancy 304, 305
Holon, definition of 21
Home Depot 237, 241
HOME Investment Partnership Program 222
Honeywell Corporation 278
Horseshoe crabs 297
Housing 27, 86-87, 220-22, 238
Housing and Urban Development, US Department of (HUD) 86, 211, 215, 220-22, 242, 279
HUBZone Empowerment Contracting 231, 242
Human resources
  and networks 175-76
  role of 65-66
Human rights 28
Hydroponics 211, 218

IBM 200
Imagination, role of 66
Impact Energy Controls Corporation 278
India 148, 345-48, 349, 352
Indian Health Service (IHS) 232
Indian tribes 216, 221, 223, 226, 229, 233, 240
Indo-Dutch Project on Alternatives in Development 345
Indonesia 79, 348-49, 352
Industrial Assessment Center 227, 242
Industrial clusters 61-62, 93-94, 165-66, 249, 250-52
Industrial ecology
  analysis of 30-50
  business systems, role of 57-60
  definition of 51-52
  human resources, role in 65-66
  industrial network theory, role in 61-63
  scavenger/decomposer functions 157, 329
  self-organisation in 53-55
  social process, role of 55-57
Industrial Ecology, Material and Energy Flows, Interagency Working Group on (US) 51-52
Industrial Estate Authority of Thailand (IEAT) 343-45
Industrial metabolism, India 345-48
Industrial parks, traditional
  creating partnerships in 319
  management of 164-65
  see also Burnside; Eco-industrial parks
Industrial symbiosis
  and eco-industrial parks, comparison 18-19
  Kalundborg, Denmark 259, 270-75
  preconditions for 274
  Red Hills EcoPlex, comparison with 313-14
Industrial Technologies, Office of (OIT) 227, 242
Information, management systems
  and government, role of 79-80, 87-88
  and networks 176
Infrastructure 85, 202
  installing 150-52
  next-generation 202
  pipework 133-34, 141, 151, 274
Innovation 147
Institute for Communication and Analysis of Science and Technology (ICAST) 345, 352
Insurance 146-47, 207-8
Intellectual property 147
Interface Inc. 208

Intermodal Surface Transportation Efficiency Act (ISTEA) 81, 85
International Human Dimensions Programme on Global Environmental Change 56
Internet 26, 176
Intervale Food Center, Burlington, VT 83, 201, 210-11, 218, 237
Intervale Foundation 211, 237
Inventions and Innovation programme 227, 242
Investment
  green 23, 209
  returns on 22-23, 202, 243-57, 298
    Cape Charles Sustainable Technology Park 298
    at Kalundborg 273-74
    at Phillips Eco-Enterprise Center 285, 286
    and sustainable design 247-48, 285, 286
  socially responsible (SRI) 209
  see also Finance; Private sector
Iron sludge 348
ISO 14000 series 22, 60, 84, 178, 186-87
ISO 14001 187-99
Italy 166

Japan 71, 352
Jobs 103-104, 236, 298, 315, 324, 338
  in remanufacturing industry, US 41
  see also Workers
Johnson Graduate School of Management 163
Joint Venture Silicon Valley Network (JVSV) 165

Kalundborg industrial symbiosis, Denmark
  case study of 270-75
  and eco-industrial parks, comparison 18-19
  investment, payback/return on 273-74
  and Red Hills EcoPlex, comparison 313-14
  success of, evaluating 258-59
Kauffman Center for Entrepreneurial Leadership 282
Kellner-Rogers, Myron 54
Kiuchi, Tachi 64
Koestler, Arthur 21, 54
Krugman, Paul 93

Labour practices 76
Land
  assembly 202-3, 294
  clearance for development 149-50, 151
  see also Brownfield redevelopment; Sites
Landfills 95, 104
  building on 47
  mining 46-47
  reclamation of 298-99
  use of, reducing 75, 156, 262, 323-24, 348
Land-use planning 24, 27, 85, 186, 205
  see also Legal issues; Zoning
Lawton, Geoff 333
Leadership
  importance of 200-201, 236, 274
  teams 182-83
Leadership for a Changing World programme 236, 241
Leadership in Energy and Environmental Design (LEED) 124, 143, 295
Learning, looped, concept of 64
Learning organisations 63-64
Lease rate incentives 234
Leather 347
Leatherwood, Thomas 132

Legal issues 138-47, 191-92
LHB Engineers and Architects 279
Lighting 328-29
Lignite 307-8
Loan guarantee programmes 87, 240-41
  see also Finance; Grants/loans
Location, importance of 201-202, 243-46
Location-efficient mortgages (LEMs) 238, 241
Logistics, integrated 173
Londonderry Ecological Industrial Park, New
    Hampshire
  case study 300-306
  covenants on 60, 266, 303-304, 305
  developers of 201, 202, 265-66, 301, 304, 305-306
  funding 305-306
  local community 300-301
  power generation 301-302
  success, evaluation of 264-66
Los Angeles Department of Water and Power (LADWP)
    80
Louisiana Army Ammunition Plant
  re-use of, case study 330-40
Lowe, Ernest 106, 182
Lowitt, Peter 98, 99
Lucent Corporation 53

Makower, Joel 23
Malaysia 352
Management
  and ecological principles 23-24
  and environmental policy 189-91
  of industrial parks, traditional 164-65
  and networks 165-81
  by property owners' associations (POAs) 140
  structures of 182-84
  systems, need for 163-64
  type of, creating 139-40
Man-kuen Chak, Almaz 64
Manufacturing Extension Project (MEP) 87
Manufacturing industry 26-27
  see also Products; Remanufacturing
Markets
  creating/expanding 173-74, 206
  failures in, correcting 71-75
  waste exchange
    see Waste
Maryland, USA 85
Maskell, P. 107
Materials
  construction 203, 273, 277-78, 281
    concrete 145-46, 262, 298, 313, 314, 348
    plasterboard 273, 312, 313
  consumption of, USA 34, 36
  exchange
    see Waste
  flows of 36, 52, 169-70, 317, 336-38
  raw, renewable 25, 44, 159
  'smart' 39, 45
  use and re-use, hierarchy of 37, 46
  waste of 17, 34
  see also Design; Products; Recycling; Re-use; Waste
Matthews, John 21
McDonough, William 21, 207
McGalliard, Tad 163
McKenna, Terence 45
McKinsey and Company 25
McNeil Biomass Electrical Generating Station 211

Merkle, Ralph 45
Metals 38, 42, 347
Methanol 318-19
Miller Composting Corporation 329
Mines
  acid mine drainage 151
  lignite 307-308, 311
Minnesota State, USA 234, 237, 279, 284
  see also Green Institute; Phillips Eco-Enterprise Center
Minority Business Development Agency (MBDA) 87
Mississippi Co-operative Extension Service 316
Mississippi Development Authority 314, 315
  see also Red Hills EcoPlex
Mississippi State University 316
Mitsubishi Electric 64
Molecular reconstruction 44-45
Mollison, Bill 333
Monitoring/measurement 196
Monsanto Co. 139
Mortgages, location efficient (LEMs) 238
Musnikow, Judy 163

Nanochemistry 44-45
Naroda Industrial Estate, India 347-48
National Aeronautics and Space Administration (NASA)
    87
National Ambient Air Quality Standards (NAAQS) 142
National and Local Technical Assistance programmes
    217
National Association of Counties (US) 70
National Center for Eco-Industrial Development 9
National Endowment for the Arts 216, 233, 242
National Industrial Competitiveness through Energy,
    Environment and Economics programme (NICE3)
    227, 242
National Institute of Standards and Technology (NIST)
    87
National Oceanic and Atmospheric Administration
    (NOAA) 71, 84, 210, 242
  funding by 215, 219-20, 290, 299
National Science Foundation (NSF) 53, 87
National Telecommunications and Infrastructure
    Administration (NTIA) 87
National Transit Administration 85
Natural Step, The 34, 65
Nature Conservancy, The 299
Nepal 352
Netherlands 75
Net metering 134, 146
Networks/networking
  and business clusters 61-62, 93-94, 165-66, 249,
      250-52
  and communities 180-81
  eco-industrial, possible areas of 59
  and eco-industrial parks 18-20, 166-81
    structure of 185
  and energy 170-71
  and environment, health and safety 177-79
  and human resources 175-76, 179
  and information/communication 176-77
  and marketing 173-74
  and materials 169-70
  in Philippines 342-43
  and production processes 179-80
  in Silicon Valley 165-66
  social 107
  of subcontractors 179

theory of 61-63
and transport 171-73
see also Kalundborg; Triangle J
New Hampshire, USA
see Londonderry
New Public Works programme 233
New towns 92
New Urbanism 27, 114-15, 141, 249
and economic development 92-93
New York State Energy, Research and Development Authority (NYSERDA) 233
'Next Industrial Revolution, The' 207
NICE3 (National Industrial Competitiveness through Energy, Environment and Economics programme) 227, 242
NIMBYs 252, 256
Non-attainment New Source Review (NNSR) 142
Non-profit organisations 237, 320-21
North, Jonathan 235
North American Coal Corporation 315
North American Free Trade Agreement (NAFTA) 209
Northampton County, USA 288-89
 Sustainable Development Action Strategy 263-64, 290
 see also Cape Charles
North Carolina
 see Triangle J
North Carolina State University 317
Northwest Louisiana Commerce Center (NWLCC)
 case study 330-40
 funding of 330, 331, 337-38
 and permaculture 330, 333-35, 340
 and wood products industry 336-38
Nova Scotia, Canada
 Solid Waste-resource Management Strategy 323-24
 see also Burnside Industrial Park
Nova Scotia Power Inc. 325-26
Novo Nordisk/Novozymes 272-73, 314

Objectives, environmental 192-93, 196
Ocean and Coastal Resource Management, Office of (OCRM) 219
Omnitility concept 132-33
Opportunities, business
 locating/identifying 244-46
 new 157
Orimulsion fuel 272, 313
Ozone, stratospheric 149, 261

Packard, Vance 39
Pakistan 352
Paper
 production 130, 346-47, 349, 351
 recycled 74, 239
 waste 323-24
Pario Research, USA 253, 255
Parks, eco-industrial
 see Eco-industrial parks
Partnerships, creation of 319
 see also Networks
Patents 147
Pauli, Gunter 32
Pax World Funds 209
Performance indicators 192-93, 196
Permaculture 330, 333-35, 340
Permits 141-43, 176, 178, 192, 343
Petersen, Astrid 163

Petrochemicals, processing of 152-53
Pharmaceuticals industry 272, 273, 297
Philippine Board of Investments 342-43, 352
Philippine National Oil Company 342, 343
Philippines 71, 341, 342-43, 349, 352
Phillips, Wendell 277
Phillips Coal Company 307-308, 310, 315
Phillips Eco-Enterprise Center, Minneapolis
 case study of 276-87
 community development 277-78, 282-83
 construction of 203, 281
 design of 56, 279-81, 284-85
 and environmental justice 277
 funding for 234, 237, 279, 284
 and Green Institute, founding of 277-78
 investment, payback/return on 285, 286
 operation of 281-83, 285-87
 origins of 205, 277-79
 planning/marketing 278, 283-84
 ReUse Centre 40, 203, 277-78
 success of, evaluating 263
 see also Green Institute
Phillips Petroleum 218, 310
Photovoltaics
 see Solar power
Piaget, J. 63
Pickering Inc. 314, 316
Pipework 133-34, 141, 151, 274
Planned Densification 253, 254-57
Planners
 guidelines for 160-61, 253-57
 local, role of 97-99
Planning charrettes 56, 109, 293
Planning programme (EDA) 217
Plasterboard 273, 312, 313
Plastics 25
Policies
 environmental 84, 189-91, 192
 to facilitate financing 237-41
 to support EID 81-88
 on transport 81, 84-85, 95-96
Policy, Planning and Evaluation, Office of (OPPE) 224-25
Pollard, Patricia 209
Pollutant Release and Transfer Registry 79
Pollution
 air 74-75, 81, 95, 142, 158
 charges 75
 definition of 33
 and distributed generation 129
 emission credits, trading 74-75
 and environmental management systems 186-87, 196
 impacts of 47-48
 monitoring 196-97
 permits 141-43, 176, 178, 192, 343
 prevention (P2)
  in India 347-48
  policies on 84
 water 75, 151
 see also Insurance; ISO 14000 series; ISO 14001; Waste
Polychlorinated biphenyls (PCBs) 148-49
Porter, Michael 62, 86, 93
Port of Cape Charles
 see Cape Charles
Poultry processing 311
Poverty 28, 33, 86, 222, 232, 289
 see also Communities, development; Equity

Power generation   83, 128-35, 151
  at Cape Charles   293
  co-generation   128, 129, 171, 211, 262
  combined-cycle gas   301-302, 305
  distributed generation   129, 141
  fluidised-bed   307-308, 313
  funding for, federal   226-29
  and green building design   301-302
  at Kalundborg   272, 313
  at Londonderry EIP   301-302
  net metering   134, 146
  pollution from   129
  at Red Hills EcoPlex   307-309, 313, 315
  from tyres   338
  waste from   145-46, 272-73, 311-15
  see also Energy
Power Technologies, Office of (OPT)   227, 242
Prairies   143, 281
President's Council on Sustainable Development (PCSD)
  9, 70, 79, 216, 243, 259, 291
  eco-industrial development, definition of   19, 100, 259
Prevention of Significant Deterioration (PSD)   142
Private financing   202, 234-37
Private-sector developers   202, 301, 304, 305-306
  see also Londonderry
Private Sector Participation in Managing the
    Environment (PRIME)   71, 342, 352
  industrial ecology module   342-43
Process mapping   124-26
Production processes   179-80
Product-Life Institute   50
Products
  disassembly/remanufacturing of   40-41, 156-57
  durability/functionality of   39-40
  life-extension loops   50
Profitability   102-103
  see also Investment, returns on
Programme for Pollution Control, Evaluation and Rating
    (PROPER)   79
Projects
  attracting finance for   206-10
  creating viable   201-205
Property owners' associations (POAs)   140
Property value, increasing   246-48
ProVento America Inc.   297
Public, involvement of   100-11
  see also Communities
Public financing   211-15, 292-93
  federal   215-33
  state   233-34
Public goods, provision of   75-80
Public–private partnerships   76, 77, 87, 237
Public Works programme (EDA)   217-18

Quality systems   179
Quantum Connection Eco-Park™   249-50, 253-57

Rail transport   81, 151, 282
Rainforests   64
Ramaswamy, Ramesh   345, 346
Real estate, value of   243-57
Recycling   41-43, 95, 156-58
  encouraging   75, 161
  in-process industrial   42-43
  viability of   42, 74
  and waste-to-energy   46, 130
  see also Re-use; Waste

Red Hills EcoPlex, Mississippi
  case study   307-16
  funding   217, 307, 312, 314, 315
  and Kalundborg, comparison   313-14
  power generation at   307-309, 313, 315
Refrigerators   40, 41
Regulations/enforcement
  beyond-compliance standards   176
  and environmental improvement   94-95
  and environmental management systems   191-92
  government   72-74
Rejeski, David   79
Relationships, eco-industrial   144-45
Remanufacturing   40-41, 156-57
Remediation
  brownfield sites   94, 149
  costs, insurance for   147, 204, 223-24
  soil   273
Renewal communities, HUD   222
Repair/re-use/remanufacture
  design for   40-41
Reporting   79, 178
Research EIT   352
Research funding   78, 87
Research Triangle Park, NC   87, 251
Reservations, American Indian   232
Resource engineering   126
Resource Recovery Centre   203
Resources
  consumption of   28, 34, 36
  management of   38-48, 158-62
  see also Energy; Entropy; Equity; Materials; Re-use;
    Waste
Response and Restoration, Office of (ORR)   219-20
Return on assets (ROA)   22-23
Re-use   40-41, 156-58, 277-78, 281
  adaptive   203
  deconstruction of buildings for   40, 156, 203, 278
  designing for   40-41
  of Louisiana Ammunition Plant   330-40
  of materials   40-41, 156-58, 281
    hierarchy of   37, 46
  see also Design; Materials; Products; Recycling
ReUse Centre, Minneapolis   40, 203, 277-78
Revolving loan funds (RLFs)   219, 223-24, 240
Rideaway Vans   305
Risk
  and capital providers   207-208
  insurance/mitigation of   145-47
Riverside Eco-Park
  see Intervale Food Center
Robèrt, Karl-Henrik   65
Robins, R.   46
Rocky Mountain Institute   261
Rouse Forum   91
RPP International   124, 352
Rural Business Co-operative Service   229
Rural development   25-26, 229-30, 232, 288-90, 300-301
Rural Development mission, USDA   229
Rural Utilities Service (RUS)   229-30

Saratoga Technology Energy Park (STEP)   233
Scale, importance of   21
Scavenger/decomposer functions   43, 157, 329
Schlarb, Mary   100, 103
School for Resource and Environmental Studies   159

School to Work Act   86
Scientific and Environmental Associates   298
Search conferences   109
Section 7(a) loan guarantees   230
Section 108 loan guarantees, HUD   221
Section 7(m) Microloans   231
Self-organisation   182, 340
 and engineered systems, comparison   53-55
Senge, Peter   54-55
Seshasayee Paper and Board Ltd   346-47
Shellfish
 see Fish
Shopping centres   282-83
ShoreBank Corporation   236, 240, 242
Silicon Valley, US   165-66, 168
Simpler Way, A   54
Singapore   352
Sites
 assembly of   202-203, 294
 clearance/preparation of   149-50, 160, 203
 historic/culturally significant   204-205, 283
 Quantum Connections plan for   253-57
 selection of   201-202, 243-46
Slaughterhouses   42
Small Business Administration, US (SBA)   86, 216, 230-31, 242
Smart Growth   81, 85, 91-92, 114-15, 205, 206, 238
Snell, Robin   64
Social capital   106-107
Social Investment Forum   209
Socially responsible investment (SRI)   209
Social networks   107
Social processes   51-66
 see also Communities
Software tools   224-25
Solar power   80, 129, 131-32, 228, 229, 233
 architecturally integrated systems   295, 298
Solar Rooftop Incentive programme   80
Southern Paper Company   309, 310
Sponsors, corporate   237
Sprawl, reducing   204, 238, 254-57
Springfield, MA   103
Sri Lanka   352
Stahel, Walter   50
Stakeholders
 government as   70
 involvement of   304
 see also Communities
Stamps, D.   53-54
State Energy Program (SEP)   228, 242
State funding sources   233-34
Statoil refinery, Kalundborg   272-73, 314
Steel   47, 262
Stonyfield Farms   300, 305
'Strategies for Financing Eco-industrial Parks'   235
Stuff   31
Subcontractors   179
Subsidies   76, 79, 80-81
 see also Grants; Public financing
Success, evaluating   258-67
Sugar   346-47, 349, 351
Sulphur   272, 313, 314
Surveys   109-10, 318, 338
Sustainable America: A New Consensus   216
Sustainable America: New Public Policy for the 21st Century   206
Sustainable Corporation, The   57

Sustainable Design and Development LLP   265-66, 301, 304, 305-306
Sustainable development
 and industry   15-18, 302-303
Sustainable Development, Centre of Excellence in (website)   80
Sustainable Development, President's Council on (PCSD)   9, 70, 79, 216, 243, 259, 291
 and EID, definition of   19, 100, 259
Sustainable Development and Intergovernmental Affairs, Office of   220
'Sustainable Finance: Seeking Global Financial Security'   208
Sustainable Jobs Fund (SJF)   236, 242
Sustainable Londonderry
 see Londonderry
Sustainable technology   296
 see also Cape Charles; Technologies
Sweden   65
Swedwood Canada Ltd   328
Symbiosis Institute   314
Szreter, S.   106-107

Taiwan   352
Tamil Nadu, India   347
Targets   192-93, 196
Taxes   80-81, 84, 86, 162, 213
 Asia   343, 351
 credits   86, 143, 221-22, 224, 233-34, 238-39
Tax increment financing (TIF)   86, 214, 234
Taxpayer Relief Act   224
TEA-3   81
Technologies
 biological   211, 297
 clean   206, 233, 297-98, 338
 detrivore   43
 sustainable, at Cape Charles   295-98
Telecommunications
 companies, buildings for   153-54
 factories, wastes from   155-56
 provision of   146
Temporary Assistance to Needy Families (TANF)   86
Tenants, advantages for   102-103
Tennessee Valley Authority (TVA)
 see Red Hills EcoPlex
Tents, Bedouin   40
Texas, USA   218, 308
 See also Brownsville
Texas Engineering Extension Service (TEEX)   218
Texas Industries   262
Textiles   346
TFM Show 2003   253
Thailand   71, 341, 343-45, 352
Tibbs, Hardin   48, 53
Tirupur, India   346
Toxics Release Inventory, US   56, 79
Toxic substances   79, 156
Tractebel Power   218, 307, 310, 315, 316
Trade   209-10
 foreign, zones   210, 241, 265
Trademarks   147
Traffic sheds   96
Training/education   76, 78, 85-86, 104, 176, 179, 181
 for brownfield redevelopment   102, 224, 282
 of developers   116-17
 environmental   194, 224, 282
 by Green Institute   203, 224, 282

INDEX   **383**

Transport
    and commuting   95-96, 172-73
    emissions from   81, 95
    and location efficiency   238, 241
    planning   96, 171-73
    policies on   81, 84-85, 95-96
    rail   81, 151, 282
    research in   228
    subsidised   81
Transportation, US Department of   85
Transportation Equity Act for the Twenty-first Century (TEA-21)   81, 85
Transportation Technologies, Office of (OTT)   228, 242
Triangle J Council of Governments (TJCOG)
    case study   317-21
    eco-industrial networking   20, 95, 139, 317-20
    funding   320-21
Trist, Eric   61
Tyres   338

UN Development Programme (UNDP)   342, 352
UN Environment Programme   352
UN International Human Dimensions Programme   16
University of Albany   233
University of Hong Kong   261
University of Kaiserslautern   347, 352
University of Louisiana Lafayette   338
University of Minnesota   279, 282
University of North Carolina   317
University of Virginia School of Architecture   291
Urban and Rural Community Economic Development programme   232
Urban development   25, 112-27
    see also Communities; Developers; New Urbanism; Smart Growth
USA
    ammunition facilities, re-use of   330-40
    Baltimore, MD   56, 79
    Burlington, VT   237
        see also Intervale Food Center
    California   80, 134, 145, 165-66
    Chattanooga, TN   79, 139, 163, 237
    EID in, history of   70-71
    federal programmes   84-87
    funding, federal sources of   215-33
    industrial policy of   69, 70-71, 74
    Louisiana
        see Northwest Louisiana Commerce Center
    Maryland   85
    materials, consumption of   34, 36
    Minnesota   234, 237, 279, 284
        see also Green Institute; Phillips Eco-Enterprise Center
    Mississippi State
        see Red Hills EcoPlex
    New Hampshire
        see Londonderry
    New York   60, 200, 233, 261
    North Carolina   20
        see also Triangle J
    public involvement policy   102
    Texas   217, 218, 308
        see also Brownsville
    Virginia   109, 210, 233, 288-89
        see also Cape Charles
US Clean Air Act 1990   74-75

US Department of Agriculture (USDA)   84, 242
    funding by   210, 216, 229-30, 338
US Department of Commerce (US DOC)   71, 78, 80, 84, 85, 87, 215-20, 242
US Department of Defense   240
US Department of Energy (US DOE)   70, 78, 200, 210, 242, 299
    funding by   215, 226-29, 307, 312
US Department of Health and Human Services (HHS)   216, 231-33, 242
US Department of Housing and Urban Development (HUD)   86, 211, 215, 220-22, 242, 279
US Department of Interior (DOI)   84
US Department of Transportation   85
US Department of Treasury   86
US Economic Development Authority   307
US Environmental Protection Agency (US EPA)   41, 52, 68, 71, 75, 242
    Energy Star programme   84, 276
    funding by   78, 210, 215, 222-26, 318
    on procurement policies   239
    public involvement policy   102, 111
    Smart Growth initiative   78, 91
    waste regulation by   146
US Green Building Council (USGBC)   124, 143, 295
US National Academy of Engineering   34
US Small Business Administration (SBA)   86, 216, 230-31, 242
US Toxics Release Inventory   56, 79

Valentec Systems Inc.   330, 331, 333
Value, creation of   243-57
Vehicles
    see Transport
Virginia, USA   109, 210, 233, 288-89
    see also Cape Charles
Virginia Coastal Programme (VCP)   290
Virginia Department of Transportation   210
Virginia Tech   299
Visioning, community   108-10, 290-91
Von Weizsäcker, Ernst Ulrich   28, 34, 46

Wako Chemicals USA   297
Wallner, H.P.   62
Washington Post   208
Waste   74-75, 95
    ash, coal   145-46, 272, 311-15
    construction   40, 156, 203, 277-78
    energy from   45-46, 130-31, 328-29, 338
        agricultural   346, 347
    exchange programmes/markets for   40-41, 74-75, 95, 98, 156-61, 327
        ash   145-46, 272, 311-15
        Asia   346-51
        Green Institute ReUse Center   40, 277-78
        green twinning   262
        risks of   145-46
        Triangle J, North Carolina   95, 317-19, 321
        see also Clusters; Networks
    hazardous   146-47, 156, 324, 345
        costs of   74, 139
    mining   46-47
    from power generation   145-46, 272-73, 311-15
    production of   34, 46, 95
    reduction of   35-37, 95, 158-62, 327
        management strategies for   323-24
    sugar   346-47, 349, 351

tyres 338
wood 130, 131, 336-38
see also Eco-Business Program; Landfill; Materials; Pollution; Recycling; Re-use
Waste Age (magazine) 47
Wastemakers, The 39
Water 346-47
　conservation/re-use 42, 151, 272, 274, 313-14
　desalination 297
　pollution of 75, 151, 224
　storm, management of 142-43, 281
　substitution for 328
　waste, treatment of 132, 140-41, 142, 146, 160, 301
　see also Wetlands
Web addresses 80, 241-42
Weiss, P.A. 54
Welfare Reform Law 86
Welfare to Work (W2W) tax credit 86, 221
Wetlands 150, 160, 281, 293, 296
Wheatley, Margaret 54
White House Council on Environmental Quality 63
Whole Earth Review 45
Wind power 129, 297, 298, 299
Whitman, Christie 111
Wolff, Rolf 65
Wood waste 130, 131, 336-38
Work and Environment Initiative 9, 38, 124, 163, 244, 251
　and Cape Charles 296, 299
　and NWLCC eco-industrial park 334, 336
　and TJCOG project 320

Workers
　health of 175, 177-79
　and networking 175-76, 179
　productivity of 281
　training 76, 78, 85-86, 104, 176, 179, 181
　　environmental 194, 224, 282
　see also Jobs
Workforce Investment Act (WIA) 86
Workshops, grants for 220
Workstage 255
World Bank 51
World Resources Institute 28
Worldwatch Institute 159
World Wide Web 176

Xerox
　Palo Alto Research Center 45, 63
　remanufacturing by 40-41, 156

Young, Annie 277

Zoning 140-41, 205, 256-57
　foreign-trade 210, 265